Advanced Modeling of Thermodynamic Energy Components and Systems

Advanced Modeling of Thermodynamic Energy Components and Systems presents intermediate-complexity models for energy system components. It addresses a critical gap in studying energy systems under off-design conditions, examining how systems behave when ambient temperature or load conditions change.

Using Thermoptim software's systemic approach, which graphically links component icons and encapsulates equations within modules, this book focuses on cycle architecture and component interactions rather than equation manipulation. It demonstrates how component technologies (dynamic versus positive displacement compressors, various heat exchanger types) fundamentally affect system behavior and provides tools to model these distinctions accurately.

This book equips mechanical and energy engineering researchers, professionals and graduate students with methods bridging technological characteristics and thermodynamic models.

Renaud Gicquel served as a Full Professor and later a Distinguished Professor at PSL University, France, directing its Centre for Energy Studies (1987–2004). A graduate of École des Mines de Paris, France, with a doctorate from Paris VI University, France, Dr. Gicquel's passion for integrating technology with thermodynamics education led him to develop Thermoptim, an innovative software widely used for energy system simulation and teaching, and author over 80 papers and three books. He has also created three MOOCs, democratizing access to advanced energy concepts globally. Dr. Gicquel's career spans numerous notable international roles: Special Assistant to the UN Secretary General for the Conference on New and Renewable Energy Sources (1980), senior positions at France's Ministry for Research and the National and CNRS, and co-founding the Mediterranean Energy Observatory (1986).

Advanced Modeling of Thermodynamic Energy Components and Systems

Volume 1

Renaud Gicquel

CRC Press
Taylor & Francis Group
Boca Raton London New York

CRC Press is an imprint of the
Taylor & Francis Group, an **informa** business

Designed cover image: Siemens Energy; Renaud Gicquel

First edition published 2026
by CRC Press
2385 NW Executive Center Drive, Suite 320, Boca Raton FL 33431

and by CRC Press
4 Park Square, Milton Park, Abingdon, Oxon, OX14 4RN

CRC Press is an imprint of Taylor & Francis Group, LLC

ISBN: 978-1-032-99786-5 (hbk)
ISBN: 978-1-032-99788-9 (pbk)
ISBN: 978-1-003-60601-7 (ebk)

DOI: 10.1201/9781003606017

Typeset in Times
by codeMantra

Access the Support Material: www.routledge.com/9781032997865

To my wife and lifelong companion,

To our three daughters

To our grand children

Contents

Disclaimer

This book is an educational text designed to teach advanced thermodynamic modeling principles and methods. While it emphasizes technological realism and practical applications, it is not intended to serve as an industrial design handbook.

The thermodynamic data, constants, and correlations used throughout this book are selected for pedagogical clarity and may differ from those found in certified industrial databases. They should not be used as reference values for professional engineering calculations.

The Thermoptim software used is a pedagogical tool that facilitates the learning of complex concepts. The examples and exercises in this book were developed and validated using Thermoptim versions 2.52, 2.62, 2.72, and 2.82. Results obtained with other versions may differ. In all cases, Thermoptim's results are illustrative and do not match the precision or validation required by professional engineering design tools used in industrial practice.

Readers engaged in actual system design should therefore consult appropriate industrial software and professional engineering resources.

The author and publisher accept no liability for any consequence arising from the use of the methods, data, or software results presented in this book in a professional or industrial context.

Additional Resources

Numerous digital resources for teaching energy systems have been progressively compiled into a portal called Thermoptim-UNIT (www.thermoptim.org), whose content is freely accessible with few exceptions, such as solutions to selected exercises and problems.

This book's website (https://server.s4e2.com/crc/amtc/) provides access to various supplementary resources, including:

- Links to digital versions of selected figures
- Updated footnote links in case of changes
- Application examples of developments presented in specific chapters, available as guided explorations or examples of Thermoptim models with their associated files
- Each chapter opens with a landing page featuring an overview and abstract, along with a dedicated Table of Contents and chapter-specific resources.
- Theoretical and methodological supplements, either on the portal or dedicated pages

Since the complete list of resources was not finalized at the time of publication, please refer to the website during your reading to discover all available materials. Some resources will be reserved for registered book purchasers.

Preface

ENGINEERING THERMODYNAMICS: ADVANCED MODELING OF ENERGY SYSTEMS AND NUCLEAR CYCLES

In nearly all courses of thermodynamics applied to energy systems worldwide, the chosen approach does not take into account the technological characteristics of components such as compressors or heat exchangers. As a result, it does not allow the study of the behavior of these systems when their operating conditions vary, for example, when the ambient temperature changes or when the load they are subjected to evolves. This is known as the study of the operation of energy systems under off-design conditions.

Moreover, the thermodynamic cycles that are studied are generally only caricatures of those implemented in reality, as learners do not have the tools that would allow them to model them correctly. This is particularly true for Thermodynamic Cycles of Nuclear Power Plants (TCNPPs), some of which are relatively complex due to the peculiarities of this energy source.

The originality of this set of books lies in addressing these two subjects, which are not covered in the thermodynamics textbooks available today. This gap has several causes:

First, curricula have gradually been simplified over the past decades, limiting the time available for in-depth studies, notably the study of the operation of energy systems under off-design conditions.

Second, this reduction in curricula has mainly affected technological issues, with most teachers favoring theoretical approaches with which learners are more familiar.

Finally, the degree of difficulty of these subjects is significantly higher than merely learning the basics of thermodynamics, which are already relatively complex.

As a result, these modeling difficulties are evaded in most courses.

In our book *Energy Systems: A New Approach to Engineering Thermodynamics*, published in 2021, we presented a method for learning thermodynamics as applied to energy systems that is simplified in terms of equations. This approach is designed to meet the needs of learners with limited backgrounds in mathematics and physics, and who do not wish to engage heavily in modeling.[1]

The cycle studies in these three books are conducted using a simulation software called Thermoptim, which allows for the modeling of a wide range of systems while faithfully representing technological reality.

Modeling is performed systemically by graphically linking icons that represent components. Since the equations are encapsulated within the modules corresponding to each component, learners don't need to constantly rewrite them. This allows them more time to focus on the architecture of the cycles, the interactions between different components, and so on. Consequently, this approach fosters a way of reasoning for learners that is distinctly different from the traditional methods used in most thermodynamics textbooks.

The first volume elaborates on the analysis of the behavior of the basic components and explains how customized components can be modeled and coupled to Thermoptim through a mechanism named external classes. Several case studies show how these developments can be put in practice in simple cycles.

The second volume shows how they can be used to model complex thermodynamic cycles coupled to nuclear reactors.

Building on a more advanced scientific level than the 2021 edition, this new set of books complements the previous work on two key fronts, addressing important aspects that were intentionally omitted in 2021 to maintain a simplified approach.

While preserving the same systemic approach, this new series expands on the earlier work by exploring more in-depth modeling, which was previously limited to the basics, and by thoroughly examining TCNPPs.

Structured in two volumes, it equips readers with the tools needed to master these timely and, as we will demonstrate, highly accessible subjects with the right resources.

Advanced Modeling of Thermodynamic Energy Components and Systems

The objective of the first volume is to present a modeling approach for the main components used in energy systems (pumps, compressors, expansion devices, heat exchangers, combustion chambers, etc.) that is detailed enough to account for their key technological characteristics (a dynamic compressor does not behave like a positive displacement compressor).

At the same time, this approach is suitable for integration into a global system that can involve dozens of such components. These models offer intermediate complexity, positioned between general models that do not consider technological specificities and highly detailed finite element models, which completely describe geometry but are not suitable for systemic analyses.

Chapters 1 and 2 provide a theoretical background to introduce the notations that will be used later.

Chapters 3 and 4 are dedicated to the modeling of different components (pumps, compressors, expansion devices, heat exchangers, combustion chambers), taking into account their technological specificities.

Chapters 6–10 address the behavior of systems under off-design conditions.

Advanced Modeling of Thermodynamic Cycles of Nuclear Power Plants

Recent events have sparked renewed interest in nuclear energy, particularly the so-called Small Modular Reactors, which are much smaller (between 30 and 340 MWe) than the large-scale pressurized water reactors that make up the majority of the world's nuclear fleet (from 900 to 1,650 MWe).

The second volume of this set of books is devoted to the study of most TCNPPs.

The first chapter explores the evolution of applied thermodynamics pedagogy, highlighting two main teaching paradigms that effectively equip students to be fully operational by the end of the course. The developments presented in this book demonstrate how to bridge the gap between these complementary approaches.

Chapter 2 explores the physical phenomena involved in nuclear reactors and present the main types of current and future nuclear reactors. In this chapter, we provide only a brief presentation of the associated thermodynamic cycles, indicating only their type and capacity, the pressure and temperature levels reached, and the efficiencies achieved. Further details are provided in the rest of this book.

Chapters 3–5 mainly deal with the water-steam cycles that are the most used or considered for nuclear power plants. They show learners how they can use powerful tools such as the pinch method or exergy balances built thanks to exergy structures to improve the performance of these cycles.

Chapter 6 is dedicated to cycles involving gas turbines and those with supercritical CO_2.

Chapter 7 begins by presenting combined cycles, which usually combine a gas turbine cycle and a steam cycle. Then, it introduces technologies that could constitute a significant application for nuclear reactors, namely seawater desalination and large-scale hydrogen production, and it concludes with cogeneration installations that have the advantage of leading to the best overall efficiencies.

Chapter 8 presents 11 case studies relating to TCNPPs.

AI Acknowledgments

Claude was used for linguistic revision purposes. ChatGPT, Grok, Perplexity, Gemini, Mistral and Claude were used to determine translations of technical terms from French to English.

About the Author

Prof. Renaud Gicquel has published several books:

- In 1992, *Introduction aux Problèmes* Énergétiques *Globaux* (Economica). A revised version was published in 2013 (in collaboration with May Gicquel) by Presses des Mines and in English by CRC Press Balkema under the title *Introduction to Global Energy Issues.*
- In 2001, *Systèmes* Énergétiques, Volumes 1 and 2 (Presses des Mines). A revised edition was published in 2009 with a third volume on advanced cycles and innovative low-environmental-impact systems. A new version was published in 2023 by Presses des Mines. The 2009 revised version was translated into English and published in 2012 by CRC Press Taylor & Francis as *Energy Systems: A New Approach to Engineering Thermodynamics.* A second edition was published in 2021.

Renaud's special interest and passion is combining thermodynamics and energy system education with modern information technology tools. To this end, he has developed various software packages to facilitate the teaching and learning of applied thermodynamics and the simulation of energy systems, including Thermoptim software, distributed by S4E2 (https://server.s4e2.com/jml/index.php/en).

CAREER

Renaud Gicquel is now retired. The main part of Prof. Gicquel's career took place at École des Mines de Paris (PSL University, France), where he led the Centre for Energy Studies from September 1987 to December 2004. Appointed as a Full Professor in July 1986 and later as a Distinguished Professor, he taught applied thermodynamics, energy system modeling, and global energy issues. He has developed three MOOCs on these topics.

He was trained as an engineer at École des Mines de Paris, France, and earned his doctorate in engineering from Paris VI University. He started his career in 1975 as a researcher on solar energy at École des Mines, before being appointed as a Special Assistant to the Secretary General of the United Nations Conference on New and Renewable Sources of Energy in New York in 1980. He then became the Head of Multilateral Issues in the International Affairs Service of the Ministry for Research and Technology in Paris in 1982. From 1983 to 1985, he was an Adviser for International Issues at the National Center for Scientific Research. In 1986, he co-founded the Mediterranean Energy Observatory with the late Michel Grenon. In 1990, Dr. Gicquel created the ARTEMIS group in Nantes, a research body for thermal energy research, together with the University of Nantes and ISITEM (now Polytech). He also served as a Deputy Director at École des Mines de Nantes for 3 years beginning in 1991.

List of Symbols

A	area (m^2)
a	hydrogen available for combustion
$\mathbf{A_c}$	cross-sectional flow area (m^2)
$\mathbf{B_o}$	Boiling number
C	concentration
C	speed (m/s)
C	fluid velocity (m/s)
$\mathbf{C_s}$	speed of sound (m/s)
$\mathbf{C_p}$	molar heat capacity at constant pressure (J/(mol K))
$\mathbf{c_p}$	specific heat capacity at constant pressure (J/(kg K))
$\mathbf{C_v}$	molar heat capacity at constant volume (J/(mol K))
$\mathbf{c_v}$	specific heat at constant volume (J/(kg K))
D	diameter (m)
$\mathbf{d_h}$	hydraulic diameter (m)
e	excess air
e	thickness (m)
f	friction coefficient
f	surface factor
F	molar flow rate (mol/s)
F	molar Helmholtz energy (J/mol)
f	mass Helmholtz energy (J/kg)
G	molar Gibbs energy (J/mol)
g	mass Gibbs energy (J/kg)
g	acceleration due to gravity (m/s)
H	molar enthalpy (J/mol)
h	mass enthalpy (J/kg)
h	heat transfer coefficient (J/(K m^2))
h	calorimetric coefficient (J/(kg bar))
h	height difference (m)
$\mathbf{h_f}$	enthalpy of formation (J/mol)
i	van't Hoff factor of the solute
I	current density (A)
k	polytropic coefficient
k	dissociation rate
$\mathbf{k_m}$	mixing entropy (J/(mol K))
K	kinetic energy (J/kg)
$\mathbf{K_{eb}}$	ebullioscopy constant K (kg/mol)
$\mathbf{K_p}$	equilibrium constant
$\mathbf{K_0}$	cone or Stodola constant
l	calorimetric coefficient (J/m)
L	enthalpy of evaporation (J/kg)
L	length (m)

Le	Lewis number
M	molar mass (kg/mol)
Ma	Mach number
m	mass (kg)
n	number of moles
N	Rotation speed (tr/mn)
$\mathbf{N_o}$	Avogadro's number $6{\cdot}10^{23}$ (mol/g)
$\mathbf{N_u}$	Nusselt number
NTU	number of transfer units
P	pressure (Pa or bar)
Pr	Prandtl number
p	wetted perimeter
P	power (W)
Q	heat (J/kg)
q′	specific enthalpy (moist gas) (J/kg)
R	universal constant 8.314 (J/(mol K))
R	ratio of heat capacity rates of a heat exchanger
r	ideal gas constant (J/K)
r	compression ratio
R	richness
RR	relative roughness
$\mathbf{R_e}$	Reynolds number
S	molar entropy (J/(mol K))
s	mass entropy (J/(kg K))
S	surface (m^2)
$\mathbf{S_c}$	Schmidt number
T	thermodynamic temperature (K)
t	temperature (°C or K)
t	dry bulb temperature (°C)
t′	wet bulb temperature (°C)
tr	dew point (°C)
u	specific internal energy (J/kg)
U	molar internal energy (J/mol)
U	peripheral speed (m/s)
U	heat transfer coefficient (W/(K m^2))
v	volume (m^3)
v	specific volume (m^3/kg)
V	volume (m^3)
V	molar volume (m^3/mol)
V	voltage (V)
$\mathbf{v_{spec}}$	specific volume (m^3/kg)
W	specific mechanical energy (J/kg)
w	specific humidity kg water/kg of dry air
W	power density (W/kg)

x length (m)
x vapor quality
x mole fraction
x mole fraction in liquid phase (mixtures)
$\mathbf{x_h}$ specific exergy (J/kg)
$\mathbf{x_q}$ specific heat-exergy (J/kg)
y mass fraction
y mole fraction in vapor phase (mixtures)
z global mole fraction (mixtures)
z altitude (m)
d exact differential
δ differential form
Δ variation

INDICES

a related to admission
c corrected
c related to compressor
c cold
cc related to combined cycle
d related to expansion
dg related to the dry gas
g related to the gas turbine
g related to the gas phase
h hot
i related to inlet
l related to liquid phase
mm related to the moist mixture
p related to polytropic
p related to products
r related to discharge
r related to reactants
s related to saturation
s related to isentropic
s related to the output
sat related to saturation
t total
t related to the turbine
v related to the steam cycle
vap related to the vapor

GREEK SYMBOLS

a	Baumann coefficient
β	losses
δ	differential
δ	Rateau coefficient
Δ	specific diameter
ε	degree of reaction
ε	relative humidity (RH)
ε	effectiveness
ε	bleed fraction
ε	initial expansion ratio
ε	roughness m
ϕ	heat flux (J/kg)
γ	ratio of thermal capacities
η	efficiency, effectiveness
φ	flow factor
λ	calorimetric coefficient (J/m)
λ	volumetric efficiency
λ	air factor
λ	thermal conductivity (W/(K m))
Λ	power factor
μ	calorimetric coefficient (J/bar)
μ	Rateau coefficient
μ	viscosity (kg/(s m))
μ_i	molar chemical potential (J/mol)
ω	acentric factor
ω	rotation speed (rd/s)
л	heat created by irreversibilities (J/kg)
л$_i$	internal pressure ratio
л	osmotic pressure (Pa or bar)
θ	Carnot factor
Θ	ratio of the turbine inlet temperature at the compressor inlet temperature
ρ	density (kg/m)
ρ	volumetric compression ratio
τ	specific shaft work (J/kg)
τ	hydrogen utilization rate
τ	optical efficiency
τ	recirculation rate
Ω	coefficient of work
ψ	enthalpy factor

Acronyms

AICVF	Association des Ingénieurs et Techniciens en Climatique, Ventilation et Froid
ASHRAE	American Society of Heating, Refrigerating and Air Conditioning Engineers
BTU	British Thermal Unit
COP	Coefficient of Performance
FDB	Fully Developed Boiling
GUI	Graphical User Interface
IP	Intermediate Pressure
HC	Hydrocarbons
HP	High Pressure
HHV	Higher Heating Value
HVAC	Heating, Ventilation, and Air Conditioning
JANAF	Joint Army Navy Air Force
LL	leaving losses
LMTD	logarithmic mean temperature difference
LNG	Liquefied natural gas
LP	Low Pressure
NPSHA	net positive suction height available
NPSHR	Net Positive Suction Head Required
NTU	Number of Transfer Units
NVG	Net Vapor Generation
NUSCLE	Nuclear Secondary Circuit Lite Emulator
ONB	Onset of Nucleate Boiling
RPDC	Rotary Positive Displacement Compressor
RR	relative roughness
SHF	sensible heat factor
SVG	Scalable Vector Graphics
SHR	sensible heat ratio
TPS	thermodynamic property server
UA	product of the exchanger surface A by the overall heat transfer coefficient U

Conversion Factors

To Obtain	Multiply	By	Reverse Factor
Meter (m)	Inch (in)	2.540000E-02	3.937008E+01
Centimeter (cm)	Inch (in)	2.540000E+00	3.937008E-01
Meter (m)	Foot (ft)	3.048000E-01	3.280840E+00
Square meter (m^2)	Square inch (in^2)	6.451600E-04	1.550003E+03
Square centimeter (cm^2)	Square inch (in^2)	6.451600E+00	1.550003E-01
Cubic meter (m^3)	Cubic foot (ft^3)	2.831685E-02	3.531466E+01
Cubic meter (m^3)	Cubic inch (in^3)	1.638706E-05	6.102376E+04
Meter per second (m/s)	Foot per second (ft/s)	3.048000E-01	3.280840E+00
Meter per second (m/s)	Inch per second (in/s)	2.540000E-02	3.937008E+01
Meter per second (m/s)	Kilometer per hour (km/h)	2.777778E-01	3.600000E+00
Meter per second (m/s)	Mile per hour (mi/h)	4.470400E-01	2.236936E+00
Kilometer per hour (km/h)	Mile per hour (mi/h)	1.609344E+00	6.213712E-01
Radian per second (rad/s)	rpm (revolution per minute) (r/min)	1.047198E-01	9.549292E+00
Gram (g)	Ounce (oz)	2.834952E+01	3.527397E-02
Kilogram (kg)	Pound (lb)	4.535924E-01	2.204622E+00
Kilogram per cubic meter (kg/m^3)	Pound per cubic foot (lb/ft^3)	1.601846E+01	6.242797E-02
Kilogram per cubic meter (kg/m^3)	Pound per cubic inch (lb/in^3)	2.767990E+04	3.612730E-05
Pascal (Pa)	Pound-force per square foot (lbf/ft^2)	4.788026E+01	2.088543E-02
Pascal (Pa)	Pound-force per square inch (psi) (lbf/in^2)	6.894757E+03	1.450377E-04
Pascal (Pa)	Bar (bar)	1.000000E+05	1.000000E-05
Pascal (Pa)	Atmosphere, standard (atm)	1.013250E+05	9.869233E-06
Watt (W)	British thermal unit per hour (Btu/h)	2.930711E-01	3.412141E+00
Watt (W)	British thermal unit per second (Btu/s)	1.055056E+03	9.478170E-04
Watt (W)	Ton of refrigeration (12,000 Btu/h)	3.516853E+03	2.843451E-04
Joule per kilogram kelvin [J/(kg·K)]	British thermal unit per pound degree Fahrenheit [Btu/(lb·°F)]	4.186800E+03	2.388459E-04
Watt per square meter (W/m^2)	British thermal unit per square foot hour [$Btu/(ft^2{\cdot}h)$]	3.154591E+00	3.169983E-01
Watt per square meter (W/m^2)	British thermal unit per square foot second [$Btu/(ft^2{\cdot}s)$]	1.135653E+04	8.805507E-05
Joule per square meter (J/m^2)	British thermal unit per square foot (Btu/ft^2)	1.135653E+04	8.805507E-05

(Continued)

(*Continued*)

To Obtain	Multiply	By	Reverse Factor
Joule per cubic meter (J/m^3)	British thermal unit per cubic foot (Btu/ft^3)	3.725895E+04	2.683919E-05
Joule per kilogram (J/kg)	British thermal unit per pound (Btu/lb)	2.326000E+03	4.299226E-04
Joule per kelvin (J/K)	British thermal unit per degree Fahrenheit (Btu/°F)	1.899101E+03	5.265649E-04
Watt per square meter kelvin [$W/(m^2 \cdot K)$]	British thermal unit per hour square foot degree Fahrenheit [$Btu/(h \cdot ft^2 \cdot °F)$]	5.678263E+00	1.761102E-01
Watt per square meter kelvin [$W/(m^2 \cdot K)$]	British thermal unit per second square foot degree Fahrenheit [$Btu/(s \cdot ft^2 \cdot °F)$]	2.044175E+04	4.891949E-05
Joule (J)	Watt hour (W·h)	3.600000E+03	2.777778E-04
Joule (J)	British thermal unit (Btu)	1.055056E+03	9.478170E-04
Joule (J)	Foot pound-force (ft·lbf)	1.355818E+00	7.375621E-01
Kelvin (K)	Degree Celsius (°C)	T/K = t/°C + 273.15	T/°C = t/K – 273.15
Degree Celsius (°C)	Degree Fahrenheit (°F)	t/°C = (t/°F – 32)/1.8	t/°F = 1.8 t/°F + 32
Kelvin (K)	Degree Fahrenheit (°F)	T/K = (t/°F + 459.67)/1.8	T/°F = 1.8 t/K – 459.67
Kelvin (K)	Degree Rankine (°R)	T/K = (T/°R)/1.8	T/°R = 1.8 T/K
Degree Celsius (°C)	Kelvin (K)	t/°C = T/K – 273.15	T/K = t/°C + 273.15

These values have been obtained from NIST Guide to the SI.[2]

NOTES

1 https://direns.minesparis.psl.eu/Sites/Thopt/en/co/secondEditionEnerSyst.html
2 http://physics.nist.gov/Pubs/SP811/appenB9.html.

Introduction

INTRODUCTION

As indicated by its title, this book addresses energy systems, that is, energy conversion technologies considered as systems based on assemblies of elementary components coupled together.

Over the past 25 years, we have developed a novel approach to teaching thermodynamics applied to energy conversion, now used in more than 120 higher education institutions at both undergraduate and graduate levels (engineering schools, universities) as well as in vocational training.

The change in educational paradigm we have introduced is based on a shift in the knowledge acquired by students. The derivation of equations describing the thermodynamic processes undergone by fluids is drastically reduced, with calculations being performed by a simulator such as Thermoptim (www.thermoptim.org) without learners needing to know the details. They devote most of their time, on the one hand, to learning technologies, and on the other hand, to reflecting on the architecture of both conventional and innovative thermodynamic cycles, graphically building and configuring models of various energy technologies.

The new teaching method has the distinction of being simultaneously much simpler than conventional approaches for introducing the discipline and much more useful for advanced students, who can progress further in their studies thanks to a powerful and open modeling environment. They can work on real-world complex innovative cycles currently being studied in laboratories and companies.

In our previous work, *Energy Systems: A New Approach to Engineering Thermodynamics*, published in 2021, we proposed a learning method based on reducing mathematical formalism without sacrificing scientific rigor, designed to meet the needs of learners with limited backgrounds in mathematics and physics who do not wish to engage in advanced modeling.

Although it limited the theoretical difficulties, the 2021 edition covered a very broad field, including a variety of cycles to which the developments in this book can be applied: conventional as well as innovative power plants, gas turbines, reciprocating engines, Stirling engines, combined cycles, cogeneration, refrigeration cycles, renewable energy conversion, combustion, Generation I–IV nuclear energy conversion, evaporation, desalination, fuel cells, oxycombustion, air conditioning, etc.

In this two-volume set, of which this is the first volume, we complement the simplified approach highlighted in 2021 by enhancing the presentation of component models, particularly regarding their equation formulation.

After a review of thermodynamic fundamentals, this volume delves into the analysis of the behavior of the most commonly used components and explains how customized components can be modeled and coupled with Thermoptim through a mechanism called external classes. It then addresses a particularly challenging

DOI: 10.1201/9781003606017-1

issue: the sizing of components and the study of systems under off-design conditions. Several case studies illustrate how these developments can be applied to simple cycles.

Let us specify that we will only focus on the components that have a significant impact on the thermodynamic performance of the systems studied, as going further would take us too far from our main concerns. Therefore, we will not particularly discuss ejectors, deaerators, or chemical water treatments.

This book provides its readers with advanced methods allowing them to analyze these systems, as well as guidance in selecting equations if they wish to build their own models.

I.1 OBJECTIVES OF THIS BOOK

For beginners, the major advantage of **Thermoptim** is that it helps them model complex energy systems simply, without having to write equations or programs. It relieves its users of many problems, including computational ones, and enables them to conduct analyses that they could not pursue otherwise, especially when starting out. When learning the discipline, it becomes possible to focus on a qualitative approach, with the calculations required for quantitative studies being performed by the software. Note that the name of this tool has a double meaning: it does allow one to optimize thermodynamic systems, but above all it allows one to learn this discipline with optimism[1].

Under these conditions, energy system operation can be studied using a completely new pedagogy, where only a small number of elements must be considered: first, those used to graphically describe the systems studied, and second, those used to configure the simulator components.

When approached this way, it is not necessary, at least initially, to consider the details of the equations that enable calculation of the processes involved: Thermoptim automatically handles this. Learning the discipline is limited to these basic concepts and their implementation in the software. The memorization effort and cognitive load required of beginners are greatly reduced, allowing them to focus their attention on understanding basic phenomenological concepts and their practical implementation.

Thus, Thermoptim permits the calculation of energy systems from simple to complex, without writing a single line of code.

Please note that, despite its emphasis on technological realism, this book is a manual intended for learning the discipline and not an industrial handbook. The same applies to the Thermoptim software, whose results do not match the precision of professional tools used by industrial engineering offices.

However, the ambition of this work extends beyond that: it also aims to provide readers with sufficient information so they can create custom models, either by introducing specific component models not available in the Thermoptim core, which involves showing them how to derive their equations, or by using the software's advanced features.

An additional objective of the first volume is to present a modeling approach for the main components used in energy systems (pumps, compressors, expansion devices, heat exchangers, combustion chambers, etc.) that is detailed enough to account for their key technological characteristics (a dynamic compressor does not behave like a positive displacement compressor).

In summary, we pursue two objectives in this book:

- Establish the constitutive equations representing the operation of key technologies, and provide readers who want to model these themselves with sufficient guidance to do so, including using the external class mechanism that allows them to add their own components or sets of algorithms to Thermoptim for controlling complex projects.
- Provide readers with methodological guidance, simple or advanced, for analyzing the systems they study. In this spirit, significant attention is devoted to exergy methods, which are increasingly regarded as among the best suited to perform optimization studies, as they can account for both the quantity of energy involved and its quality.

Pedagogically, this book and Thermoptim can be used in various contexts, under both a traditional approach (neo-behaviorist or objectivist) and a more recent constructivist approach. We hope it will help colleagues overcome difficulties they may face in their practice.

For proponents of a conventional approach, Thermoptim particularly allows enrichment of the classical presentation through accurate simulations and enables students to study multiple real-world examples. In a constructivist approach, this book and software additionally serve as tools allowing students to work independently to explore a broad field and analyze very open topics:

- For beginners, it provides a structured environment that reduces cognitive load while acquiring the vocabulary and basic concepts encapsulated in the screens. Once this vocabulary is learned, cooperative learning with peers and teachers is strengthened.
- Very quickly, it becomes possible to work on realistic problems rather than oversimplifications (such as conventional internal combustion engines with perfect air as the working fluid). In addition, during internships, students using Thermoptim are fully operational in the companies where they work, which is highly motivating for them and leads them to engage more thoroughly.
- For experienced users, Thermoptim enables them to study very complex systems (for example, Areva Framatome used Thermoptim to optimize combined cycles and cogeneration thermodynamic cycles coupled vérifier with high-temperature nuclear reactors; see Sections 7.1.3 and 7.4.3 of Volume 2).

Readers will understand that we seek above all to make the study of energy systems as accessible as possible, demonstrating very concretely how realistic models can be developed to represent them. This focus has led us to deliberately limit our discussion of scientific issues to key points, especially regarding the fundamentals of thermodynamics and heat transfer. As a result, this book sacrifices generality for ease of use: practical know-how is privileged over theoretical knowledge itself. Readers interested in further developments may refer to documents cited in the bibliography.

Experienced engineers will find a coherent body of theory and practice through which they can become quickly operational, without having to personally engage in solving equations or developing a computer modeling environment.

1.2 CONTENT OF THIS BOOK

The overall structure of this book is as follows:

- A structured approach to modeling energy conversion technologies with Thermoptim is first presented. After reviewing thermodynamic fundamentals, we establish the main equations that enable calculation of the behavior of basic components used in most energy technologies (compressors, turbines, combustion chambers, throttling devices, etc.). Chapter 3 contains the essential elements for understanding how components that come into play in energy conversion technologies can be modeled.
- Chapter 4 addresses the challenging topic of heat exchanger modeling.
- Chapter 5 deals with Thermoptim extensions that have been introduced particularly to conduct component sizing and off-design operation studies, which are developed in Chapters 6 through 9. It explains how to model different components while accounting for their technological specificities.
- Chapter 10 presents a series of case studies illustrating how such studies can be carried out with Thermoptim. This chapter is illustrated by examples of cycles modeled with the software, providing readers with models whose structure or settings they may customize as they wish to perform various simulations.

For the bibliography, we have made a specific choice: we only include references that we have actually used in our writing work, as well as some general books. We have deliberately omitted exhaustive bibliographic lists to avoid overwhelming the reader with a mass of references that are now easily accessible through modern databases.

Many digital resources for teaching energy systems have been gradually collected into a portal called Thermoptim-UNIT (www.thermoptim.org), whose content is freely accessible, with few exceptions, including solutions to some exercises and problems.

In addition, this book's website (https://server.s4e2.com/crc/amtc/) provides access to various supplementary resources, including:

- Links to digital versions of selected figures
- Updated footnote links in case of changes
- Application examples of developments presented in specific chapters, available as guided explorations or small Thermoptim projects with their associated files
- Theoretical and methodological supplements, either on the portal or dedicated pages

Since the complete list of resources was not finalized at the time of publication, please refer to the website during your reading to discover all available materials. Some resources will be reserved for registered book purchasers.

I.3 CONCLUSION

This work introduces an innovative pedagogical and methodological approach for the modeling and analysis of energy systems using the Thermoptim simulator.

By replacing heavy mathematical formalism with graphical modeling and interactive simulation, it allows learners to focus on understanding technologies and system architectures. Designed for both beginners and advanced users, it establishes a bridge between theory and practice by providing a unified framework for the study of real-world energy conversion technologies.

This book combines fundamental thermodynamics, component modeling, and system optimization within a constructivist learning environment that fosters autonomy, realism, and critical analysis.

NOTE

1 A demo version of Thermoptim is freely available (https://server.s4e2.com/jml/index.php/en/thermoptim-en/download), allowing users to view existing projects and diagrams and to create small ones in order to become familiar with the tools. Naturally, it does not provide access to all software features. In particular, it does not allow saving projects, diagrams, or output result files, although it does enable the loading of projects of any size. The 2.82 demo version is sufficient to work with all the examples presented in this book.

1 Presentation of the Approach

INTRODUCTION

The approach proposed in this book is based on the observation that thermodynamics is much simpler in qualitative than in quantitative terms.

To explain it, we first briefly introduce the operation of a gas turbine, an engine whose principle is among the simplest. This will allow us to address the core concepts by introducing fundamental elements needed for studying energy technologies: the fluids involved, the processes they undergo and the corresponding components, and finally assemblies of these components. It will demonstrate the relevance of a twofold methodological approach.

We then show that this methodology involves two complementary approaches: the analytical approach used to represent each component, and the systems approach for defining the internal architecture of the system.

The subsequent sections explain how the proposed methodology can be implemented and present the Thermoptim primitive types that enable modeling of energy technologies according to this framework.

1.1 A TWO-LEVEL METHODOLOGY

1.1.1 Physical Phenomena Taking Place in a Gas Turbine

The gas turbine, also called combustion turbine, is a heat engine that is currently very popular, given its excellent performance (efficiency greater than 35% when used alone, and 55% or more in combined cycle).

This machine is composed of three elements (Figure 1.1.1): a compressor, a combustion chamber, and a turbine.

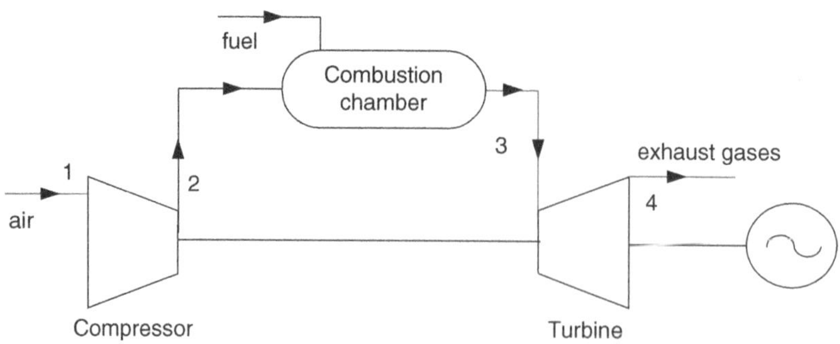

FIGURE 1.1.1 Sketch of a gas turbine.

DOI: 10.1201/9781003606017-2

This very simple example is not representative, by far, of the operation of all internal combustion engines. It can, however, provide a first indication of the complexity of the phenomena that take place, and therefore the knowledge necessary to calculate them:

- The various components of the machine are coupled. Their couplings arise on the one hand from the fluid passing through them, and on the other hand from the mechanical connection between the compressor and turbine shafts.
- Working fluids are gas mixtures: initially air and fuel, and then combustion gases. In each case, they can be considered as ideal gases, whose thermodynamic energy properties depend only on temperature.
- In other power plants such as steam power plants, the working fluid alternates between liquid and vapor states. The ideal gas model is no longer sufficient and must be replaced by much more complex real fluid models, with energy properties depending on both pressure and temperature.
- The compression and expansion phases are of crucial importance in the engine's operation, because energy conversions between the fluid and the drive shaft occur during these phases.
- The combustion reaction is highly complex, still imperfectly understood, but can nevertheless be approached by different methods that calculate the energies involved and provide insight into the origin of pollutants.
- Finally, the working fluids pass through the various engine components, and understanding their flow conditions requires advanced concepts in fluid mechanics (which are not discussed in this book).

In summary, the study of an energy technology such as a gas turbine faces a dual challenge:

- Fluids that pass through its various components follow relatively complex behavior laws and undergo fundamental processes whose analysis can be difficult (non-linear laws, combustion, etc.).
- Components are coupled together, so they cannot be calculated indepen dently.

1.1.2 Energy Technologies: Component Assemblies

Like the gas turbine, energy technologies are assemblies of components through which thermodynamic working fluids flow and undergo transformations of varying complexity.

To analyze these systems, a classical thermodynamic schematic diagram is generally used, sometimes referred to as a process flowsheet, representing the various components and their interconnections. This diagram is often close to the actual physical schematic, but for a thorough thermodynamic analysis, it is preferable to reason in terms of **functional structure** rather than being limited to the physical arrangement, as the same set of mechanical parts can serve several successive functions. For example, in a diesel engine, the assembly of the cylinder, cylinder head, and piston is first used to compress the incoming air, then to ensure combustion, and finally to expand the combustion gases while producing work.

Note, however, that in the example shown, the connections between components are very simple: the outlet of the compressor is the inlet of the combustion chamber, and the outlet of the latter is the inlet to the turbine. This simplicity of the interconnection network is general and suggests adopting a dual approach in the study of energy technologies, separating on the one hand the description of the interconnection network between components, and on the other hand the analysis of their internal behavior. This procedure has many advantages as discussed later in this book.

The approach proposed in this book is thus based on the observation that thermodynamics is much simpler in qualitative than in quantitative terms.

Since energy technologies are assemblies of components through which thermodynamic fluids flow and undergo various processes, we greatly simplify matters if we adopt a dual approach, beginning by separating the representation of the overall system, usually quite simple, from the study of its components taken individually.

The overall representation is very useful at the qualitative level: it is visual and helps us understand the role played by each component in the complete system. When learning the subject, it is essential to fully assimilate the design principles of these technologies. Once one has in mind the internal structure of an engine or refrigerator, the study of the behavior of one of its components is made easier because we understand its place in the whole and how it contributes to the overall performance.

If one has a proper graphical environment such as the Thermoptim diagram editor (Figure 1.1.2), the internal structure of the system can be described very easily. Figure 1.1.2 shows all the constituent elements of the gas turbine: the three connected components (compressor, combustion chamber, and turbine) and the fluid inlets and outlets, represented by small rectangles topped with a dot, called process points.

This provides a qualitative representation, very meaningful for the engineer, which can be quantified by setting the thermodynamic properties of the different components and then calculating them. This qualitative representation has furthermore the distinction of being largely independent of the assumptions adopted for calculating the various components: it is an invariant of the system. Thus, a given system architecture can be calculated with various component models.

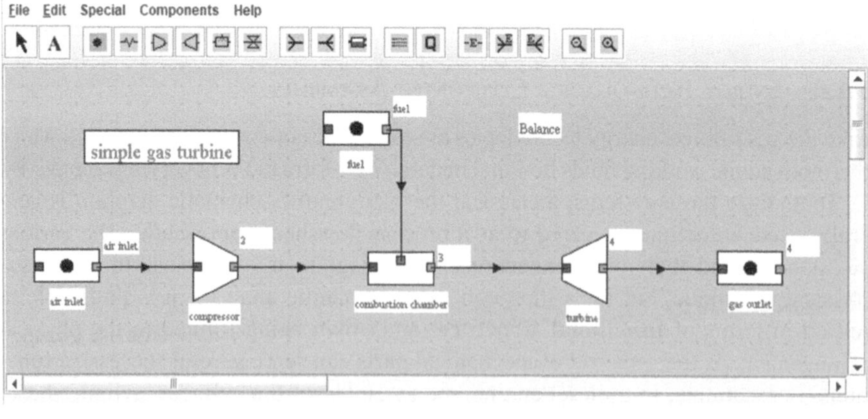

FIGURE 1.1.2 Gas turbine in Thermoptim diagram editor.

This dual modeling approach involves two methods often presented as mutually exclusive but which are in this case complementary and mutually enriching:

- The analytical method used to represent each component by a number of characteristic parameters, coupling variables, and a set of appropriate equations.
- The systems perspective regarding the definition of the internal architecture of the system (choice of components, description of their relationships).

Not only does such a framework significantly simplify the modeling process and facilitate subsequent use and maintenance of the model, but most importantly it secures its construction by automating the establishment of linkages between the various component elements and ensuring consistency. This point is especially important when the system under study comprises a large number of components.

This methodology is moreover the only way to focus efforts on innovative cycle studies, which is now a research field both fascinating and essential for the future.

The numerous innovative ideas proposed in recent years—such as the design of CO_2-free cycles—demonstrate the vitality of this field of research, which plays a crucial role in the education of future engineers.

1.2 PRACTICAL IMPLEMENTATION OF THE DOUBLE ANALYTICAL-SYSTEMS APPROACH

Incidentally, we must use the term "model" in two different senses, which the context will usually distinguish without difficulty: component models are used to calculate or analyze the internal behavior of a component, while system models take the form of an assembly of component models.

The originality of this work lies in its ability to make this advanced modeling accessible, thanks to a rigorous systems approach supported by powerful methodological tools.

It is notably based on the Thermoptim software, designed to faithfully and pedagogically represent complex energy systems. This book is therefore aimed at a broad audience.

1.2.1 Use of the Thermoptim Software

The Thermoptim core was built to represent the most common components: compressors, turbines, combustion chambers, expansion valves, and heat exchangers. Since their thermodynamic models are almost universally accepted, it was possible to program them once and for all.

The software includes a second family of components, which take the form of external Java classes. These components (e.g., solar collectors, cooling coils, cooling towers, fuel cells, chemical reactors, etc.) are easily customizable by users, who thus have the opportunity to define themselves the phenomenological and behavior models they wish to use.

Among these classes, we should mention the possibility of creating controller modules that coordinate the sequence of calculations performed by Thermoptim. These controllers are particularly used to solve coupled equation sets, especially using an optimization library called minPack. We will see several examples of this in the case studies.

To implement the dual analytical/systems approach that we advocate, we must operate in two corresponding stages:

- Begin by having a set of component models, either by using existing ones or by developing new ones.
- Assemble them to create the model of the overall system.

The first step is essential and constitutes a prerequisite for implementing the systems approach. For the user, the situation differs depending on whether a sufficiently realistic model is already available in the Thermoptim core or among the external classes:

- When all the required components are already available, the modeling task mainly concerns the systems phase—assembling pre-built component models and configuring them as needed.
- When some component models must be developed, the external class development environment allows users to create these models and integrate them into the software.

Once the component models are available, building the model of a thermodynamic system with Thermoptim is very simple:

- We first make a qualitative description by representing it graphically as a set of components connected by links representing the fluid pipes or heat exchangers.
- The defined model is then quantified by configuring the various primitive types involved, in order to calculate its performance, making use of predefined templates.

The diagram editor facilitates the qualitative stage: in a first step, the user provides only the minimum information necessary for the logical definition of the project they build (implicitly the types of components they select and explicitly their name and exit point, the substance associated with them, as well as the value of the flow rate through them). Then, when they interconnect these components, some of this information is automatically propagated from upstream to downstream (for example, the inlet point of the downstream component becomes the same as the outlet point of the upstream one). This small amount of basic information provided to the graphical modeler corresponds to the systems description of the technology studied.

When this step is complete, it becomes possible to transfer the diagram components to the simulator to create the required primitive types, with a default configuration of their thermodynamic properties. Quantification of the desired model can then be made by refining this configuration, with each element of the simulator being easily displayed by double-clicking either on the corresponding component in the diagram editor or on a table row of the main simulator project screen.

Once the element parameters are configured and calculated, the results can be directly displayed in the diagram editor, which thus becomes a true synopsis of the installation, or plotted as a cycle in an interactive thermodynamic diagram.

1.2.2 Functional and Exergy Structures

Once this work has been completed, the *functional structure* of the system under study is established, providing a basis for calculating its overall performance.

However, this structure does not make it possible to assess irreversibilities. To do so, an *exergy structure* can be defined, which analyzes the distribution and evolution of exergy flows within the system.

The exergy structure describes how exergy is supplied, transformed, recycled, and lost. It provides two particularly useful tools for guiding system optimization:

- **The exergy diagram**, which graphically illustrates the inputs, conversions, recirculations, and losses of exergy.
- **The exergy balance**, which is a table showing the exergy resources and products associated with each functional component, enabling the calculation of exergy efficiency and the magnitude of irreversibilities, both in absolute terms and as percentages.

The functional and exergy structures thus offer two complementary conceptual perspectives, while their respective diagrams serve as their graphical representation.[1]

These two structures are dual: the functional structure describes the physical organization and the processes, while the exergy structure evaluates exergy efficiency and irreversibilities, offering a complementary perspective focused on energy quality.

1.3 THERMOPTIM PRIMITIVE TYPES

When analyzing the most common energy technologies, we observe that the number of elements of which they are composed is relatively limited, and that in most cases each component for a given phase of operation may be called mono-functional, as it exchanges energy according to a preferred mode, either as purely thermal (heat transfer) or by converting mechanical energy into thermal energy or vice versa.

Consequently, the functions that the various components are required to perform can be grouped into relatively few categories and calculated independently from each other, forming what one might call the primitive types basis which constitutes the core of Thermoptim. It is sufficient to represent a large number of energy technologies, including those used in most conventional cycles. As we have already indicated, and as we will have the opportunity to demonstrate repeatedly in this work, the external class mechanism allows us to supplement this basis by adding other components to the software, particularly multi-functional ones.

The case of heat exchangers is somewhat unusual because it is clear that the calculations of the cooling of one fluid and the heating of the other must be conducted in a coupled manner. Special treatment will be reserved for this as discussed below.

1.3.1 Component Modeling

The second part of this book is mainly devoted to the presentation of component models, which we have grouped into three types:

- Phenomenological models constructed on the basis of an analysis of the physical functioning of the component. Their parameters then have physical significance. Their goal is the study of thermodynamic cycles.
- Empirical or behavior models, often of the black-box type, which have the advantage of representing the overall input/output component behavior, and can especially be used to study its operation in off-design mode. Their parameters usually have no physical significance.
- Component sizing models which provide access to detailed internal design of the components. The more accurate of these models are specialized tools that take into account both theoretical aspects (mechanics, thermodynamics, fluid mechanics, and heat transfer), and technological (manufacturing processes, surface treatments, and material selection), economic (costs), or environmental constraints. They are sophisticated multidisciplinary tools that typically equip only manufacturers.

A compromise must be found between the desired accuracy and the complexity (and cost) of modeling. Given the objective of this book, we limit ourselves to a steady-state representation of phenomena, and we do not generally seek to obtain detailed models of how a particular component operates, which can be extremely complex. Only in Chapters 6–10 do we provide some guidance on how to approach the component sizing and off-design study of systems.

As we shall see, the models we use are often mixed, being partly phenomenological and partly empirical.

For example, the study of heat exchangers (see Chapter 4) shows that a model of type (ε, NTU) is quite sufficient to study the insertion of such a component in a system, but it only provides access to the product UA of the heat transfer coefficient U and the exchange area A: it is therefore an incomplete phenomenological model. If it is associated with an empirical law giving the evolution of the value of UA as a function of the two fluid flows through the exchanger, it can be used as a model for analyzing the response of a specific exchanger to changes in its operating conditions. It can also become a component sizing model if it is completed by the set of functions for determining the value of U, as explained in Chapter 7.

Some additional definitions relating to models deserve to be made. The models we are interested in involve four different concepts:

- Parameters characterizing the component: some are purely geometric, such as a cross-sectional flow area or an exchange area, while others have thermodynamic significance, such as an isentropic efficiency. The thermodynamic properties of fluids are a special category of parameters.
- Input variables represent the constraints imposed on the component, such as flow rate, temperature, or pressure. These are exogenous, determined outside the component.
- Internal variables are used to determine the component state. These will often be state variables in the thermodynamic sense of the term.
- Output variables are the model results, such as temperatures, compositions, etc.

To study the insertion of a component in a system, it is usually not necessary to know the details of its internal operation. In most cases, and this is the case with most technologies that interest us here, a phenomenological model may represent a given component by a small number of characteristics and its coupling variables (inputs and outputs). These characteristics depend on the type of component and the technology.

For example, a positive displacement compressor will be characterized by a volumetric compression ratio, a volumetric efficiency, and an isentropic or polytropic efficiency, while a heat exchanger will be characterized by a number of transfer units. The coupling variables are pressures, temperatures, flow rates, etc.

We will focus on understanding the overall performance of components, which is sufficient for our purpose, committing ourselves to identifying expressions of characteristics representative of their operation. By limiting our discussion to models that are as simple as possible while remaining thermodynamically relevant, we will not achieve completeness. Readers interested in further developments will find them in the literature.

1.3.2 Thermoptim Primitive Types

The list of functional elements that may appear in the main energy conversion technologies corresponds to the concepts implemented in Thermoptim, which are detailed below.

The following list is a basis of primitives sufficient for studying a large number of energy technologies:

- You must first be able to represent the intensive properties of the fluids used and calculate their state for various values of pressure, temperature, etc. The concepts of *substances* and *points* allow us to do this.
- These fluids undergo processes that can be grouped into several major categories, the most common being: compression, expansion, throttling or flashes, combustion, and heat exchanges. It is in processes that the mass flow rates involved are specified, and therefore where extensive properties can be calculated.
- Fluids pass through components, forming more or less complex networks that we must be able to describe. The processes above are a portion of these circuits.

To complement these, we must make use of *nodes* (dividers, mixers, and separators).
- When two fluids exchange heat together, they form *heat exchangers*, components coupled with two processes that cannot be calculated separately.

1.3.2.1 Substance Thermodynamic Properties

The representation of thermodynamic properties of substances is of course a necessity. It requires, first, adequate fluid models and, second, data to represent the fluids used.

Every substance exists in at least one of the three phases: solid, liquid, or gaseous. When the pressure is sufficiently low and the temperature sufficiently high, it is reasonable to consider that the substance behaves like an ideal gas whose heat capacity, internal energy, and enthalpy depend only on temperature (not on pressure).

Thermoptim includes numerous data needed to calculate the thermophysical state of fluids, which can be either pure substances or gas mixtures. Their composition can change due to either mixing or combustion reactions.

It defines three categories of substances: pure ideal gases, compound ideal gases, and condensable vapors or real fluids (without mixtures). The perfect gas corresponds to the particular case of ideal gases whose heat capacity is temperature independent.[2]

1.3.2.2 States of Fluids: Points

Once you have a representation of substance properties, it becomes possible to calculate the state of a fluid as a function of relevant variables, such as pressure, temperature, enthalpy, etc.

For this, Thermoptim defines points. A point represents a substance particle and determines its intensive state variables: pressure, temperature, heat capacity, enthalpy, entropy, internal energy, exergy, quality, absolute or relative humidity, etc. A point is identified by a name and the name of the substance associated with it.

To calculate it, you must:

- Either enter at least two of its state variables, typically pressure and temperature for open systems, or volume and temperature for closed systems.
- Or automatically determine it by using one of the processes defined below.

1.3.2.3 Processes

In thermodynamics, a process is defined as the evolution of a system or substance from one equilibrium state to another. A process therefore associates two points as defined previously, an inlet point and an outlet point. In addition, it indicates the mass flow rate involved and thus allows one to calculate extensive state variables and notably to determine the variation of energy involved during the process.

The most common processes have been modeled and are directly accessible. Knowing the state of the fluid at the process inlet, Thermoptim can solve either the direct problem or the reverse problem. In the first case, knowing the characteristics of the process, it computes the state at the end of evolution and the energies involved, and updates the outlet point. In the second case, it identifies the parameters of the selected process that allow it to reach the defined outlet point state.

FIGURE 1.3.1 Nodes.

Processes can be of several types: compression, expansion, combustion, throttling or flash (pressure drop without work production, as in a valve or a filter), heat exchange, and water vapor/gas mixtures (the latter includes six different categories of evolutions). They are presented in detail in Chapter 3.

In each case, various characteristics of the process must be specified, for example, in compression, its isentropic or polytropic efficiency, as well as the type of calculation required (open or closed system, direct or inverse calculation, etc.).

A cycle can thus be described as a set of points connected by processes. To the extent that the mass flow rate of fluid is the same in all the evolutions, processes and points are sufficient. If this is not the case, it may be necessary to describe at least partially the network of involved fluids using the nodes defined below.

1.3.2.4 Nodes

Nodes can describe the network elements that constitute mixture, division, or separation of fluids. In a node, several fluid branches are interconnected to form a single stream (Figure 1.3.1).

If this is a mixer, the various branches join to form a single stream. The mass flow rate of the main stream is equal to the sum of those of the branches, and the enthalpy balance is used to calculate the enthalpy and temperature of the mixture.

If it is a divider, the main stream is divided into several branches for which the flow rates must be specified. Temperature and enthalpy are conserved.

The logical definition of a node is by (1 – n) association of processes: one process is the main stream, and n processes correspond to the branches. The processes themselves being connected to points, and the latter to substances, updates of the fluid state are made automatically.

A third type of node also exists: the separator, which is a somewhat special divider that receives as input a vapor in two-phase state and separates the liquid and gas phases.

1.3.2.5 Heat Exchangers

Heat exchangers are components that combine two streams, one that warms up and the other that cools, whose evolutions are coupled and cannot be computed independently. The simplest definition of a heat exchanger would require indicating which two processes it couples.

Thermoptim can design a heat exchanger, that is to say calculate the value of the UA product of its exchange surface A by its heat exchange coefficient U, if you indicate what constraints are imposed on the flow rates and temperatures (for instance, minimum pinch, set effectiveness value).

The software can also, once the design is completed (i.e., for a given value of the UA product), calculate, for off-design operating conditions of the flow rates or inlet temperatures, what the exchanger behavior is, that is to say what the outlet temperature values are.

1.3.2.6 Components of the Diagram Editor

Thermoptim is equipped with a diagram editor (Figure 1.1.2) that can qualitatively describe the system studied. It has a palette with the various available components (heat exchangers, compressors, turbines, combustion chambers, mixers, dividers, etc.), and a working panel where these components are placed and interconnected by links. This environment offers particularly interesting user-friendliness for viewing and controlling the connections of large systems.

If we note, first, that a point is always associated with a process and a substance with a point, and, second, that a heat exchanger in fact defines a coupling between two exchange-type processes, we see that the description of a system can be made solely from processes, nodes, and their interactions.

It is this property that is used in the diagram editor, whose palette includes only processes and nodes, which are then connected using two types of links: oriented links representing a fluid line, i.e., essentially a point and a flow rate, and non-oriented links symbolizing exchanger connections.

The great usefulness of the diagram editor is that you can build a visual representation of the systems studied from a very small number of elements, which need only be understood and known in order to model many energy technologies. Once this description is made, you simply have to appropriately configure the Thermoptim primitive types involved. The modeling work is thus divided into two distinct and complementary phases, the qualitative and quantitative. The tight coupling between the diagram editor and Thermoptim simulator makes it easy to set the parameters and quantify the model.

When the elements are configured and their calculation is performed, the results can be directly displayed in the diagram editor (Figure 1.3.2), which behaves like a synoptic view of the system, and represented as a cycle in the interactive charts (Figure 1.3.3).

1.3.2.7 Projects

The set of primitive types for describing an energy technology (substances, points, processes, nodes, heat exchangers) is known as a project in the package. The structure of Thermoptim can be summarized by the graph in Figure 1.3.4. A project can be easily manipulated through appropriate interfaces. All results can be exported to a text file and can be post-processed by a word processor or a spreadsheet.

The complete set of equations corresponding to the project can be generated automatically, allowing them to be solved in other environments if the user wishes (Figure 1.3.5).

To facilitate archiving and reuse of different projects generated, a number of rules and project management tools have been defined. In particular, a project can receive a name and description to characterize it succinctly. Then, as it is clear that the same diagram can correspond to several projects, each with a different setting, and that a

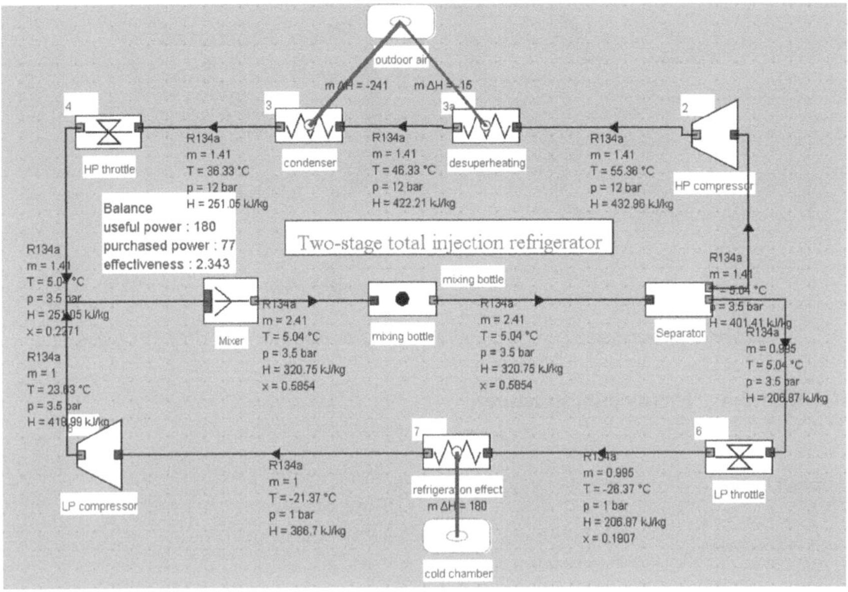

FIGURE 1.3.2 Viewing cycle point states.

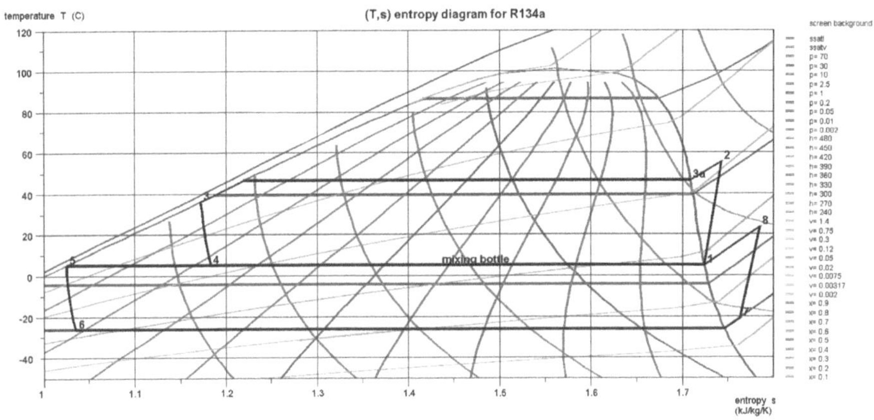

FIGURE 1.3.3 Two-stage R134a refrigeration cycle in the entropy chart.

project cannot in principle have several diagrams, it is possible to associate a diagram name with a project.

In this way, when a diagram is displayed in the graphical modeler, you can search among the list of projects associated with it and display it in a private library.

On this page of this book's website, you will find some guidance on how to work with Thermoptim by opening existing work files.[3]

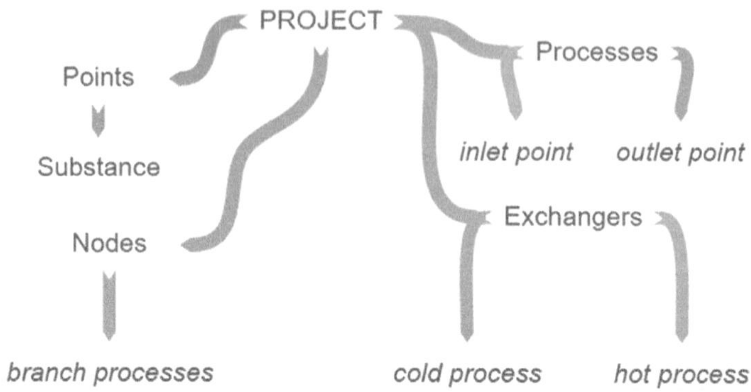

FIGURE 1.3.4 Thermoptim structure.

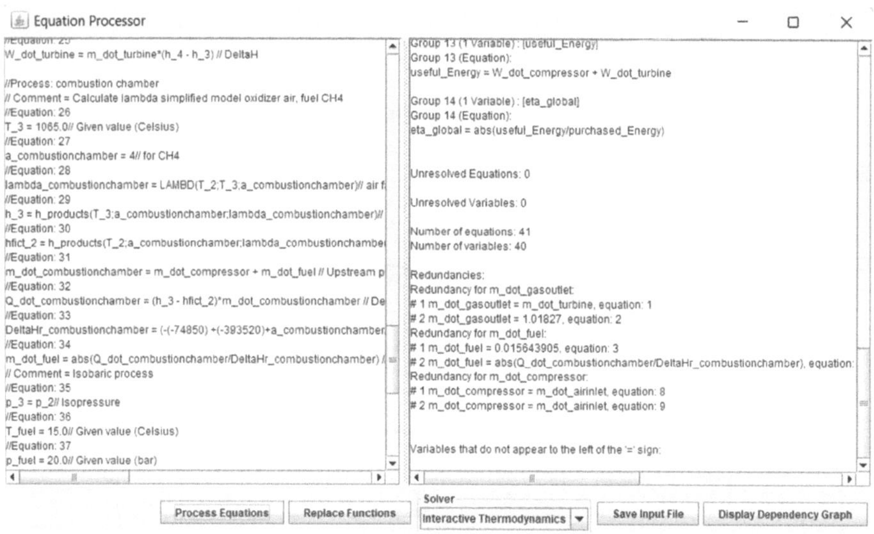

FIGURE 1.3.5 Gas turbine equations generated by Thermoptim.

1.3.3 Thermoptim Assets

The preceding sections allowed us to introduce the main Thermoptim primitive types.

This very small number of elements is sufficient to model a wide range of energy technologies. It is very remarkable that a dozen widgets (including links) are sufficient to represent so many. The modeling of these technologies, which is generally considered difficult, can be greatly simplified when you have a tool like Thermoptim, with the diagram editor allowing one to obtain a qualitative description of the model that is particularly meaningful for an engineer and that is then relatively easy to quantify.

Once a cycle model is built in Thermoptim, it is possible to visualize the state of the various points in many thermodynamic diagrams: (T,s) entropy, (h,p) enthalpy, (h,s) Mollier, or exergy.

For a deeper understanding of these topics, two notes expand on the pedagogical use of Thermoptim: the first elaborates on the new educational paradigm it enables[4], while the second addresses its theoretical foundations[5].

Thermoptim appears as an applied thermodynamics package whose main assets are:

- It provides a coherent analytical and systems modeling environment. For a relatively small initial investment in understanding the underlying logic, itself quite natural and intuitive, it enables a rigorous analysis process leading to significant productivity gains. The models developed can easily be documented, archived, and reused or modified.
- Its primitive type basis is an analytical grid for energy technologies allowing one to highlight the component interrelationship graph. With a little experience, the modeler obtains a representation that is both faithful to reality and calculable by the software. This step is greatly facilitated by the graphical editor.
- Since fluid properties are calculated automatically, the modeler can obtain very accurate results while being relieved of many computational problems that would otherwise be avoided only by resorting to oversimplified assumptions.
- The description of the problem is made using natural concepts of the engineer or physicist, with primitive types having very clear physical meaning. The modeler can concentrate on the physical analysis of phenomena, the mathematical translation, and numerical and computer-generated models being provided by the tool in a transparent manner.
- Modelers do not write a single line of code or solve an equation themselves, Thermoptim processing the project description submitted to it. Users are thus guaranteed to obtain realistic results quickly without the risk of programming errors. If they want to make calculations not provided in the tool, they can process (e.g., with macros) the text file of project results.
- However, once the model is built, Thermoptim is able to generate all the corresponding equations, which can be exported to be solved in other environments.
- Cycles can be easily visualized in the interactive thermodynamic diagrams.
- Thermoptim includes two powerful tools designed to facilitate the optimization of complex energy systems: the exergy structure editor and the pinch method.
- The package is an open environment, written in Java. If necessary, the basis of primitive types can be extended to represent additional components.

1.4 CONCLUSION

This chapter introduced a comprehensive two-level methodology for the analysis of energy systems, based on the observation that thermodynamics is simpler qualitatively than quantitatively.

The approach distinguishes between analytical methods for individual component representation and a systemic perspective for the overall architecture. The functional structure describes the physical arrangement, while the exergy structure evaluates the energy quality.

The implementation relies on Thermoptim, combining a diagram editor for qualitative description and a simulator for quantitative analysis. The software utilizes a basis of primitive types (substances, points, processes, nodes, heat exchangers).

This dual analytical-systemic approach significantly simplifies the modeling process while ensuring consistency, allowing focus on the development of innovative cycles essential to future energy challenges.

You will find on this book's website[6] links to additional information about Thermoptim.

NOTES

1 It should be noted that exergy structures are only available in versions 2.6 and 2.8 of the software.
2 In this book, we distinguish between an **ideal gas**, which obeys the equation of state PV = nRT, and a **perfect gas**, defined as an ideal gas with a constant specific heat. It should be noted that in much of the English-language literature, particularly in North America, the term "ideal gas" is often used for both concepts.
3 https://server.s4e2.com/crc/amtc/resources/first-steps-thopt/.
4 https://www.s4e2.com/download/peda/PedagogicalParadigmEquations.pdf.
5 https://www.s4e2.com/download/peda/PedagogieSystemesEnergetiquesThermoptim-en.pdf.
6 Available at: https://server.s4e2.com/crc/amtc/index.html.

BIBLIOGRAPHY

Gicquel, R. (2021). *Energy systems: A new approach to engineering thermodynamics,* 2nd ed. CRC Press. ISBN 978-0367545129.

Gicquel, R. (2022). *Thermoptim reference manual*, Volume 1. Mines ParisTech, English ed., Available online: https://direns.minesparis.psl.eu/Sites/Thopt/en/res/refThopt1US.pdf

2 Thermodynamics Fundamentals

INTRODUCTION

This chapter presents the fundamental thermodynamics knowledge required to study energy technologies.

One ambition of this book, as previously stated, is to propose a much simpler approach to learning applied thermodynamics than traditional teaching methods. This emphasis on simplicity has led us to limit our presentation of thermodynamics to only the foundations required for our purpose.

After demonstrating how to express energy exchanges during a process, a review of the first two laws of thermodynamics provides the theoretical basis for conducting most subsequent calculations. The concept of exergy is then introduced.

These sections provide definitions of many concepts that are essential for calculating energy technologies, even when using Thermoptim. Examples include state functions such as internal energy, enthalpy, and entropy (Sections 2.3 and 2.4), the properties of matter, and the various diagrams used to represent fluids (Section 2.6).

2.1 BASIC CONCEPTS AND DEFINITIONS

Thermodynamics is the science that deals with energy and its transformations between different forms—mechanical, thermal, chemical, and electrical energy.

Presenting thermodynamics in a general way is particularly challenging, since the set of definitions, postulates, and demonstrations must apply to all these forms of energy.

However, for the energy systems discussed in this book, we can in practice limit ourselves to the first two forms—mechanics and heat transfer—except for moist mixtures and combustion reactions.

Accordingly, except when calculating moist mixtures and combustion (each receiving special treatment), we always assume that the substances involved have invariant composition. We therefore do not include chemical variables, which significantly simplifies the formalism. Moreover, the systems we study are always non-electrified.

This approach, which is common practice in power engineering, will occasionally lead us to deviate from strict thermodynamic formalism. What we lose in generality will be offset by gains in ease of use.

At this stage, it is essential to define or recall several fundamental concepts needed later: thermodynamic systems, variables and equations of state, phase equilibrium, and reversible and irreversible processes. These concepts are necessary to compute the evolutions experienced by thermodynamic fluids. It is indeed necessary to introduce variables and functions connected by equations that can only be determined if

DOI: 10.1201/9781003606017-3

the environment is considered continuous and homogeneous—that is, if these functions are sufficiently regular in the mathematical sense. In particular, we focus only on macroscopic quantities.

2.1.1 Open and Closed Systems

A thermodynamic system is a quantity of matter isolated from its surroundings by a real or imaginary boundary. This system is called closed if it does not exchange matter with the outside through its boundaries; otherwise, it is called open. Note that this notion of system differs from the one introduced earlier to characterize an energy technology as an assembly of components. This vocabulary difficulty is unavoidable given common practice, but the context generally makes it easy to distinguish between the two meanings of the term.

Beginners are often confused by the distinction between closed and open systems, the latter representing a new concept for them, since undergraduate instruction typically considers only closed systems (to avoid accounting for matter exchange at boundaries).

In a diesel or gasoline engine, the valves are closed during compression, combustion, and expansion, isolating the mass of gas between the piston, jacket, and head from the outside. The processes occurring inside the engine must be calculated as closed systems. In a combustion gas turbine, the compressor, combustor, and turbine are driven by a continuous flow of gas. Matter is transferred at the entrance and exit of each component. The changes must then be calculated as open systems.

There is a certain paradox in that calculations are generally easier for open systems than for closed systems, although the former involve matter exchange with the outside. The reason is that pressure is usually given in open system calculations, whereas it depends on many factors in a closed system.

For example, combustion in a gas turbine (open system) occurs at constant pressure (neglecting small pressure drops), whereas in a diesel or gasoline engine (closed system), the pressure varies greatly during the process. Therefore, if one wants to remain realistic, it is preferable to decompose the process into several stages: for example, first at constant volume, then constant pressure, and finally constant temperature.

To further complicate matters, it is natural (but incorrect) to confuse the concepts of open and closed systems in the thermodynamic sense just introduced with their meaning when considering the entire cycle traversed by fluids in a thermal machine.

For example, a refrigerator compression cycle is a closed cycle (Figure 2.1.1)—otherwise, the refrigerant would be lost to the atmosphere, which would be both costly and environmentally harmful—while each of its components taken separately (except possibly for displacement-type compressors) is an open system through which a steady-state flow of refrigerant passes.

The machine operates as a closed cycle involving several components, each working as an open system.

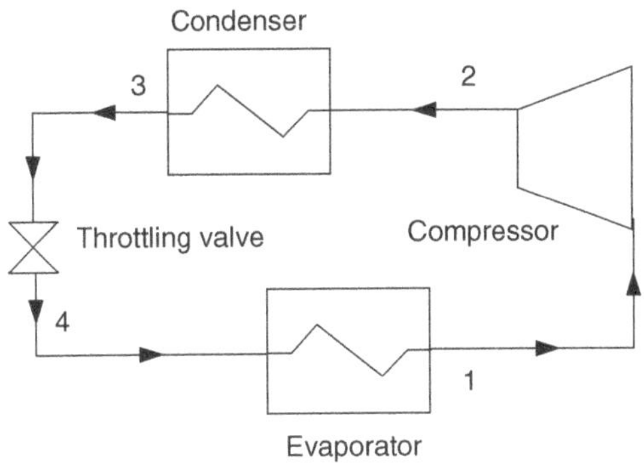

FIGURE 2.1.1 Compression refrigeration cycle.

2.1.2 State of a System, Intensive and Extensive Quantities

The notion of a system's state represents "the minimum information necessary to determine its future behavior." State variables (temperature, pressure, etc.) are all physical quantities (or thermodynamic properties) necessary and sufficient to fully characterize a system at a given moment. Several sets typically meet this definition. We will see later that for a single-phase unit mass, two quantities suffice to determine all others. These results ensure the existence of equations relating each state variable to two independent quantities: $v = f(P, T)$. Such relationships are called equations of state, which are fundamental in practice.

Depending on the problem, the following pairs are generally used: (pressure, volume), (pressure, temperature), (temperature, volume). A state function is a quantity whose value depends only on the state of the system, not its history.

A physical model involves variables representing the state of a system, which is *a priori* a function of time and position. These quantities can be grouped into two broad classes:

- **Intensive quantities** such as pressure, temperature, or specific enthalpy, which are independent of the amount of matter considered.
- **Extensive quantities** such as mass, enthalpy, or entropy, which depend on the mass of the system.

An intensive quantity relates the condition at a point in the medium to a reference condition at another point or in another medium. For example, temperature is defined relative to zero or to the triple point of water.

An extensive quantity is additive: if a system comprises several phases, the extensive quantities characterizing it equal the sum of those of its component phases. Mass and total enthalpy are extensive quantities.

Note that multiplying the phase mass by certain intensive quantities such as specific enthalpy or entropy yields extensive quantities such as enthalpy or entropy.

Recall that a control volume is a region of space with real or imaginary boundaries, thus defining the system of interest (thermodynamic in this case). An example of a control volume is given in Figure 2.3.1. It is bounded by the walls of the fixed and mobile machine M and two immaterial geometric surfaces named A_1 and A_2 at the initial instant t_0, and B_1 and B_2 at t_0+dt.

In the example in Figure 2.3.1, the control volume is considered a closed system, which is followed in its movement over time, from A1 to B1 and A2 to B2. In other cases, control volumes can define open systems. The masses they isolate may then vary depending on mass flow rates crossing their boundaries.

In general, the component models we use are established by writing the laws of continuity and conservation of extensive variables: mass, energy, entropy, etc., with balance sheets established for well-chosen control volumes.

Conservation laws are written in the general form:

$$\left\{\begin{array}{c}\text{accumulation}\\\text{in the control}\\\text{volume}\end{array}\right\}=\left\{\begin{array}{c}\text{inward transport}\\\text{by the}\\\text{surface}\end{array}\right\}-\left\{\begin{array}{c}\text{outward transport}\\\text{by the}\\\text{surface}\end{array}\right\}$$
$$+\left\{\begin{array}{c}\text{transfer}\\\text{through}\\\text{the surface}\end{array}\right\}+\left\{\begin{array}{c}\text{generation}\\\text{in the}\\\text{volume}\end{array}\right\}-\left\{\begin{array}{c}\text{consumption}\\\text{in the}\\\text{volume}\end{array}\right\} \tag{2.1.1}$$

For conservative quantities, such as mass, internal energy, and enthalpy, the last two terms in the equation vanish. For non-conservative quantities, such as entropy or the number of moles of chemical species in reaction, it may be necessary to introduce one or the other.

Building a model means translating that expression into a set of semantic equations and specifying the initial and boundary conditions.

2.1.3 Phase, Pure Substances, Mixtures

A phase is a continuous medium having the following three properties:

- It is homogeneous (implying uniform temperature).
- Velocity at each point is zero in a suitable reference frame.
- It is subject to no external force at a distance (uniform pressure).

As is well known, matter exists in three phases: solid, liquid, and gas. A thermodynamic system may consist of a single pure substance or multiple substances. In the latter case, the mixture is characterized by its molar or mass composition. Each component of the mixture may be present in one or more phases. If the components and their phases are uniformly distributed throughout the volume defined by the system boundaries, the mixture is homogeneous; otherwise, it is heterogeneous. The properties of a mixture obviously depend on its homogeneity.

The concept of phase plays a very important role in practice, because we always assume in what follows that **any physical system decomposes into a set of phases**.

Thus, note that the equations we establish are theoretically valid only for one phase, and there are risks in using them outside this assumption.

2.1.4 Equilibrium, Reversible Process

A phase is said to be in static equilibrium, or simply in equilibrium, if:

- Pressure and temperature are uniform in space.
- All state variables are constant over time.

We define a reversible process between two equilibrium states 1 and 2 as a fictitious process having the following two properties:

- It is sufficiently slow in all respects (velocity, heat and matter transfer, etc.) so it can be considered a continuous equilibrium state.
- It is the common boundary of two families of real processes, one leading from 1 to 2, and the other from 2 to 1.

A process is said to be irreversible in the following two cases:

- The reverse process is not feasible without substantial equipment modification (mixing, combustion, etc.).
- It contains a source of irreversibility such as friction or viscosity.

2.1.5 Temperature

The concept of temperature can be introduced in several different ways. We content ourselves here with the operational definition given by F. Fer (1970), which is based on two propositions:

- We know how to construct a thermometer device whose physical properties are, under well-defined operating conditions, a function of a single variable called temperature.
- We know how to obtain a physical system such that when a thermometer is immersed in it, its indication remains constant over time and independent of its orientation and location.

We say the temperature of the medium equals that of the thermometer and that the medium is in thermal equilibrium.

It is possible to link this presentation with axiomatic definitions of temperature deduced, for example, from the Second Law. A rigorous and comprehensive presentation of this notion, however, is beyond the limits we have set for this book.

Furthermore, this introduction is quite intuitive and its practical use generally does not pose particular problems.

2.1.6 Symbols

We generally write state variables in lowercase (except temperature in Kelvin and pressure) as well as intensive state functions (referred to unit mass), and capitalize extensive quantities (referred to total mass).

Following established practice, however, an exception is made for kinetic energy K, the work of external forces W, and heat exchanged with the outside Q, which are written in capitals even when expressed in mass units, and for useful work τ, which is always written in lowercase.

We convert from one to another by multiplying by the mass m (for closed systems) or the mass flow rate $\dot{m}$ (for open systems).

$$\text{volume } v = \frac{V}{m} = \frac{\dot{V}}{\dot{m}}\left(m^3/kg\right) \quad \text{internal energy } u = \frac{U}{m} = \frac{\dot{U}}{\dot{m}}\left(J/kg\right)$$

$$\text{entropy } s = \frac{S}{m} = \frac{\dot{S}}{\dot{m}}\left(J/kg/K\right) \quad \text{enthalpy } \quad h = \frac{H}{m} = \frac{\dot{H}}{\dot{m}}\left(J/kg\right)$$

For convenience, we will refer, whenever possible, to the unit mass of fluid.

In this book, since we use extensive quantities only rarely, to avoid burdening the notation, we also use capital letters for molar quantities, which we occasionally employ. When this occurs, we will specify it.

Recall that the kilomol (1 kmol) is the amount of a substance whose mass equals its molar mass M in kilograms: 1 kmol of O_2 has a mass of 32 kg.

If M* is the mass of a molecule, $M^* = M/N_0$, where N_0 is Avogadro's number:

$$N_0 = 6 \times 1{,}023 \text{ molecules/g}$$

In the normal state (0°C, 1 bar), the volume occupied by 1 kmol equals 22.414 m^3.

2.2 ENERGY EXCHANGES IN A PROCESS

Thermal machine components are traversed by fluids that are most often gaseous or liquid. During the processes they undergo, these fluids exchange energy with the outside or among themselves in two forms: mechanical, traditionally denoted W, and thermal, denoted Q. Given their practical importance, we first study these energy exchanges.

Consistent with our preference for the simplest possible presentation of thermodynamics fundamentals, we limit ourselves in what follows to steady-state systems, because even though the results we establish generalize quite easily, a heavier formalism must be adopted to be rigorous.

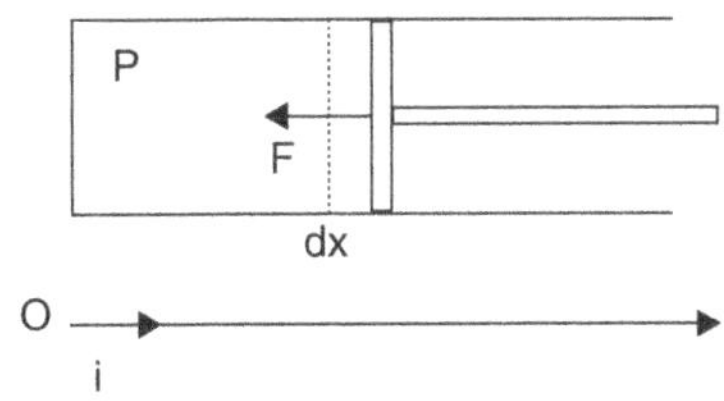

FIGURE 2.2.1 Sketch of a cylinder-piston assembly.

2.2.1 Work δW of External Forces on a Closed System

Consider a closed single-phase system. External forces acting on it are usually limited to the action of gravity on the fluid mass and pressure on its boundaries.

In heat engines, the work of gravity is in most cases negligible compared to that of so-called "contact" actions. To illustrate, the work provided by a mass of 1 kg of water falling from a height of 100 m equals 980 J, while that of the same mass of steam at 500°C isentropically expanded from 100 bar to 1 bar equals 983 kJ, or 1,000 times greater.

Consequently, in thermal machines, the work of mass forces is most often negligible compared to that of pressure.

Consider a fluid at rest (Figure 2.2.1), at uniform pressure P enclosed in a cylinder whose walls are all fixed except a piston capable of moving in one direction. Now exert a force $\vec{F}$ on the piston to move it slowly enough (so that at any time the system can be considered a phase).

Consider an axis $(O,\vec{i})$ directed from the bottom of the cylinder $(x = 0 \Rightarrow V = 0)$. The force being exerted by the system on the outside, $\vec{F} = -\|F\|\vec{i}$.

Let A be the piston section, and $\vec{dx} = dx\,\vec{i}$ its motion.

Obviously $\|F\| = PA$, and work $m\delta W_A$ received by the mass m of fluid is:

$$m\,\delta W_A = \vec{F}\cdot\vec{dx} = -\|F\|dx \quad \vec{i}\cdot\vec{i} = -PA\,dx = -PdV$$

dV being the volume change of the mass, equal to mdv, therefore:

$$\delta W_A = -Pdv \tag{2.2.1}$$

In compression, the volume change is negative and the work received is positive. We denote this work with index A to indicate it is exercised on the surface of the system considered.

This expression, which generalizes easily to a closed system of any form, is valid only if the system remains in static equilibrium, i.e., if the following assumptions are true:

- The pressure remains uniform across the system.
- The fluid remains at rest.

Note that formula (2.2.1) implies that work received by the system is positive, and work it provides is negative. By convention, we generalize this result by counting energy received by a system as positive, and energy it provides to the outside as negative.

During a reversible process making a closed phase pass from state 1 to state 2, the mass work of contact actions has the value:

$$W_A = -\int_1^2 P dv \tag{2.2.2}$$

If gravity work is not negligible, one can easily show that it is expressed as:

$$dW_v = -g dz \tag{2.2.3}$$

where $g > 0$ is the acceleration of gravity, and z is the altitude of the point, counted positively upwards. We denote such work with index v to indicate it is exercised in the volume of the system considered.

We then have:

$$W = W_A + W_V \tag{2.2.4}$$

2.2.2 Heat Transfer

Consider a simple fluid mass within which no friction occurs. δQ is the quantity of heat exchanged with the outside and received by the unit of fluid mass in an infinitesimal process.

An important experimental fact, the basis of compressible fluid thermodynamics, is that δQ is a differential form of the fluid mass state, called a calorimetric equation.

For example, in the case of a mono-variant system (equilibrium between phases during a change of state), the calorimetric equation depends only on the quality x, the ratio of vapor mass to total mass (liquid + vapor) in the case of vaporization or condensation:

$$\delta Q = L dx$$

where L is the change of state enthalpy.

More generally, the calorimetric equation that connects δQ to the simple fluid (bivariant) state variables can take three well-known equivalent canonical forms:

$$\begin{aligned} \delta Q &= c_p dT + h dP \\ \delta Q &= c_v dT + l dv \\ \delta Q &= \mu dP + \lambda dv \end{aligned} \tag{2.2.5}$$

Coefficients c_p, c_v, h, l, μ, and λ are called equilibrium fluid calorimetric coefficients (note that h is **not** enthalpy, which will be introduced later). They are linked by differential relations that can be obtained without significant difficulty, but which are not of interest here. It suffices to know that the calorimetric properties of a fluid are determined by knowing two of these six coefficients.

Subsequently, we will use only coefficients c_p and c_v, which are respectively called specific heat capacity at constant pressure and specific heat capacity at constant volume, formerly known as "specific heat."

For a solid or liquid, $c_p \approx c_v$, and $\delta Q \approx c dT$

For an ideal gas, depending on whether we are dealing with an open system operating at constant pressure, or a closed one operating at constant volume, $\delta Q = c_p\, dT$ or $\delta Q = c_v\, dT$. For a real gas, use one of the canonical forms above.

The above equations are valid only if certain conditions are met:

- First, the fluid mass must be completely homogeneous, i.e., comparable to a phase.
- Second, we assumed there was no irreversibility inside or at the boundary of the fluid mass. If any exists, the relationship becomes:

$$\delta Q < c_p\, dT + h\, dP \quad \text{We will then write:}$$
$$\delta Q = c_p\, dT + h\, dP - \delta\pi \tag{2.2.6}$$

$\delta\pi$, essentially positive term, has a very simple physical meaning: it is the heat generated by mechanical friction within the fluid. Although it differs fundamentally from heat received from outside, it produces the same effect. Its meaning will be clarified during the presentation of the Second Law of thermodynamics (Section 2.4).

By convention, therefore, we denote δQ as the heat exchanged with the outside, counted positively if received by the system, and $\delta\pi$ as the heat dissipated by internal friction if any. In practice, it is important to distinguish these two forms of heat; otherwise, serious reasoning errors can occur. In particular, processes without heat exchange with the outside, called adiabatic, are such that $\delta Q = 0$, whether or not they are the seat of irreversibility—that is, whether $\delta\pi$ is zero or not.

2.3 FIRST LAW OF THERMODYNAMICS

Having established expressions for calculating the mechanical and thermal energy exchange between a fluid mass and its surroundings, we can now formulate the First Law of thermodynamics. As usual, we begin with the expression for closed systems, which is well known, and then generalize it for open systems.

The First Law, also known as the equivalence principle or the principle of energy conservation, states that the energy contained in an isolated system or in a system evolving through a closed cycle remains constant, regardless of the processes it undergoes. The various forms that the energy of a system can take—mechanical energy, thermal energy, potential energy, kinetic energy, etc.—are all equivalent to each other under the First Law.

2.3.1 Definition of Internal Energy U (Closed System)

Any closed physical system is characterized by a scalar U, which depends exclusively on state variables and satisfies the following relation for any real process:

$$\Delta U + \Delta K = W + Q \tag{2.3.1}$$

where K is the kinetic energy of the system, W is the work done by external forces expressed for the total mass of the system and given by relation (2.2.4) $W = W_A + W_V$, and Q is the heat exchanged by the system with its surroundings during the process under consideration.

U is an extensive quantity called the internal energy of the system. U+K is sometimes referred to as the total energy of the system. For a phase of mass m, U = mu, where u is the specific internal energy.

Recall that the state variables that define a physical system in the most general case fall into four broad categories:

- mechanical variables (position or deformation);
- temperature;
- chemical variables;
- electrical variables.

Only the first two categories will be considered in most applications we address, and the third will be used for fluid mixtures and combustion reactions. The fourth category will not be addressed in this book.

Note that many authors distinguish the work of gravity (W_v) from the work of pressure forces (W_A) in the expression of the First Law, and therefore express it in the equivalent form:

$$\Delta U + \Delta K + m g \Delta z = W_A + Q$$

For an infinitesimal control volume, equation (2.3.1) becomes:

$$dU + dK = \delta W + \delta Q \tag{2.3.2}$$

Physically, the First Law derives from experimental evidence that, regardless of the processes undergone by a given system, the sum W+Q depends only on the initial and final states. It follows that W+Q is a state function of the system.

Mathematically, the First Law states that while δW and δQ are not exact differentials, their sum is an exact differential equal to the sum of kinetic energy changes and a state function: the internal energy.

2.3.2 Application to a Fluid Mass

We will see later that in the case of a single fluid phase, the physical state of the system, expressed in mass quantities, is characterized by variables P, v, T, connected by the equation of state. u is a function of these variables, or more precisely of two of them.

Let us apply the First Law to an infinitesimal reversible process of a fluid, neglecting the action of gravity. During this process, the kinetic energy remains constant and equal to zero, which implies $dK = 0$. Moreover, $\delta W = -Pdv$ and δQ is expressed as:

$$\delta Q = du + Pdv \tag{2.3.3}$$

By identification with the calorimetric equation (2.2.5), we obtain: $c_v = \left(\frac{\partial u}{\partial T}\right)v$, which is often used as a definition of c_v.

The previous equation shows that for constant-volume heating without mechanical friction, the heat exchanged with the surroundings equals the change in the fluid's internal energy:

$$Q_1^2 = u_2 - u_1$$

This relationship forms the basis for calorimetric determinations of u.

2.3.3 Work Provided, Shaft Work τ

Industrial operations generally occur continuously, with each component (turbine, pump, valve, etc.) permanently receiving and discharging matter. When operating conditions are stabilized, possibly periodically, the operation is called "steady-state" or "permanent." As mentioned previously, these devices must be analyzed as open systems, and the previous expression, valid only for closed systems, must be generalized.

The reasoning principle is to follow the evolution of a closed control volume and calculate the work of external forces on all its boundaries, distinguishing between cross-sections of the fluid, fixed walls (which obviously neither produce nor receive work), and moving walls such as blades, where shaft work τ is exerted.

In the most common case (Figure 2.3.1), we can assume that the component operates between two large chambers where the fluid is in equilibrium. The upstream (1) and downstream (2) states are defined by their pressure and temperature, assumed to be uniform despite the extractions and additions due to suction and discharge. For example, a gas turbine compressor draws in air and discharges it into the combustion chamber, where the pressure is substantially uniform.

In its passage from (1) to (2), each unit mass of fluid receives shaft work τ from the moving walls. Knowledge of this quantity is fundamental, since its product with the mass flow rate gives the output power (neglecting mechanical losses in bearings and transmission).

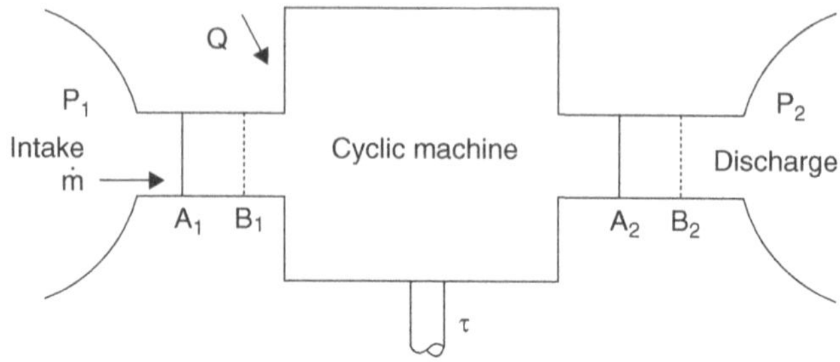

FIGURE 2.3.1 Periodic machine.

It can easily be shown that, per unit mass, a component performing any type of processing provides work τ algebraically equal to the work of pressure forces calculated in a closed system W_A, increased by the transfer work, i.e., the variation of the product Pv:

$$\tau = W_A + P_2 v_2 - P_1 v_1 = W_A + \Delta(Pv) \tag{2.3.4}$$

In general, $P_2 v_2 - P_1 v_1 \neq 0$, and τ differs from W_A.

2.3.3.1 Demonstration

The machine, traversed by a flow rate, operates in an unsteady but periodic state, meaning all its components, including the fluid mass within it, are periodically in the same state (note that steady-state operation is obtained from the periodic regime by letting the period tend to zero). Consider the closed system consisting of the fluid mass initially contained in the control volume bounded by the fixed and movable walls and sections A_1 and A_2, respectively, located in the inlet and outlet chambers.

After a period dt, the control volume has moved and is now bounded by sections B_1 and B_2, also in the inlet and outlet chambers, respectively. The machine has been traversed by the fluid mass $\dot{m}\,dt$, which is the common mass of sections $A_1 B_1$ and $A_2 B_2$ (by conservation of mass flow), and has provided work on the fluid $\tau \dot{m}\,dt$ (by definition).

The work of external forces W dt equals the sum of the work of gravity and the work of pressure forces exerted on the four boundaries of the machine: A_1, A_2, moving walls, and fixed walls.

On the fixed walls, $\delta W_{A4} = 0$.

On A_1, reasoning similar to that in Section 2.2.1 (the machine cannot be treated as a single phase) shows that:

$$\delta W_{A1} = -P_1\, dV_1 = -P_1(-\dot{m}\ dt)v_1 = P_1 v_1 \dot{m}\ dt$$

On A_2, in the same way, $\delta W_{A2} = -P_2\, v_2\, \dot{m} dt$

By definition, the shaft work is that of pressure forces on the moving walls:

$$\tau \dot{m}\, dt = \delta W_{A3}$$

We thus have:

$$\delta W = \delta W_A + dW_v = \delta W_{A1} + \delta W_{A2} + \delta W_{A3} + \delta W_{A4} + dW_v$$

$$\delta W = W\,dt = P_1\,v_1\,\dot{m}\,dt - P_2\,v_2\,\dot{m}\,dt + \tau\,\dot{m}\,dt + dW_v$$

$$\tau = W_A + P_2\,v_2 - P_1\,v_1 = W_A + \Delta(Pv)$$

This concept of shaft work is far from trivial. In practical terms, however, it poses no particular problem. In all open system compressions and expansions, it is the shaft work that must be considered in calculations.

We will later introduce another concept, that of useful energy, used to distinguish it from purchased energy in calculating machine efficiency.

Note that in the previous demonstration, we made no restrictive assumption about the nature of the processes within the machine itself. We simply recognized that the pressure was uniform and constant over time at the inlet and outlet of the machine. The relationship obtained thus applies only under this condition, regardless of the intermediate processes, whether reversible or not.

2.3.3.2 Special Case of a Reversible Process

If the process is reversible, $W_A = -\int_1^2 P dv$

$$\tau = -\int_1^2 P dv + P_2\,v_2 - P_1\,v_1$$

We recognize the expression for integration by parts, hence:

$$\tau = \int_1^2 v dP \qquad (2.3.5)$$

All these relationships can be represented graphically in the Clapeyron diagram, which corresponds to the (v, P) plane with v on the x-axis and P on the y-axis (Figure 2.3.2). The area between the curve 1 → 2 representing the process and the

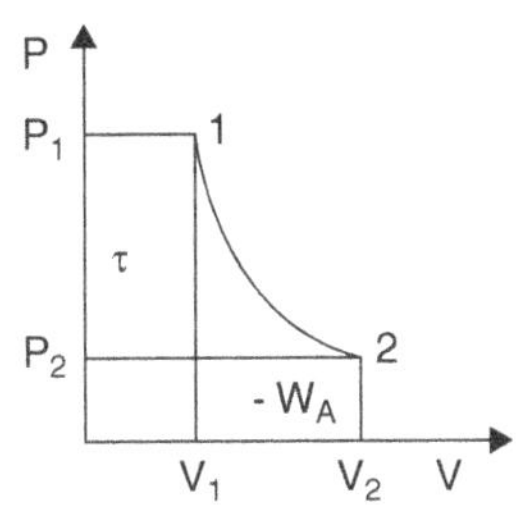

FIGURE 2.3.2 Compression process in a (P,v) diagram.

x-axis equals the opposite of the work of external forces W_A, while the area between the curve and the y-axis equals the shaft work τ.

Note that these areas depend on the path of the curve $1 \rightarrow 2$, and therefore that mere knowledge of the initial and final states is not sufficient to determine W_A and τ.

Mathematically, this follows from the important fact that the differential expressions $-Pdv$ and vdP are not exact differentials. Physically, it is explained by the involvement of a form of energy other than mechanical work, namely thermal energy.

If two different reversible processes starting from the same initial state 1 and ending at the same final state 2 do not correspond to the same shaft work, it is because the fluid does not exchange the same amount of heat with the surroundings during these two processes.

To distinguish them well, we denote differential forms with a small $\delta(\delta W_A, \delta\tau)$ and exact differentials with d.

2.3.4 Shaft Work and Enthalpy (Open Systems)

Let us return to equation (2.3.1) and express it in terms of τ instead of W for the total mass of the system. We obtain:

$$\Delta U + \Delta K = W + Q = W_A + W_v + Q = \tau - \Delta(PV) - mg\Delta z + Q$$

which can be written:

$$\Delta(U + PV) + \Delta K + mg\Delta z = \tau + Q$$

Let us define enthalpy as the state function $H = U + PV$ and specific enthalpy as $h = u + Pv$.

The enthalpic form of the First Law is therefore written in mass variables (note that K, τ, and Q are now expressed in mass quantities):

$$\Delta h + \Delta K + g\Delta z = \tau + Q \tag{2.3.6}$$

For an infinitesimal control volume, this equation becomes:

$$dh + dK + gdz = \delta\tau + \delta Q \tag{2.3.7}$$

Transposing the reasoning of Section 2.3.2, we find that the calorimetric equation is written as:

$$\delta Q = dh - vdP \tag{2.3.8}$$

and $c_p = \left(\frac{\partial h}{\partial T}\right)_P$, which is often used as a definition of c_p.

2.3.5 Establishment of Enthalpy Balance

We have seen that in industrial processes, operations usually occur continuously, i.e., in open systems. It is therefore the enthalpic expressions of the First Law that apply most often.

Although familiar to almost everyone, experience shows that its implementation may present some difficulties. It suffices, however, to apply the conservation law of extensive energy quantities (shaft work, heat exchanged with the surroundings Q, enthalpy flow of the fluids involved) to an appropriate control volume, which most often is simply the system defined by the boundaries of the component being studied. Given the First Law, enthalpy is a conservative quantity, and equation (2.1.1) is simplified, with generation and consumption terms disappearing, to become:

$$\left\{\begin{array}{l}\text{accumulation}\\ \text{in the control}\\ \text{volume}\end{array}\right\} = \left\{\begin{array}{l}\text{inward transport}\\ \text{by the}\\ \text{surface}\end{array}\right\} - \left\{\begin{array}{l}\text{outward transport}\\ \text{by the}\\ \text{surface}\end{array}\right\} + \left\{\begin{array}{l}\text{transfer}\\ \text{through}\\ \text{the surface}\end{array}\right\}$$

If we assume that the system operates in steady-state mode, which is usually the case, there is no net accumulation in the control volume, and this equation reduces to:

$$\left\{\begin{array}{l}\text{inward transport}\\ \text{by the}\\ \text{surface}\end{array}\right\} - \left\{\begin{array}{l}\text{outward transport}\\ \text{by the}\\ \text{surface}\end{array}\right\} + \left\{\begin{array}{l}\text{transfer}\\ \text{through}\\ \text{the surface}\end{array}\right\} = 0$$

In most cases, the enthalpy balance is determined when the boundary flows of heat, useful work, and enthalpy are known.

Consider, for example, the case of an adiabatic axial compressor for a gas turbine, which will be calculated in more detail in Chapter 3. It draws air at rest from the atmosphere and discharges it into the combustion chamber of the machine.

The control volume in this case corresponds to the geometry of the compressor, bounded upstream, and downstream by its connecting flanges.

The mass balance is straightforward to establish because, denoting the compressor inlet with index 1 and the outlet with index 2:

$$\dot{m}_2 = \dot{m}_1 \text{ in steady-state}$$

Since the machine is adiabatic, there is no heat exchange with the surroundings, and therefore Q = 0. At both suction and discharge, the air velocity is low and its kinetic energy is negligible. The enthalpy balance for this machine is very simple: $\Delta h = \tau$. The enthalpy imparted to the fluid equals the work received on the shaft.

If the compressor is cooled, the balance is slightly more complex. If the heat exchanged with the surroundings is $Q<0$, we have: $\Delta h = \tau + Q$. The enthalpy imparted to the fluid equals (in absolute value) the difference between the work received on the shaft and the heat extracted.

Now let us consider a slightly more complex example involving closed systems: combustion in a diesel engine, modeled as a succession of two elementary combustions (called mixed cycle)—first at constant volume, then at constant pressure.

The mass balance is now somewhat more complicated, with m_c being the mass of fuel injected during each combustion phase:

$$m_2 = m_1 + m_c$$

The control volume in this case is the entire assembly: cylinder head, cylinder, and piston. Since the engine is cooled by water, the combustion is not adiabatic.

Since the first combustion phase occurs at constant volume, it involves no work ($W=0$), and the energy balance should be written here $\Sigma \Delta U_i = Q$, where the summation is performed over all species present, assumed to be ideal gases. A first difficulty arises here because this equation requires that the internal energy of the gases in the chamber be calculated taking into account the state of the combustion reaction, i.e., considering changes in chemical variables.

However, in industrial practice, it is very rare to proceed as indicated above: the internal energy (or enthalpy) of an ideal gas is generally defined relative to the standard reference of 1 bar and 298 K for a given composition.

To overcome this difficulty, it is customary, although formally inconsistent with the First Law, to reintroduce somewhat artificially into the balance equations a term for volumetric power generation, which corresponds to the heat Q_r (> 0) released by the combustion reaction. As we indicated earlier in this chapter, this approach allows us to ignore chemical variables and thus significantly simplifies the formalism.

For the constant-pressure combustion phase, the piston is in motion and produces shaft work W that must also be included in the balance.

2.3.6 Application to Industrial Processes

With C denoting the velocity of the fluid, the First Law applied to a component through which a fluid passes at unit flow rate can be expressed in mass units as:

$$\tau + Q = h_2 - h_1 + \Delta K + g\Delta z = h_2 - h_1 + \frac{C_2^2 - C_1^2}{2} \tag{2.3.9}$$

This fundamental relationship explains why in industrial devices it is virtually impossible to implement simultaneously both significant mechanical work and substantial heat flux.

- Components that transfer heat from one fluid to another require large exchange areas, as their heat fluxes are proportional to these areas. Technical and economic considerations lead to the adoption of purely static devices, for example, large bundles of parallel tubes with one fluid flowing internally while the other flows externally. τ is zero due to the absence of movable walls.

- Machines that compress or expand a fluid have very compact designs for reasons of weight, size, and cost. For similar reasons, they rotate at very high speeds (several thousand revolutions per minute). Each parcel of fluid remains within them very briefly. Moreover, the heat transfer coefficients of gases have low values. Short residence time, small fluid-wall contact areas, and low transfer coefficients mean that heat exchange is minimal and that the operation of these machines is nearly adiabatic.
- There is a class of devices where both τ and Q are zero: these are static expansion devices such as valves, filters, etc. The corresponding process is called isenthalpic throttling or flash expansion.

2.4 SECOND LAW OF THERMODYNAMICS

The First Law postulates the equivalence of different forms of energy, but it does not account for an essential experimental fact: when a system interacts with its surroundings, the energy processes it undergoes can take place only in a particular direction that cannot be reversed without qualitatively changing the system.

Thus, heat flows naturally from a high-temperature substance to a low-temperature substance, but the reverse can only be achieved by using a complex thermal machine called a refrigeration machine or heat pump, or by using a Peltier electrical component.

The Second Law complements the First by introducing a function called entropy, which is used to quantitatively characterize the effects of irreversibilities occurring in a system and to explain the phenomena we have just discussed.

A rigorous and comprehensive presentation of the Second Law requires significant development due to the precautions that must be taken in formulating assumptions and explaining their connection with experimental observations, as well as the demonstrations that must be performed.

Given the objectives of this book, such treatment is not justified here, especially since for the applied thermodynamics calculations we are considering, the main practical advantages of the Second Law can be summarized in two points:

- First, entropy is, as we will show, the state function most closely related to heat Q exchanged with the surroundings (through the simple relation $\delta Q = Tds$). It therefore appears in many formal equations governing the operation of energy components.
- Second, the generation of entropy allows one to quantify all irreversibilities occurring in these components and at their boundaries, which is fundamental.

For these reasons, we have selected the presentation below, which is sufficient for our purposes and has the advantage of being concise and easy to understand. Readers interested in a comprehensive presentation of the subject may refer to the literature.

2.4.1 Definition of Entropy

To any phase are attached two quantities T and s, respectively, called "thermodynamic temperature" and "entropy," having the following properties:

- T is a function of temperature alone, independent of the system considered.
- s is a function of the system's state variables.
- In any elementary process involving heat exchange δQ with the surroundings, $\delta Q \leq Tds$, with equality satisfied if and only if the process is perfect (i.e., not irreversible).

$$ds = \frac{\delta Q}{T} + d_i s \tag{2.4.1}$$

where $d_i s$, positive or zero, is called "entropy generation."

Equation (2.4.1) can be rewritten:

$$\delta Q + \delta\pi = Tds \tag{2.4.2}$$

where $\delta\pi$ is called "uncompensated work" or "uncompensated heat." It is positive for an irreversible process and zero otherwise.

The *ex abrupto* introduction of entropy may seem surprising and arbitrary. To explain it more gradually, we will study the case of a perfect gas, whose equation of state is written:

$$Pv = rT$$

where r is a constant depending on the gas (r = 287 J/kg for air), v is the specific volume, and T is temperature in Kelvin.

This equation stems from the laws of Gay-Lussac (P/T = Const at constant volume) and Mariotte (Pv = Const at constant temperature).

The thermal properties of perfect gases result from Joule's laws, which can be stated as follows:

- First law: the internal energy of a perfect gas depends only on its temperature.
- Second law: the heat capacity at constant volume cv is a constant.

$$c_v = \left(\frac{\partial u}{\partial T}\right)_v = \frac{du}{dT}$$

$$cv = v = du / dT$$

Since $\delta Q = du + Pdv$, we have: $\delta Q = c_v dT + Pdv$

By replacing P with its value from the equation of state, we obtain:

$$\delta Q = c_v\, dT + r\, T\frac{dv}{v}$$

In this form, we immediately see that 1/T is an integrating factor, and obviously the simplest possible one. Recall that multiplying all terms of a differential form δy by an integrating factor I transforms it into an exact differential.

Since $ds = \frac{dQ}{T} = c_v \frac{dT}{T} + r\frac{dv}{v}$ is an exact differential, we obtain:

$$s = c_v \ln(T) + r \ln(v) + \text{Const}$$

The concepts of thermodynamic temperature and entropy are thus introduced naturally, with entropy being the state function most closely related to heat. There is no evidence in this approach, however, that these concepts retain their relevance for real gases. The advantage of the axiomatic approach at the beginning of this section is that perfect gas properties can then be retrieved as a special case of the general theory.

2.4.2 Irreversibility

Irreversibilities encountered in the energy devices we consider in this book, except those occurring in combustion reactions, can be grouped into two broad classes, which we present briefly:

- mechanical irreversibilities due to viscosity;
- irreversibilities due to temperature heterogeneity.

2.4.2.1 Mechanical Irreversibility

Relations (2.2.2) and (2.3.5) allow us to calculate the mechanical energy received by an open or closed system in the absence of irreversibility. When the processes are not reversible, these relations are no longer valid, and in the general case, there are no simple differential relations to determine the energy exchange mechanisms involved.

The First Law of thermodynamics allows us to solve this problem, however, as we shall now see.

Consider an open system in which an irreversible adiabatic compression occurs from state 1 to state 2′, with negligible variations in kinetic energy (Figure 2.4.1). According to the First Law, since Q=0, we have:

$$\tau = h_{2'} - h_1$$

Since h is a state function, it is reasonable to decompose the enthalpy change of the irreversible compression as the succession of two changes corresponding to the following processes:

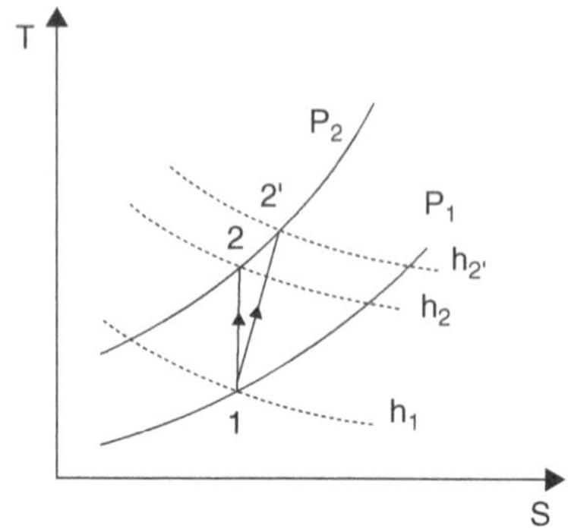

FIGURE 2.4.1 Irreversible compression.

- isentropic compression from 1 to 2 involving work $\tau_s = \int_1^2 vdP$;
- fictitious isobaric heating from 2 to 2′ with value π corresponding to irreversibilities.

$$h_{2'} - h_1 = h_{2'} - h_2 + h_2 - h_1 = \tau_s + \pi = \int_1^2 vdP + \pi$$

In conclusion:

$$\tau = \tau_s + \pi \tag{2.4.3}$$

$$\delta\tau = vdP + \delta\pi \tag{2.4.4}$$

The reasoning can be easily transposed to irreversible expansion, as well as to closed systems, for which we have:

$$W = Ws + \pi \tag{2.4.5}$$

$$\delta W = -Pdv + \delta\pi \tag{2.4.6}$$

These relations can be generalized to cases where kinetic energy or mass forces must be taken into account:

$$\delta\tau + \delta Q = dh + dK$$

$$\delta\tau = dh + dK - \delta Q$$

$$\delta\tau = dh + dK - (dh - vdP) = vdP + dK$$

$$\text{If kinetic energies are not negligible}: \delta\tau = vdP + dK$$

$$\text{If mass forces are not negligible}: \delta\tau = vdP + dK + gdz$$

$$\text{In the presence of irreversibilities}: \delta\tau = vdP + dK + gdz + \delta\pi \qquad (2.4.7)$$

These internal irreversibilities result in an entropy increase, which can be calculated by applying the Second Law.

2.4.2.2 Irreversibility Due to Temperature Heterogeneity

The second major source of irreversibilities encountered in the energy systems considered here arises from the temperature difference that must exist in practice between two bodies when they exchange heat. Indeed, for both technical and economic reasons, the exchange surfaces between these bodies are necessarily finite. It can then be shown that heat exchange is accompanied by an entropy increase at the boundary between the two bodies (at temperatures T_1 and T_2), given by the following equation:

$$d_i s = \left(\frac{T_1 - T_2}{T_1\, T_2}\right)\delta Q \qquad (2.4.8)$$

where δQ is the heat absorbed by body 2. These irreversibilities are often described as external because they occur at the boundary of the system under consideration.

2.4.3 Carnot Effectiveness of Heat Engines

Thermal engines are designed to transform heat into mechanical energy. Let us assume that the heat to be transformed is provided by an external source, the so-called hot source, whose temperature T_1 is fixed. A second source, known as the cold source, is necessary to evacuate the rejected heat. Its temperature T_2 is necessarily lower than T_1.

A heat engine connected between these two sources operates as follows: a fluid mass undergoes a simple closed process or cycle in which it exchanges heat Q_1 with the hot source and heat Q_2 with the cold source, and receives shaft work τ.

$$\text{Under the First Law, we have}: \tau + Q_1 + Q_2 = 0 \quad \text{or} \quad \tau = -Q_1 - Q_2$$

The efficiency η of the machine (often called thermal efficiency) is defined as the ratio (less than 1) of the work produced to the heat from the hot source:

$$\eta = \frac{|\tau|}{Q_1}$$

For the cycle to be fully reversible, it is necessary first that the process occurs without friction, and second that the heat exchange between the fluid and the external sources is reversible and therefore occurs with no temperature difference.

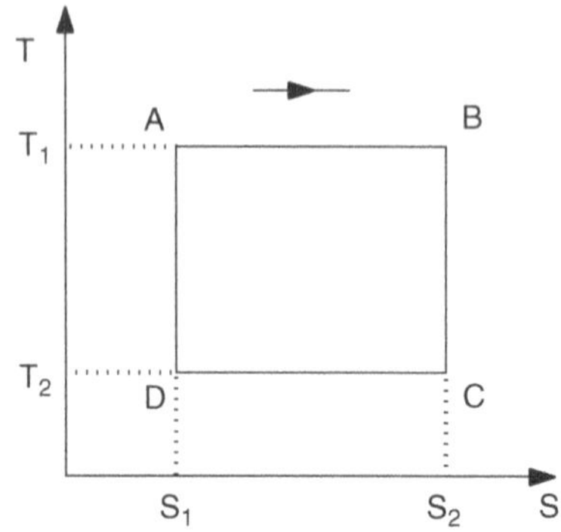

FIGURE 2.4.2 Carnot cycle.

Therefore, during the process phase where the fluid receives heat from the hot source, it will be at temperature T_1, and during the phase where it rejects heat to the cold source, it will be at temperature T_2.

During the intermediate phases, where the fluid evolves between these two temperatures, it must not receive any heat from the sources, and if no internal exchange occurs (which we assume), the process must be adiabatic and reversible, i.e., isentropic.

This leads to the Carnot cycle (named after its inventor). In the entropy diagram (T, s) (for details, see Section 2.6.1), it is represented by rectangle ABCD, traversed in the clockwise direction (Figure 2.4.2).

- AB is a segment of isotherm T_1 traversed from left to right. The fluid receives heat (isothermal expansion in the case of a perfect gas).
- CD is a segment of isotherm T_2 traversed from right to left. The fluid rejects heat (isothermal compression in the case of a perfect gas).
- BC is a segment of isentrope S_2 traversed from top to bottom (reversible adiabatic expansion in all cases).
- DA is a segment of isentrope S_1 traversed from bottom to top (reversible adiabatic compression in all cases).

The calculation of Q_1, Q_2, and η is straightforward for this cycle. Indeed, for an isotherm:

$$Q = \int T ds = T \Delta s$$

It follows that:

$$Q_{AB} = Q_1 = T_1(S_2 - S_1) \quad \text{and} \quad Q_{CD} = Q_2 = T_2(S_1 - S_2)$$

Note in passing that these two equations combine to give the Carnot-Clausius relation (2.4.9), which shows that the measurement of thermodynamic temperatures can be reduced to that of heat:

$$\frac{Q_1}{T_1} + \frac{Q_2}{T_2} = 0 \tag{2.4.9}$$

Returning to our cycle, let's calculate its performance:

$$\tau = -Q_1 - Q_2 = (T_2 - T_1)(S_2 - S_1)$$

$$\eta = \frac{|\tau|}{Q_1} = \frac{T_1 - T_2}{T_1} = 1 - \frac{T_2}{T_1} \tag{2.4.10}$$

$$\eta = 1 - \frac{T_2}{T_1}$$

Relation (2.4.10), established by Carnot in 1824, is justifiably one of the best known in thermodynamics. It is of great practical importance given its implications:

- First, with the assumptions made (reversible cycle without friction and heat transfer without temperature differences), it is easy to show that this effectiveness is the highest that can be achieved by a simple fluid heat engine operating between the two sources at T_1 and T_2. It is therefore a maximum limit, and real cycles generally have much lower efficiencies.
- Second, the effectiveness of all reversible machines operating between the two sources at T_1 and T_2 is the same and depends only on their temperatures, not on the thermodynamic fluid used.
- For these reasons, we prefer to speak of effectiveness, as some authors consider that the term "efficiency" generally used to describe this limiting cycle is not the most appropriate, and it is better to reserve it for describing the performance of a real machine compared to that of an ideal machine.
- In the foregoing, we focused on power-producing heat engines. A second class of heat engines operates in reversed cycles to transfer heat from a cold source to a hot one, using mechanical power input. This is called a refrigeration cycle or heat pump.
- The above reasoning can be transposed to the case of these reversed cycles, and it can be shown that the maximum effectiveness of these cycles (also known as the coefficient of performance) is equal to that of the reversed cycle, given by equations (2.4.11) or (2.4.12), depending on whether one considers the heat exchanged with the cold source (refrigeration cycle) or with the hot source (heat pumps):

$$\eta = \frac{Q_2}{\tau} = \frac{T_2}{T_1 - T_2} \tag{2.4.11}$$

$$\eta = \frac{Q_1}{\tau} = \frac{T_1}{T_1 - T_2} \tag{2.4.12}$$

2.4.4 Fundamental Relations for a Phase

Given equations (2.4.4) and (2.4.6) and neglecting the kinetic energies involved, the expressions of the First Law are written:

$$du = \delta W + \delta Q = -Pdv + \delta Q + \delta\pi$$

$$dh = \delta\tau + \delta Q = vdP + \delta Q + \delta\pi$$

Substituting $(\delta Q + \delta\pi)$ with Tds in these equations, we immediately obtain the following identities, which are fundamental in practical applications:

$$du = -Pdv + Tds \quad (2.4.13)$$

$$dh = vdP + Tds \quad (2.4.14)$$

2.4.5 Thermodynamic Potentials

A more general study of the behavior of a system requires that we know two of its state functions, for example u and s, or h and s, but it may be advantageous to gather all thermodynamic knowledge about it in a single well-chosen state function.

2.4.5.1 Free Energy and Free Enthalpy

We are led to define two functions, either of which is sufficient for this purpose, called, respectively, the free energy f or Helmholtz function and the free enthalpy g or Gibbs function:

$$f(v, T) = u - Ts \quad (2.4.15)$$

$$g(P, T) = h - Ts \quad (2.4.16)$$

For example, let us show that for a substance of constant composition, knowledge of f is sufficient to determine its behavior.

We have: $df = du - Tds - sdT$, and $du = -Pdv + Tds$, which gives:

$df = -Pdv - sdT$, which yields:

$$P = -\frac{\partial f}{\partial v}, \quad s = -\frac{\partial f}{\partial T}, \quad u = f - T\frac{\partial f}{\partial T}$$

$$c_v = \frac{\partial u}{\partial T} = -T\frac{\partial^2 f}{\partial T^2}$$

Knowledge of the free energy or free enthalpy is sufficient to fully characterize a thermodynamic system. In practice, however, these state functions are little used by specialists because their physical meaning is less directly accessible.

2.4.5.2 Equilibrium Relationship for Chemical Reactions

The Second Law of thermodynamics provides the fundamentals for studying chemical equilibrium relationships.

Equation (2.4.1) can be written $dQ \le Tds$.

Since, according to (2.3.3), $dQ = du + Pdv$, we have:

$$du + Pdv - Tds \le 0 \tag{2.4.17}$$

We can rewrite inequality (2.4.17) by introducing the free energy or free enthalpy differential form:

$$df = du - Tds - sdT$$

$$dg = du + Pdv + vdP - Tds - sdT$$

We obtain:

$$df + sdT + Pdv \le 0 \tag{2.4.18}$$

At constant temperature ($dT = 0$) and constant volume ($dv = 0$), a process can only evolve in the direction of decreasing free energy, and a stable equilibrium must be a minimum of free energy:

$$df \le 0 \tag{2.4.19}$$

Furthermore:

$$dg + sdT - vdP \le 0 \tag{2.4.20}$$

At constant temperature ($dT = 0$) and constant pressure ($dP = 0$), a process can only evolve in the direction of decreasing Gibbs free enthalpy, and a stable equilibrium must be a minimum of Gibbs free enthalpy:

$$dg \le 0 \tag{2.4.21}$$

A classic example of this law concerns the equilibrium between phases in a system composed of a liquid and its vapor.

When considering a mixture of several components, the constant temperature and pressure equilibrium equation becomes, expressed in molar quantities (Gibbs-Duhem relation):

$$dG_{T,p} = \sum n_i d\mu_i = 0 \tag{2.4.22}$$

where n_i is the number of moles of the ith component, and μ_i is its chemical potential, equal for an ideal gas to its partial molar free enthalpy:

$$\mu_i = G(T, P_i) \tag{2.4.23}$$

Equation (2.4.22) is the basis for calculating chemical reaction equilibrium. We will use it in paragraph 3.5.1.4 to establish the Law of Mass Action.

2.5 EXERGY

The theory of exergy aims to develop an integrated analysis method that incorporates the first two laws of thermodynamics, thus allowing us to account for both the quantity of energy involved and its quality, which the First Law alone cannot do. Its value lies in providing a rigorous thermodynamic framework to quantify the quality of any system, whether open or closed, in steady state or not.

The first to introduce this concept was G. Gouy, who in the late 19th century defined the concept of usable energy, now renamed exergy by many authors. The basic idea is to consider that a thermodynamic system interacts with what is called its environment, which behaves like an infinite reservoir at constant temperature and pressure and fixed composition, meaning that the system studied is small enough not to disturb this environment.

The environment serves, for example, as a cold source for a power cycle or as a hot source for a refrigeration cycle. Since the environment state determines the performance of the system studied, the theory of exergy can implicitly account for it.

Within the framework of this book, we provide only a brief and steady-state (and therefore not rigorous) presentation of exergy analysis. Readers wishing to delve deeper into this subject will find a comprehensive theory, especially in the work of L. Borel and D. Favrat (2005) or Moran (1989).

2.5.1 Presentation of Exergy for a Monothermal Open System in Steady State

Let us consider a steady-state open system exchanging energy with its environment, assumed to be at uniform temperature T_0 (Figure 2.5.1).

The energy equation is given by the First Law (2.3.6):

$$\Delta h + \Delta K = \tau + Q$$

and the entropy production by the Second Law:

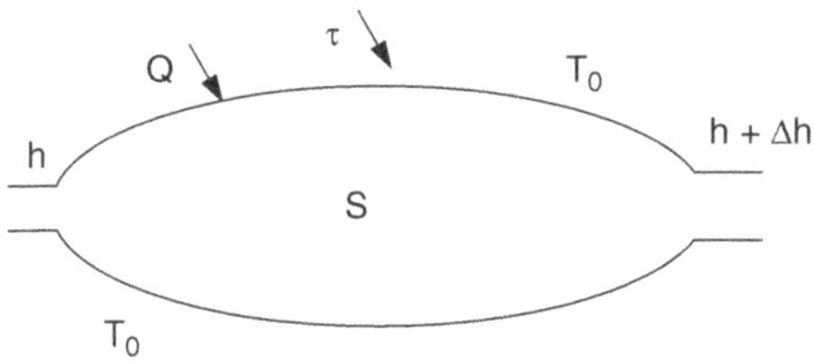

FIGURE 2.5.1 Monothermal system.

$$Q + \pi = T_0\,\Delta s \quad 0 \le \pi$$

$$\Delta s - \frac{Q}{T_0} \ge 0$$

If we assume that all heat is provided by the environment and T_0 is the temperature at the boundary of the open system, Q can be eliminated by combining these two equations, giving:

$$\tau = \Delta h + \Delta K - Q \ge \Delta h + \Delta K - T_0\,\Delta s$$

$$-\tau \le -\left(\Delta h + \Delta K - T_0\,\Delta s\right)$$

In most cases, we can neglect the kinetic energy changes ΔK.

For an open system, the function $\mathbf{x_h = h - T_0 s}$ is usually called **exergy**.

The maximum work that the system can provide equals the reduction of its exergy.

In practice, the actual work provided by the system may be less than this value. We define the exergy loss Δxhi as the difference between the maximum possible work and the actual work provided:

$$\Delta x_{hi} = (-\tau_{max}) - (-\tau_{real}) = (-\Delta x_h) - (-\tau_{real})$$

$$\Delta x_{hi} = T_0\,\Delta s - \Delta h - (-\tau_{real})$$

If the process is furthermore adiabatic:

$$(-\tau_{real}) = -\Delta h \text{ and } \Delta x_{hi} = -\Delta x_h + \Delta h = T_0\,\Delta s$$

Note that this description is not entirely satisfactory because for a closed system, the definition of exergy leads us to consider the function $\mathbf{u - T_0 s}$, so there is some ambiguity.

Given the success of the concept of usable energy introduced by G. Gouy, widely used in the Anglo-Saxon world under the name availability, we could also speak of usable enthalpy or usable energy, depending on whether one considers an open or closed system.

Beyond the question of vocabulary, we note that this new function is characterized by the introduction of a variable outside the system, T_0. As such, it is no longer a state function in the strict sense. However, since T_0 is generally assumed to be constant, exergy is sometimes presented as a linear combination of state functions.

Another remark can be made at this stage: exergy allows us, through the product of entropy by the environment temperature, to express variations of this state function in one dimension and by orders of magnitude similar to those usually encountered by engineers in energy calculations, which facilitates understanding of physical phenomena.

Finally, as discussed below, by this artifice we manage to formally eliminate from the calculations the "free" heat source represented by the environment.

Exergy can be used to evaluate the quality of changes made in real processes compared to ideal reversible ones.

2.5.2 Multithermal Open Steady-State System

Let us consider an open system that exchanges work and heat with n external sources at constant temperatures T_k and with the environment at temperature T_0 (Figure 2.5.2).

Applying the First Law gives:

$$\Delta h + \Delta K = \tau + Q_0 + \sum_{k=1}^{n} Q_k$$

and entropy production is given by the Second Law:

$$\Delta s = \frac{Q_0}{T_0} + \sum_{k=1}^{n} \frac{Q_k}{T_k} + \Delta s_i$$

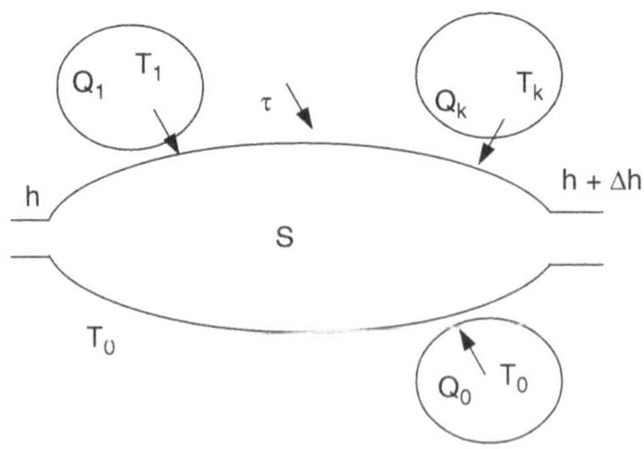

FIGURE 2.5.2 Multithermal open system.

where Δsi, positive or zero, is the entropy generation. Neglecting kinetic energy changes, we obtain:

$$\Delta x_h = \Delta(h - T_0 s) = \tau + \sum_{k=1}^{n}\left(1 - \frac{T_0}{T_k}\right)Q_k - T_0\,\Delta s_i \quad (2.5.1)$$

or, with $\Delta x_{hi} = T_0\,\Delta s_i \geq 0$:

$$\tau + \sum_{k=1}^{n}\left(1 - \frac{T_0}{T_k}\right)Q_k - \Delta x_h - \Delta x_{hi} = 0 \quad (2.5.2)$$

where τ is the work received by the system, Δx_h is the exergy variation of the fluid passing through it, and Δx_{hi} is the exergy dissipation resulting from irreversibilities. We call the term $\theta = 1 - T_0 / T$ the Carnot factor. This is the factor by which we must multiply a quantity Q of heat available at temperature T to obtain the value of its exergy. If we call useful heat (or heat-exergy according to Borel) the quantity $x_q = \theta Q$, equation (2.5.2) takes the form:

$$\begin{aligned} &\tau + \sum_{k=1}^{n}\theta_k\,Q_k - \Delta x_h - \Delta x_{hi} = 0 \\ &-\tau = \sum_{k=1}^{n} x_{qk} - \Delta x_h - \Delta x_{hi} \end{aligned} \quad (2.5.3)$$

The maximum work that an open system can provide equals the sum of the exergy-heats of the sources with which it exchanges heat, minus the change in exergy of the fluid passing through it and the exergy destroyed due to irreversibilities.

2.5.3 Application to a Two-Source Reversible Machine

Let us consider a cyclic two-source reversible machine operating between a source at temperature T_1 and the environment at temperature T_0.

Over a cycle, we have: $\Delta x_h = 0$, $\Delta x_{hi} = 0$ (reversible machine).

Applying formula (2.5.3) gives:$-\tau = x_{q1} = \left(1 - \frac{T_0}{T_1}\right)Q_1$

If $T_0 < T_1$, $x_{q1} \geq 0$: the system receives heat-exergy from the hot source at T_1 and converts it into work τ.

Here we recover the Carnot cycle, whose efficiency is $\eta = 1 - \frac{T_0}{T_1}$.

If $T_0 > T_1$, $x_{q1} \leq 0$: the system delivers heat-exergy to the cold source at T_1, drawing it from the hot source at T_0. For this, we must provide work τ. This is then a refrigeration cycle or heat pump.

2.5.4 Special Case: Heat Exchange without Work Production

In the special case where no work is produced, equation (2.5.3) can be rewritten as:

$$\Delta x_{hi} = \sum_{k=1}^{n} x_{q_k} - \Delta x_h \tag{2.5.4}$$

The exergy destroyed in the process equals the sum of the exergy-heats of the various sources minus the change in exergy of the fluid passing through the system.

2.5.5 Exergy Efficiency

Exergy allows us to rigorously define the concept of system efficiency and therefore to quantify its thermodynamic quality: it is the ratio of exergy uses to exergy resources. It is always between 0 and 1, and is much greater as irreversibilities are smaller. Exergy resources are the sum of all exergies supplied to the cycle from outside. Exergy uses represent the net balance of the cycle, i.e., the algebraic sum of exergy produced and consumed within it.

$$\eta = \frac{\text{exergy uses}}{\text{exergy resources}} \tag{2.5.5}$$

2.6 REPRESENTATIONS OF REAL FLUIDS

An increase in pressure and/or decrease in temperature can justify reconsidering the ideal gas equation. This is especially the case near the liquid–vapor equilibrium zone.

When the fluid no longer satisfies the ideal gas equation, its internal energy and enthalpy are no longer based solely on temperature.

This behavior is shown in Figure 2.6.1, for superheated steam. It is clear that the heat capacity c_p of this substance is more significantly affected by pressure at higher

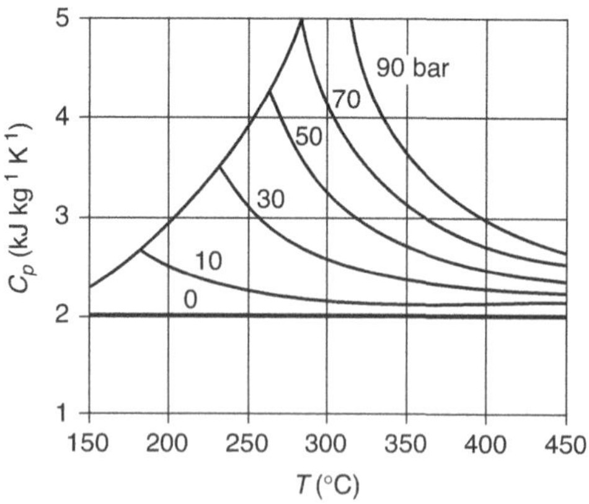

FIGURE 2.6.1 Influence of pressure on steam heat capacity.

pressures and lower temperatures (i.e., especially in the immediate vicinity of the saturation curve).

To determine the state of a real fluid, we most often use a thermodynamic diagram, a table of thermodynamic property values, or a set of equations of state covering the various necessary zones.

Traditionally, thermodynamic diagrams have been the most widely used. However, there has been a marked evolution in practice, with the development of computers making possible the direct calculation of fluid thermodynamic properties over a wide range of state variable variations. Thermoptim enables such calculations to be made precisely.

Nevertheless, even when a fluid properties computer is available, diagrams retain strong educational value because they can easily visualize real gas properties, including the liquid–vapor zone. We present them in this section.

Moreover, conventional diagrams are increasingly computerized, which allows us to combine ease of use through graphical display with numerical accuracy. Interactive diagrams of the Thermoptim family, based on the same equations as the software, are one example.

2.6.1 Thermodynamic Diagrams of Pure Substances

A pure substance of given mass being a bivariant system, its state can be represented on a two-axis diagram on which a number of iso-values or "contours" are plotted.

Among the many possible coordinate systems, we will consider five:

- The Clapeyron diagram (P, v), which gives the most direct representation of mechanical processes but is little used for studying industrial projects because it does not allow accurate reading of common energy quantities, including entropy and enthalpy.
- The (T, s) entropy diagram, which directly visualizes reversible heat transfers ($\int Tds$) and various possible irreversibilities.
- The (h, s) Mollier diagram, which is a transformation of the former, intended to directly show energy transfers in an open system, with the advantage that the enthalpy involved can be read without difficulty.
- The (h, ln(P)) diagram, with enthalpy as the abscissa and pressure as the ordinate (usually with a logarithmic scale to cover a wide range of values). This diagram is traditionally used in studying reversed cycles (heat pumps or refrigeration). Its value lies in combining the essential mechanical constraint (P) of compressors with the energy variable (h). They are also very interesting for studying power cycles.
- Exergy diagrams (or usable enthalpy diagrams) (h,x_h) and (s,x_h), with enthalpy or entropy as the abscissa and exergy as the ordinate, allow visualization of the exergy involved in processes and therefore lend themselves well to studying irreversibilities occurring in cycles.

Thermoptim provides electronic versions of these diagrams, called interactive diagrams, which can be configured at will and on which cycles can be represented.

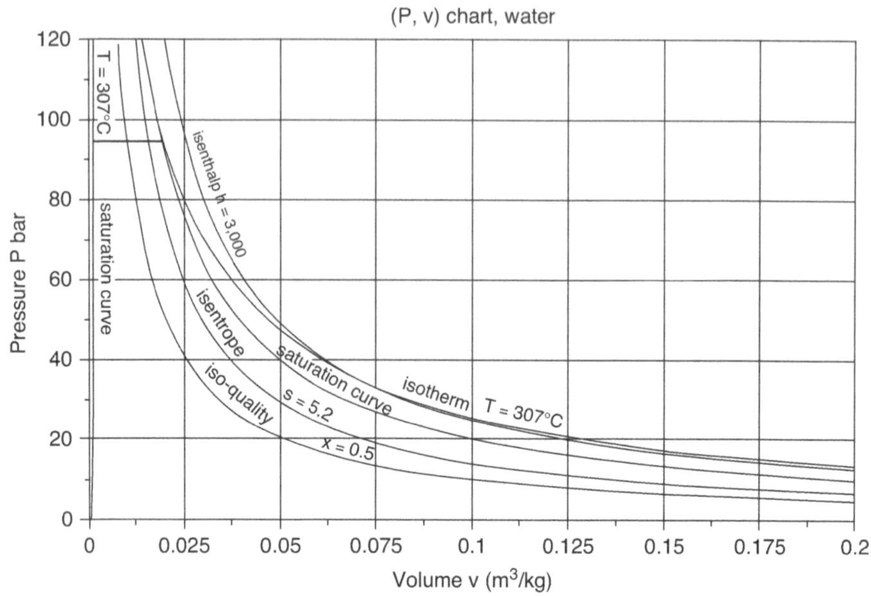

FIGURE 2.6.2 Clapeyron diagram.

2.6.1.1 Clapeyron Diagram

The Clapeyron diagram (Figure 2.6.2) represents the fluid properties with specific volume v as the abscissa and pressure P as the ordinate. In this plane, one curve is of particular importance: the saturation curve.

Its apex is the critical point (not shown in the figure as its pressure is about twice that of the ordinate scale). The left side, upward sloping, represents the beginning of the boiling process (boiling curve), and its right side, downward sloping, represents saturated vapor (dew-point curve). Below this curve is the two-phase liquid–vapor equilibrium zone, and in the rest of the plane, the single-phase fluid zone.

2.6.1.1.1 Form of Isotherms

In the "liquid" zone on the left of the diagram, isotherms have a very steep negative slope: liquids are generally incompressible.

In the two-phase domain, pressure and temperature are linked by the saturation pressure law, and isotherms are horizontal. The volume increases enormously since the gas density is much lower than that of the liquid (in the diagram of Figure 2.6.2, the abscissa scale is logarithmic).

In the zone to the right of the vaporization curve, isotherms curve downward. As the distance from the vaporization curve increases, the gas approaches the corresponding ideal gas behavior, and isotherms gradually recover to approach the equilateral hyperbolas described by the equation Pv = Const. We show an isotherm for $T = 307°C < T_c$ in Figure 2.6.2.

2.6.1.1.2 Other Iso-Value Curves

The main drawback of the Clapeyron diagram is that it does not contain common thermodynamic quantities like enthalpy and entropy, and if we equip it with isenthalpic and isentropic curves, it becomes very imprecise, as shown in the figure: the useful area of the diagram is very narrow, and isentropic, isenthalpic, and iso-quality curves have very similar shapes, making determination of their intersections difficult.

2.6.1.2 Entropy Diagram

In the entropy diagram (Figure 2.6.3), entropy is the abscissa and temperature is the ordinate. The vaporization curve again separates the plane into two parts, defining the two-phase zone and the single-phase fluid zone. The critical point is still at its maximum (C in the figure).

For some substances, the right branch of the vaporization curve (saturated vapor) presents a point of maximum entropy. This is particularly the case for certain hydrocarbons and certain chlorofluorocarbons.

One advantage of such a diagram (see Figure 2.6.4) is that any perfect cyclic process appears in the (T, s) plane as a contour (Γ) whose area A measures the absolute value of the amount of heat Q involved (or the work provided or received τ).

Indeed, $Q = \int Tds$ by definition. Since the cycle is closed, $Q + \tau = 0$, and $|\tau| = A$.

The sign convention is as follows:

- If the cycle is traversed clockwise, the work is negative, thus transferred by the fluid to the surroundings: we speak of a power cycle or a cycle operating in the forward direction.
- If the cycle is traversed counterclockwise, the opposite occurs: we speak of a refrigeration cycle or a reversed cycle.

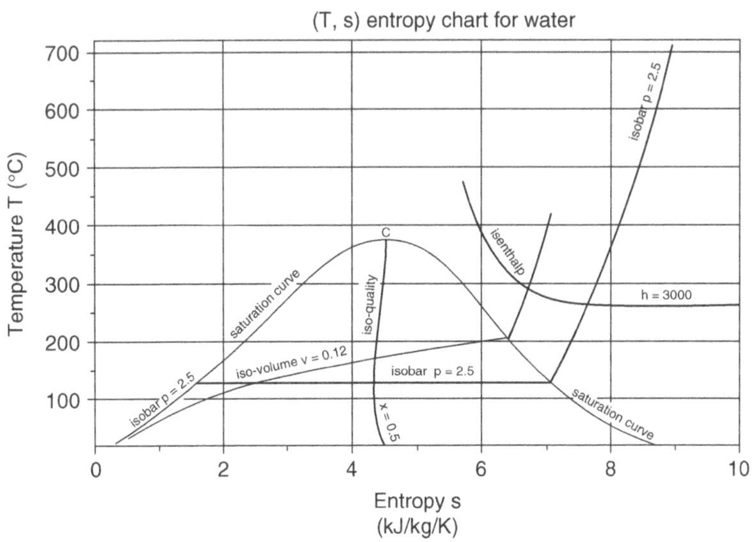

FIGURE 2.6.3 Entropy diagram.

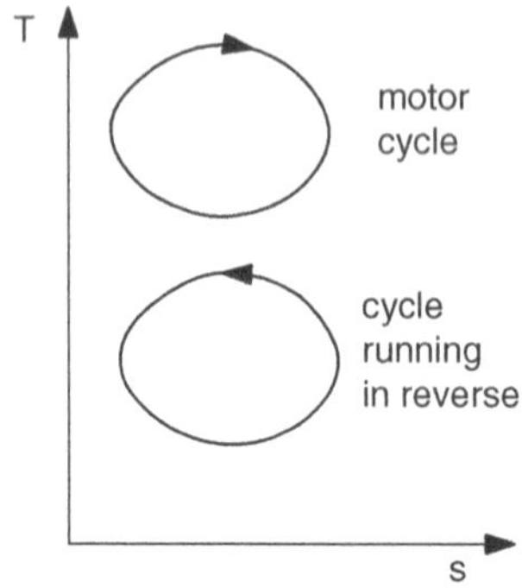

FIGURE 2.6.4 Types of cycles in entropy diagram.

2.6.1.2.1 Form of Isobars

To the left of the saturation curve, isobars curve upward. Since liquid isentropic compression has almost no effect on temperature, liquid isobars virtually merge with the ascending branch of the vaporization curve.

The diagram is very imprecise in this area, and it is preferable to use a table or a program providing thermodynamic properties along the vaporization curve.

Inside the two-phase zone, temperature and pressure are linked by the saturation pressure law, and isobars are horizontal.

To the right of the vaporization curve, curves are rising and, for an ideal gas, become exponential. They can be derived from each other by horizontal translation.

If the pressure exceeds the critical pressure, isobars are strictly ascending curves that do not intersect the vaporization curve.

2.6.1.2.2 Form of Isenthalps

Near the saturation curve, isenthalps curve downward with a steep negative slope.

As the distance from the vaporization curve increases, the gas approaches the corresponding ideal gas behavior, and isenthalps become horizontal since enthalpy is a function of temperature alone.

In an entropy diagram, a Carnot cycle consisting of two isentropic and two isothermal processes forms a rectangle.

Figures 2.6.5 and 2.6.6 provide two examples of entropy diagrams available in Thermoptim, one for helium and the other for water. The abscissa unit (entropy s) is kJ/kg/K.

2.6.1.3 Mollier Diagram

In the Mollier diagram (Figure 2.6.7), entropy is the abscissa and enthalpy is the ordinate. The advantage is that enthalpies can be directly measured on the vertical axis. The critical point C is here to the left of the vaporization curve maximum.

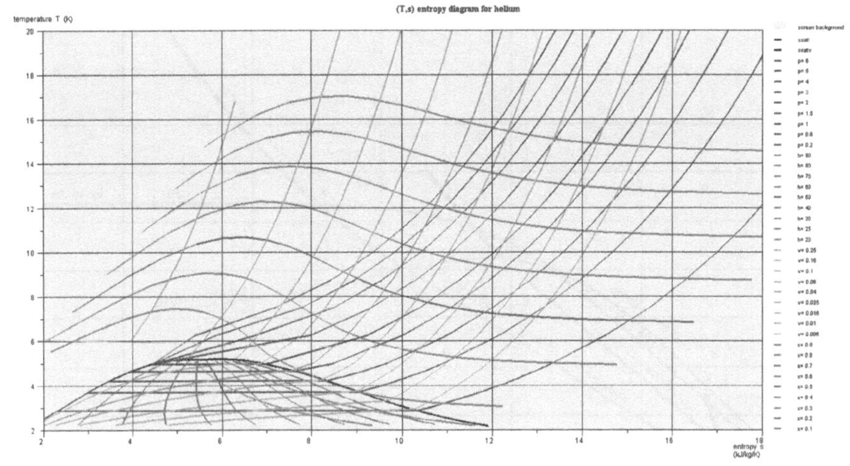

FIGURE 2.6.5 Helium entropy diagram.

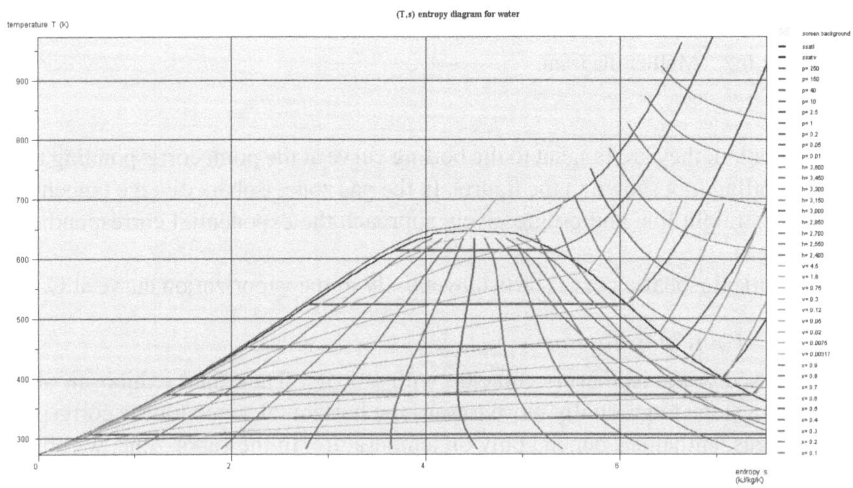

FIGURE 2.6.6 Water entropy diagram.

2.6.1.3.1 Form of Isobars

We have: dh = vdP + Tds and therefore $\left(\frac{\partial h}{\partial s}\right)_P = T$. The slope of isobars at each point equals the absolute temperature and therefore presents no angular point. These curves are strictly ascending.

In the liquid zone, they practically merge with the vaporization curve for the reasons outlined above. In the mixed zone, since pressure and temperature are linked by the saturation pressure law, isobars are straight line segments with slope T. As a first

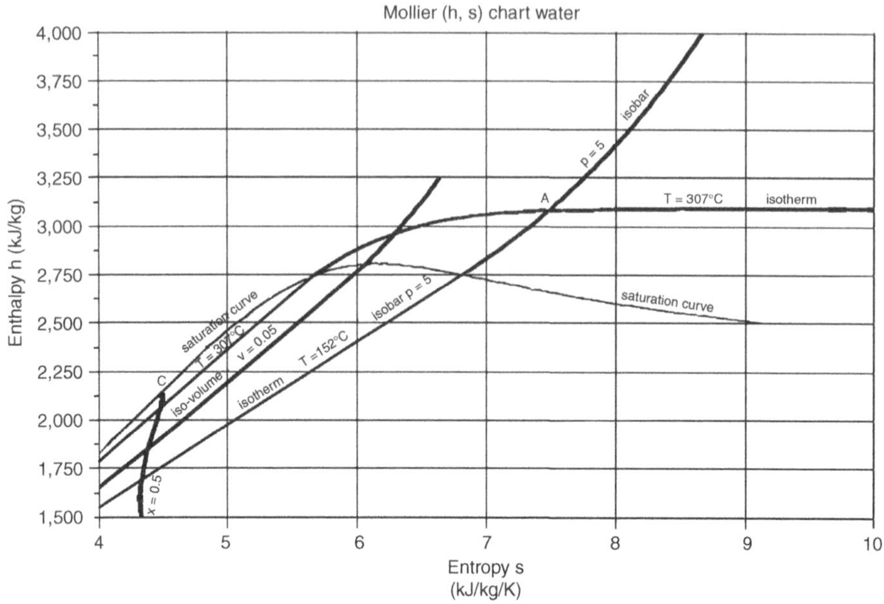

FIGURE 2.6.7 Mollier diagram.

approximation, they are tangent to the boiling curve at the point corresponding to the start of boiling (not shown in the figure). In the gas zone, isobars diverge tangentially from this straight line and progressively approach the exponential corresponding to the ideal gas.

The critical isobaric curve starts tangentially to the vaporization curve at C.

2.6.1.3.2 Form of Isotherms

In the mixed zone, isotherms coincide with isobars. They then exhibit an abrupt decrease in slope to gradually move toward the horizontal, since for the corresponding ideal gas, enthalpies depend only on temperature. In the vapor zone, we observe that isotherms (low slope) and isobars (high slope) intersect at appreciable angles, providing very accurate reading (point A in Figure 2.6.7).

2.6.1.3.3 Form of Isovolumes

It can be shown that the slope of isovolumes is slightly greater than that of isobars.

Moreover, exergy is defined by the relation $x_h = h - T_0 s$, which can be written $h = T_0 s + x_h$.

Once environment reference conditions (temperature and pressure) are defined, it is very easy to equip an (h, s) diagram with iso-exergy curves, which are lines with slope T_0.

Figures 2.6.8 and 2.6.9 show two examples of Mollier diagrams, one for water and the second for R 134a.

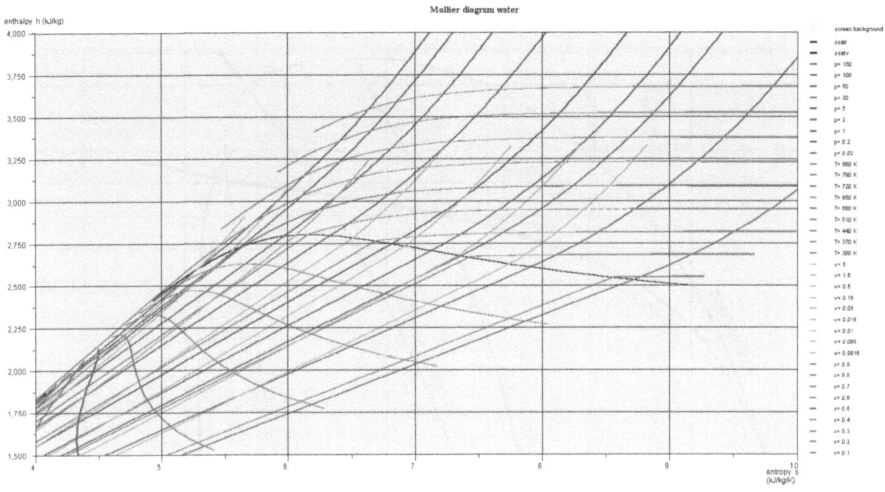

FIGURE 2.6.8 Mollier diagram (h, s) for water.

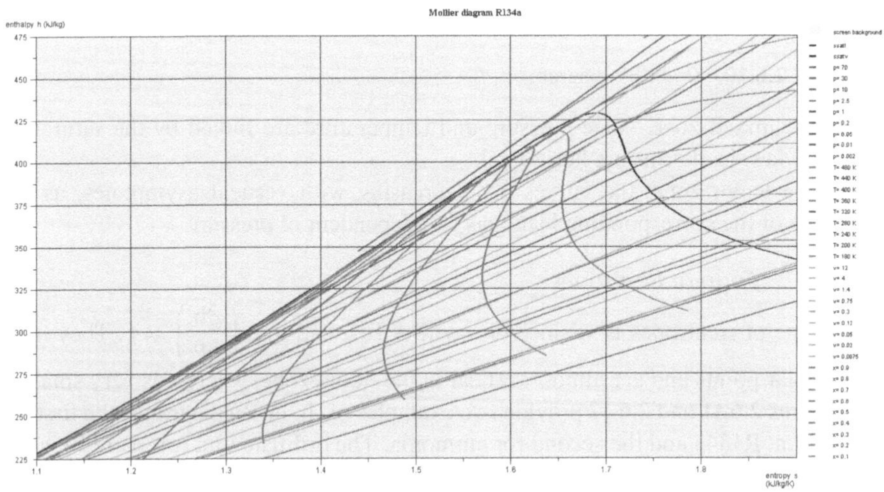

FIGURE 2.6.9 Mollier diagram (h, s) for R134a.

2.6.1.4 (h, ln(P)) Diagram

In the (h, ln(P)) diagram (Figure 2.6.10), the abscissa is enthalpy and the ordinate is pressure, usually on a logarithmic scale. This diagram is equipped with isentropic, isothermal, and isovolume curves, and in the mixed zone, iso-quality curves.

The critical point C is here at the vaporization curve maximum.

2.6.1.4.1 Form of Isotherms

In the liquid region, isotherms are practically vertical, as the liquid enthalpy is almost independent of pressure.

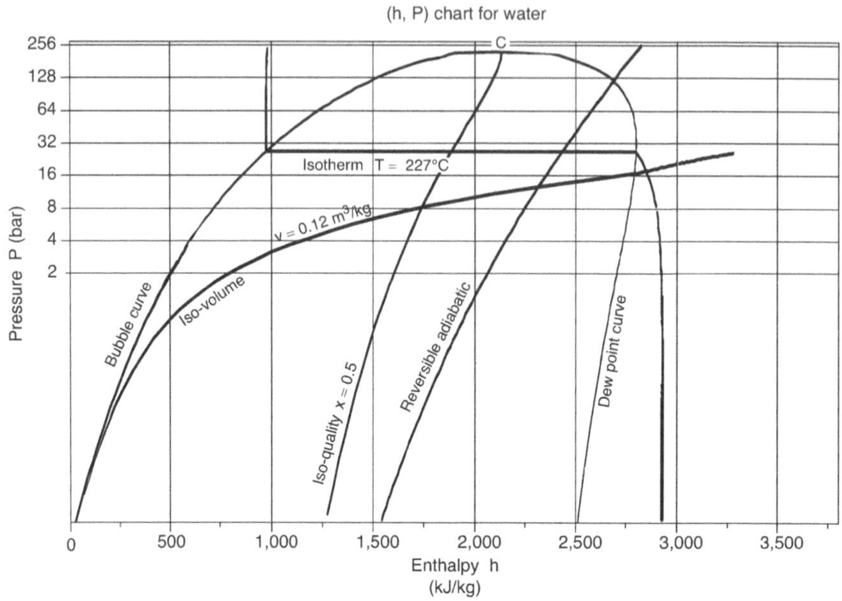

FIGURE 2.6.10 (h, ln(P)) diagram.

In the mixed zone, since pressure and temperature are linked by the saturation pressure law, isotherms are horizontal.

In the vapor zone, the curves are decreasing with vertical asymptotes, as the enthalpy of the corresponding ideal gas is independent of pressure.

2.6.1.4.2 Form of Isentropes

The slope of isentropes is v, since dh = vdP + Tds, and thus $\left(\frac{\partial h}{\partial P}\right)_s = v$. They have no angular points and are almost vertical in the liquid zone, where v is very small.

Figures 2.6.11 and 2.6.12 provide two examples of (h, ln(P)) diagrams, the first for refrigerant R134a and the second for ammonia. The ordinate unit (pressure P) is the bar, and the abscissa (enthalpy h) is kJ/kg.

2.6.1.5 Exergy Enthalpy Diagram (h, x_h)

In the exergy enthalpy diagram (h, x_h) (Figures 2.6.15–2.6.17), enthalpy is the abscissa and exergy is the ordinate.

Since $x_h = h - T_0 s$, this diagram is easily constructed from the Mollier diagram by linear combination of axes once the reference temperature T_0 is known. Given this configuration, it would be more accurate to speak of a family of exergy diagrams for a given substance (Figure 2.6.13).

As with the Mollier diagram (Figure 2.6.11), one advantage is that enthalpies can be measured directly on the x-axis. The critical point here is to the left of the vaporization curve maximum.

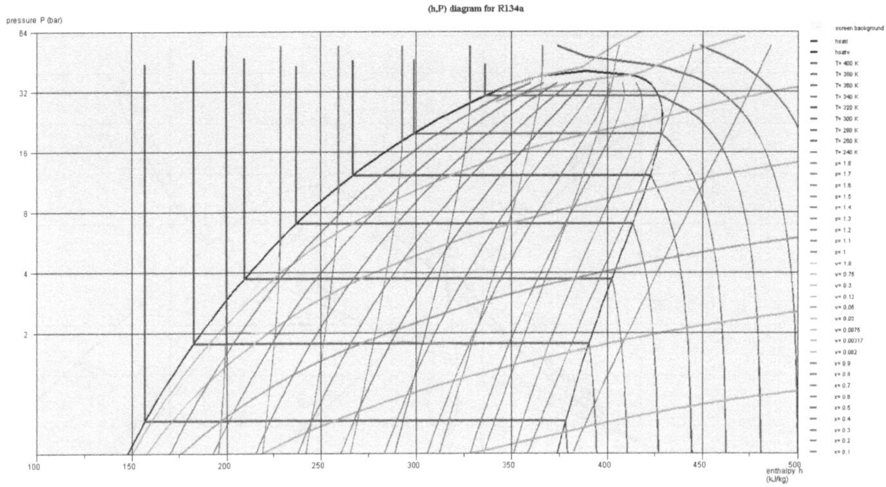

FIGURE 2.6.11 (h, ln(P)) diagram for R 134a.

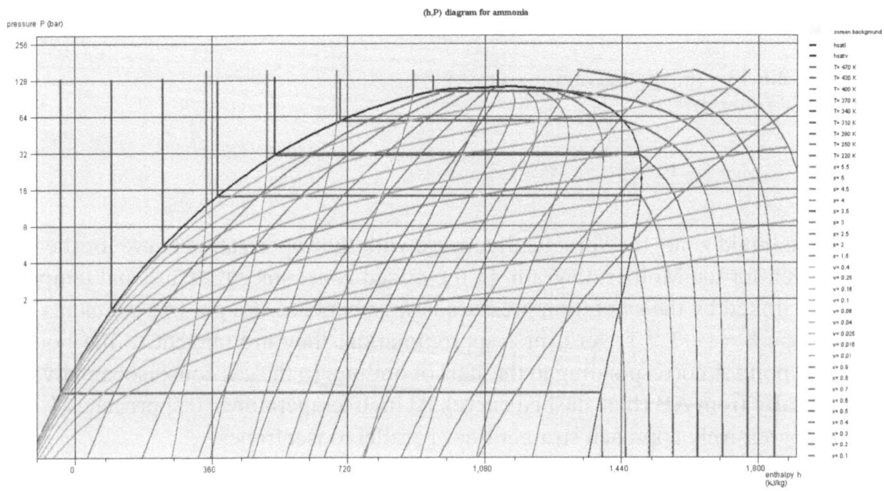

FIGURE 2.6.12 (h, ln(P)) diagram for ammonia.

2.6.1.5.1 Form of Isobars

We have: $dh = vdP + Tds$, and $dx_h = dh - T_0\,ds$, hence $dx_h = \left(1 - \frac{T_0}{T}\right)dh - \frac{T_0}{T}vdP$, and $\left(\frac{\partial x_h}{\partial h}\right)_P = \left(1 - \frac{T_0}{T}\right)$

The slope of isobars at each point equals the Carnot factor $\theta = 1 - \frac{T_0}{T}$, and therefore presents no angular point. It becomes zero for $T = T_0$, which corresponds to the minimum exergy of the isobar chosen.

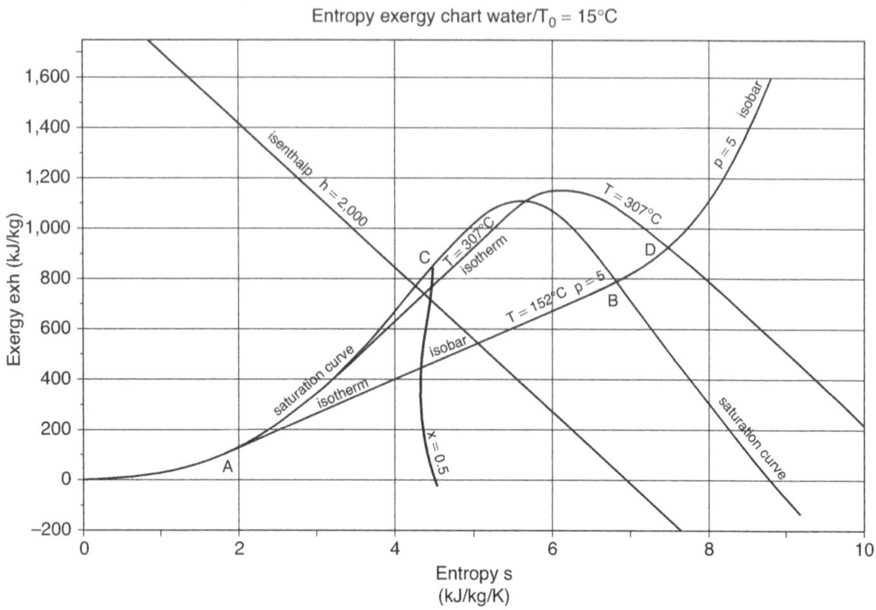

FIGURE 2.6.13 (h, xh) diagram.

In the liquid zone, they practically merge with the vaporization curve for the reasons given for the Mollier diagram. In the mixed zone, since pressure and temperature are linked by the saturation pressure law, isobars are straight line segments AB with slope $\theta = 1 - T_0 / T$. As a first approximation, they are tangent to the boiling curve at point A corresponding to the start of boiling. In the gas zone, isobars diverge tangentially from AB (blue dashed curve). At high temperatures, θ approaches 1, and they progressively approach straight lines parallel to isentropes.

The critical isobaric curve starts tangentially to the vaporization curve at C (Figure 2.6.14).

2.6.1.5.2 Form of Isotherms

In the mixed zone, isotherms coincide with isobars. Depending on the value of T_0 and temperature, they then pass through a maximum or exhibit a sudden decrease in slope to gradually tend toward the vertical, since for the corresponding ideal gas, enthalpies depend only on temperature. In the vapor zone, as shown in Figure 2.6.13, isotherms (high slope) and isobars (low slope) intersect at large angles (point D), providing very accurate reading.

The isotherm at temperature T_0 is horizontal and tangent to the minimum of the saturation curve. Generally, it is chosen as the zero scale for ordinates.

2.6.1.5.3 Form of Isovolumes

It can be shown that the slope of isovolumes is slightly greater than that of isobars.

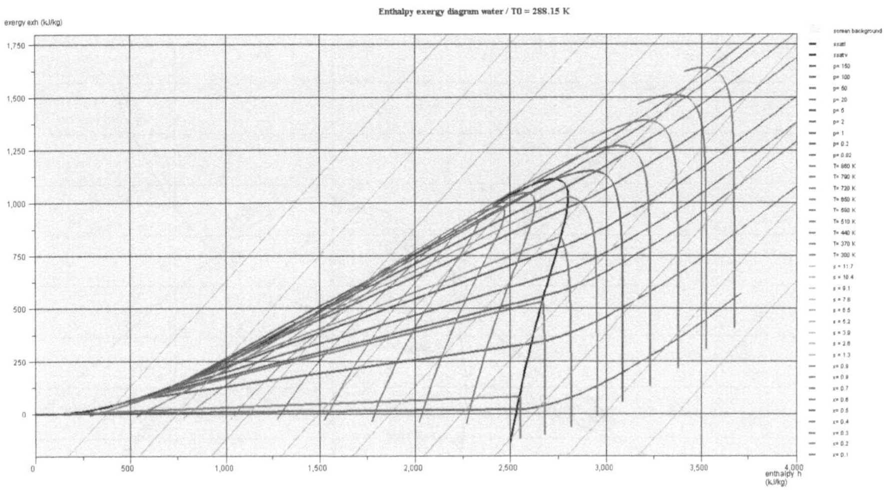

FIGURE 2.6.14 (h, xh) exergy diagram for water.

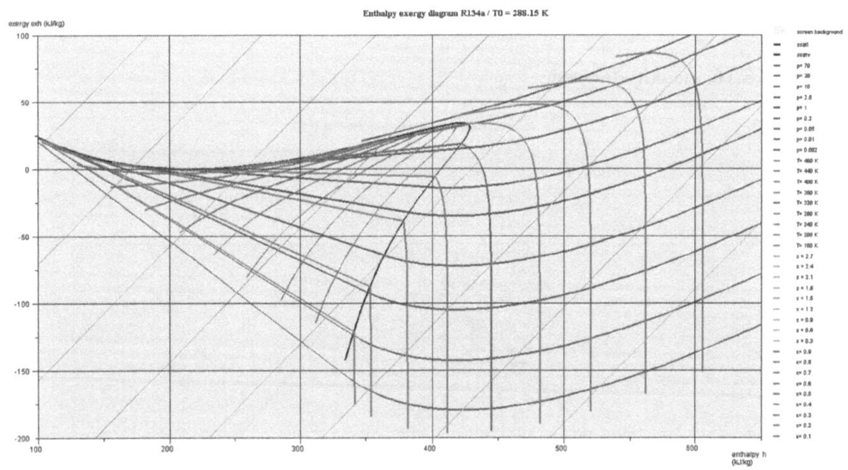

FIGURE 2.6.15 (h, xh) exergy diagram for R134a.

2.6.1.5.4 *Form of Isentropes*

Since $dx_h = dh - T_0\, ds$, $\left(\dfrac{\partial x_h}{\partial h}\right)_s = 1$, the slope of isentropes equals unity: isentropes are straight lines with a 45° slope in a vertical rectangular frame. In the example given in Figure 2.6.14, whose axes do not have the same scale, they have a slightly smaller angle.

As shown in this figure, it is very easy to identify on an exergy diagram, for a given fluid temperature, the pressure leading to maximum exergy: it is the one passing through the maximum of the corresponding isotherm.

Figures 2.6.14 and 2.6.15 show two examples of enthalpy exergy diagrams, one for water and the second for R 134a, prepared for $T_0 = 15°C = 288.15$ K.

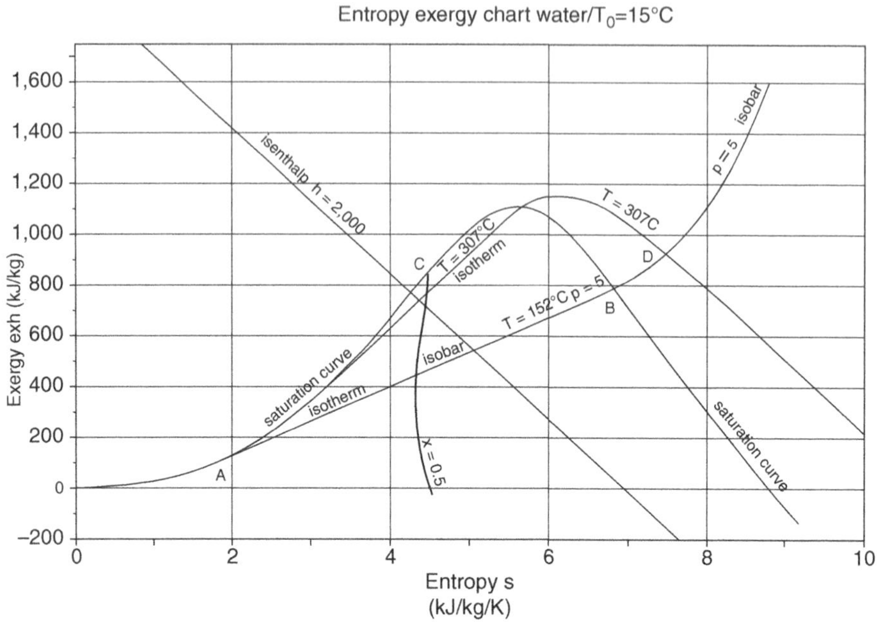

FIGURE 2.6.16 (s, xh) diagram.

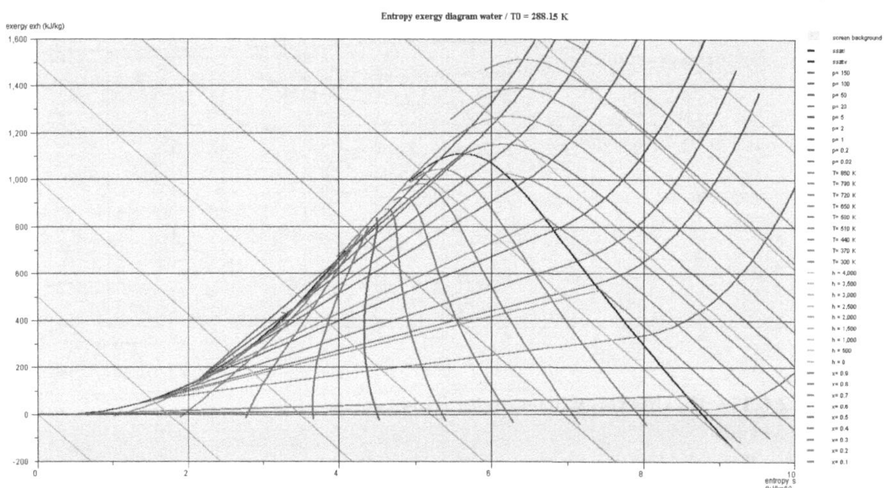

FIGURE 2.6.17 (s, x_h) exergy diagram for water.

2.6.1.6 Exergy Entropy Diagram (s, x_h)

In the entropy exergy diagram (Figures 2.6.16–2.6.18), entropy is the abscissa and exergy is the ordinate.

Like its predecessor, this diagram is easily constructed from the Mollier diagram once the reference temperature T_0 is known.

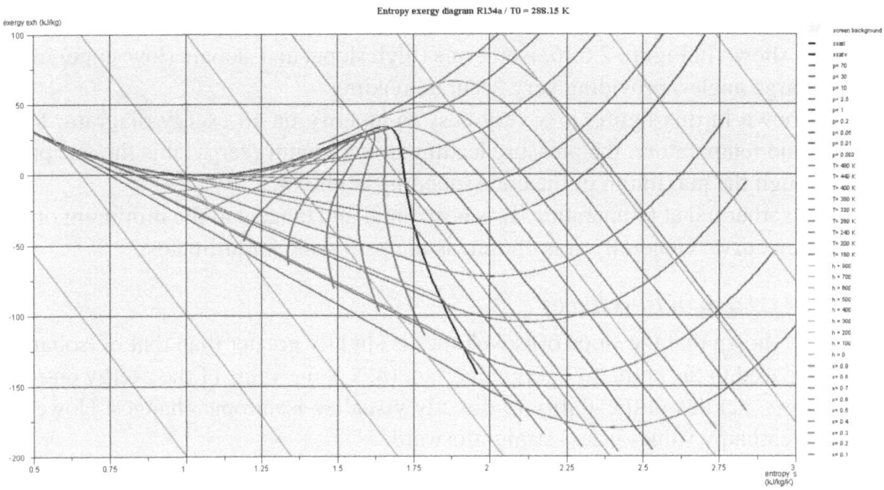

FIGURE 2.6.18 (s, x_h) exergy diagram for R134a.

The critical point here is still to the left of the vaporization curve maximum.

2.6.1.6.1 Form of Isobars

We have: $dh = vdP + Tds$, and $dx_h = dh - T_0\,ds$, hence $dx_h = vdP + (T - T_0)ds$, and $\left(\frac{\partial x_h}{\partial s}\right)_P = (T - T_0)$.

The slope of isobars at each point equals $(T - T_0)$ and therefore presents no angular point. It becomes zero for $T = T_0$, which corresponds to the minimum exergy of the chosen isobar.

In the liquid zone, isobars practically merge with the vaporization curve for the reasons stated previously. In the mixed zone, since pressure and temperature are linked by the saturation pressure law, isobars are straight line segments AB with slope $(T - T_0)$. As a first approximation, they are tangent to the boiling curve at point A corresponding to the start of boiling. In the gas zone, isobars diverge tangentially from AB. At high temperatures, they have a paraboloid appearance.

The critical isobaric curve starts tangentially to the vaporization curve at C.

2.6.1.6.2 Form of Isenthalps

Since $dx_h = dh - T_0\,ds$, $\left(\frac{\partial x_h}{\partial s}\right)_h = -T_0$, the isenthalp slope is equal to $-T_0$: isenthalps are straight lines with negative slope. Given the difference in scale, their apparent slope is close to 45°.

2.6.1.6.3 Form of Isotherms

In the mixed zone, isotherms coincide with isobars. Depending on the value of T_0 and temperature, they then pass through a maximum or exhibit a sudden decrease in slope to gradually move toward a straight negative slope parallel to isenthalps, since for the corresponding ideal gas, enthalpies depend only on temperature. In the vapor zone, as shown in Figure 2.6.16, isotherms (high slope) and isobars (low slope) intersect at large angles, providing very accurate reading.

As shown in this figure, it is very easy to identify on an exergy diagram, for a given fluid temperature, the pressure leading to maximum exergy: it is the one passing through the maximum of the corresponding isotherm.

The isothermal at temperature T_0 is horizontal and tangent to the minimum of the saturation curve. Generally, it is chosen as the zero scale for ordinates.

2.6.1.6.4 Form of Isovolumes

It can be shown that the slope of isovolumes is slightly greater than that of isobars.

Compared to the enthalpy exergy diagram (h, x_h), the value of the exergy entropy diagram (s, x_h) lies in the ability to directly visualize isentropic changes. However, reading enthalpy values is less straightforward.

Figures 2.6.17 and 2.6.18 show two examples of entropy exergy diagrams, one for water and the second for R134a, prepared for $T_0 = 15°C = 288.15$ K.

2.6.1.7 Ideal Gas Diagrams

The diagrams above are well suited to representing real fluids. When the ideal gas assumption is valid, however, most of them lose their usefulness, as isotherms or isobars become much simpler.

In practice, the entropy diagram is sufficient for most applications. Indeed, the Mollier diagram for an ideal gas is an entropy diagram whose ordinate scale has undergone an affine transformation, since $\Delta h = c_p \Delta T$. Exergy diagrams may be derived by recombination of axes, as can the (h, ln(P)) diagram.

Traditionally, computational difficulties were such that diagrams were established for a generic ideal gas, which is no longer necessary given the availability of computers. The interactive diagrams mentioned above allow construction of entropy diagrams for a very large number of gases from pure gases available in the databases (see Figure 2.6.19).

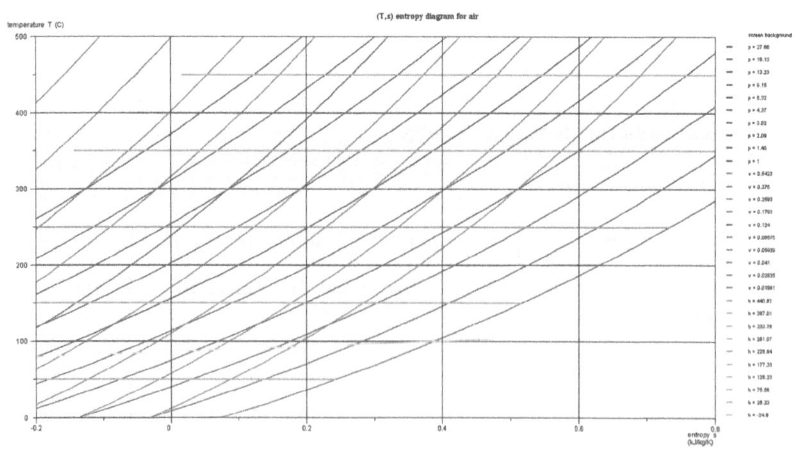

FIGURE 2.6.19 Air (T,s) entropy diagram.

2.6.2 Moist Mixtures: Properties and Diagrams

Many commonly used gases are mixtures whose composition may vary due to condensation or vaporization of one of their components. In practice, the component that changes state is mostly water, which justifies dedicating a special chapter in this book to mixtures of gases and water vapor, hereinafter referred to as moist mixtures.

Dry air in particular is rarely found in nature: it almost always contains water in the form of gas, liquid, or ice crystals. Controlling air humidity determines comfort conditions and the preservation of food and many products.

In this section, we limit ourselves to the principles for calculating thermodynamic properties of moist mixtures. While the most numerous practical applications involve air, our presentation will cover all gases for which our assumptions are valid, that is to say the vast majority of cases. For example, the humidity properties of combustion products, which almost always contain water, may be determined by the methods presented here.

2.6.2.1 Definitions and Conventions

When examining a moist mixture, we are dealing with a mixture of a non-condensable gas, which we call the dry gas, and water that may condense. Note that we will not discuss mist here; we can assume that the volume occupied by any condensed water is negligibly small compared to that occupied by the gas phase. We will therefore neglect mist. Under these conditions, for a given volume, the total mass of the gas phase can vary, while the mass of non-condensable constituents remains constant. Therefore, it is customary to refer all thermodynamic properties of the mixture to the invariant mass, which is the dry gas.

Everything proceeds as if the moist mixture were a mixture of two substances: the dry gas, whose composition is fixed, and water that may be present in one or more phases. As discussed later, it is essential for carrying out calculations to clearly specify which substance the thermodynamic quantities used relate to, and what are the reference units and origin. The notations will be chosen accordingly.

We call **specific humidity** w the ratio of the mass of water contained in a given volume of moist mixture to the mass of dry gas contained in this volume.

The water in a moist mixture can be characterized in various ways: through its specific humidity, which we just introduced, but more commonly by its mole or mass fraction, or by its partial pressure, which is of particular interest here. As long as the partial pressure of water remains below its saturation pressure at the gas temperature, the water is in the form of vapor. Otherwise, it exists at least partly in the condensed liquid or solid state. There is therefore an upper limit to the amount of water vapor that may be contained in a moist mixture. It depends on the temperature and pressure. When, at a given pressure, the temperature drops below the saturation temperature of water vapor, water vapor begins to condense as a mist or on cold walls that define the system, if they exist.

Given its practical importance, the saturated state is the reference, and we call **relative humidity** ε the ratio of the partial pressure of water vapor divided by its saturation vapor pressure at the temperature of the mixture. This ratio equals 1 (or 100%) when the vapor begins to condense. Otherwise, it is less than 1.

We call **specific enthalpy q′** and **specific volume v_{spec}** the enthalpy and volume of the moist mixture referred to 1 kg of dry gas, respectively.

2.6.2.2 Principles of Calculation

We have seen that a condensable fluid remains comparable to an ideal gas in the immediate vicinity of the liquid state when its pressure is lower than its critical pressure. When this fluid is a mixture component, this rule applies taking into account the partial pressure.

As long as the liquid state does not appear, the mixture behaves like an ideal gas. When the mixture temperature drops below the condensation temperature corresponding to the partial pressure of the condensable component in the dry gas, liquid begins to appear. In most cases, although this is not strictly true, it is reasonable to consider that this liquid phase is pure, consisting exclusively of the condensable component.

At that point, provided that equilibrium is established, experience shows that the situation is as follows:

- The vapor–liquid equilibrium relationship is satisfied by the condensable component, as if the other constituents did not exist. Its saturated vapor (index sv) partial pressure, given by the saturation pressure law, is therefore $P_{SV} = P_S(T)$.
- Dalton's law applies to the calculation of thermodynamic functions of the gas phase.
- The law of phase mixture applies to the calculation of thermodynamic functions of the liquid and gas phases (assuming that the interfacial tensions are negligible).

To calculate the thermodynamic properties of the mixture in the presence of a liquid phase, the procedure is as follows:

- Knowing the temperature of the medium, one can determine the partial pressure of the condensable component. We deduce the sum of partial pressures ($P - P_{SV}$) of the other constituents. As we know their relative molar fractions, we can completely determine the composition of the gas phase and the number of moles of the condensable component in the gaseous state. By difference, we know the number of moles liquefied.
- Dalton's law and the phase mixture law are then used to calculate all the properties of the mixture. It is clear that in this case the composition of the mixture changes as a function of the quality of the condensable component, and this must be taken into account in the calculation of thermodynamic properties.

In Thermoptim, the calculation of moist mixtures is performed using this method. It is thus possible to study the behavior of a vapor and gas mixture of any composition.

Table 2.6.1 shows the equivalence between notations used in this book and Northern America.

TABLE 2.6.1
Moist Air Nomenclature

	This Book	Northern America
Humidity ratio or specific humidity	w	W
Relative humidity	ε	Φ or RH
Dry-bulb temperature	T	T
Dew-point temperature	Tr	Td
Wet-bulb temperature	t′	t*
Specific enthalpy	q′	H
Specific volume	v′	vspec

2.6.2.3 Key Relationships

We will use indices dg to describe the dry gas, and mm for the moist mixture.

2.6.2.3.1 Case Where the Moist Mixture Composition Is Known

By definition of the partial pressure of water:

$$P_{vap} = x_{H_2O} \cdot P \qquad (2.6.1)$$

The molar mass of the dry gas is:

$$M_{dg} = \frac{1}{1 - x_{H_2O}} \sum M_{nc} x_{nc} \qquad (2.6.2)$$

nc representing the various non-condensable constituents.

The specific humidity is then by definition:

$$w = \frac{y_{H_2O}}{y_{dg}} = \frac{M_{H_2O}\, x_{H_2O}}{M_{dg}\, x_{dg}} \qquad (2.6.3)$$

If P_{sv} is the saturation vapor pressure of water:

If $P_{vap} \leq P_{sv}$ $\quad \varepsilon = \frac{P_{vap}}{P_{sv}}$

$$w = \frac{M_{H_2O}\, P_{vap}}{M_{dg}\, P - P_{vap}} \qquad (2.6.4)$$

If $P_{vap} > P_{sv}$ $\quad \varepsilon = 1$

$$w = w_{sat} = \frac{M_{H_2O}\, P_{sv}}{M_{dg} P - P_{sv}} \qquad (2.6.5)$$

$$w - w_{sat} = \text{condensate} = \frac{M_{H_2O}}{M_{dg}}\left[\frac{P_{vap}}{P - P_{vap}} - \frac{P_{sv}}{P - P_{sv}}\right] \quad (2.6.6)$$

The values of specific volume v_{spec} and specific enthalpy q′ can then be determined, the latter being calculated with an enthalpy zero at 0°C.

Generally (and this is the case in Thermoptim), the reference state for calculating gas enthalpies is taken at the standard value of 298 K. However, Heating Ventilation and Air Conditioning (HVAC) engineers customarily choose 0°C as the reference state for dry air and for saturated liquid water. This results in a mismatch between the enthalpies of moist gas as they are **commonly calculated and those represented on standard psychrometric diagrams.** This difference varies depending on the specific humidity.

In practice, since the enthalpies can be referred to three different references, one must know how to distinguish them to avoid making mistakes. In what follows, we will use as appropriate:

- the moist mixture enthalpy hmm (hmm = 0 at T = 298 K)
- the dry gas enthalpy hdg (hdg = 0 at T = 298 K)
- the moist mixture specific enthalpy q′ (q′ = 0 at T = 0°C, liquid water).

In general, unless otherwise indicated, the first is used by default, whereas in calculations specific to moist mixtures, it is usually the third form that is used.

In what follows, the index water indicates that the calculation is performed with the equations of water as a real fluid, and the index H_2O indicates that it is conducted with the equations of water treated as an ideal gas. h_{vwater} is the enthalpy of water in the vapor state, and L_{0water} represents the enthalpy of vaporization of water at 0°C.

$$q'(t,w) = h_{dg}(t) - h_{dg}\left(0^{\circ}C\right) + w h_{vwater}\left(t, P_{vap}\right) \quad (2.6.7)$$

or by making the approximation that water vapor behaves like an ideal gas:

$$q'(t,w) = h_{dg}(t) - h_{dg}\left(0^{\circ}C\right) + w\left[h_{H_2O}(t) - h_{H_2O}\left(0^{\circ}C\right)\right] + wL_{0\ water} \quad (2.6.8)$$

By introducing h_{mm}, enthalpy of the (ideal) moist mixture:

$$h_{mm}(t) = \frac{h_{dg}(t) + w h_{H_2O}(t)}{1 + w} \text{ we get:}$$

$$q'(t,w) = \left(h_{mm}(t) - h_{mm}\left(0^{\circ}C\right)\right)(1 + w) + wL_{0\ water} \quad (2.6.9)$$

Similarly, $v_{spec} = v_{mm}(1 + w)$

In the supersaturated zone, the gas is saturated with water in the vapor state, the remainder being water in the liquid state. The specific enthalpy is then given by:

$$q'(t,w) = h_{dg}(t) - h_{dg}\left(0^{\circ}C\right) + w h_{vwater}\left(t, P_{SV}\right)$$

2.6.2.3.2 *Case Where the Dry Gas Composition Is Known*

When we know the dry gas composition, the calculations are performed slightly differently, because w is then given.

We deduce, M_{dg} being the dry gas molar mass:

$$P_{vap} = P\frac{w\dfrac{M_{dg}}{M_{H_2O}}}{1+w\dfrac{M_{dg}}{M_{H_2O}}} \tag{2.6.10}$$

$$\varepsilon = \frac{P_{vap}}{P_{vs}} \tag{2.6.11}$$

Other calculations are carried out as before.

2.6.2.4 Temperatures Used for Moist Mixtures

The study of moist mixtures requires us to introduce several temperatures whose definitions are important to understand, because for a given state of the gas, their values may be substantially different.

We call **dry-bulb temperature t** (°C) or T (K) the temperature indicated by a thermometer whose sensing portion is completely dry and is placed in the moist mixture. This is the temperature in the usual sense.

We call **dew-point tr** (°C) or Tr (K) the temperature at which the first drops begin to appear when the moist mixture is cooled at constant humidity and pressure. If water condenses as ice, it is also known as frost point temperature. This is the saturation temperature of the water at partial pressure P_{vap}.

With the previous notations, $T_r = T_{sat}(P_{vap})$ if $P_{vap} \le P_{sv}$, and $T_r = T_{sat}(P_{sv}) = T$, otherwise.

We call **wet-bulb temperature t′** (°C) or T′ (K) the temperature indicated by a thermometer whose sensing element is covered with a thin film of water evaporating due to strong mixing of the gas. In practice, the bulb of the thermometer is covered with a wick soaked with water. If measured under proper conditions, this temperature is substantially equal to the adiabatic saturation temperature that would be obtained by moistening the gas to saturation in an adiabatic device by spraying water at the equilibrium temperature.

One can show that its value is given by the following equation:

$$\text{enthalpy of liquid water at } t'(\text{saturated moisture at } t' - \text{initial moisture})$$

$$= \text{enthalpy of saturated mixture at } t' - \text{enthalpy of the initial mixture at } t$$

Relative to the dry gas, the equation is:

$$h_{lwater}(t')(w_{sat} - w) = h_{dg}(t') - h_{dg}(t) + w_{sat}\,h_{vwater}(t') - wh_{vwater}(t) \tag{2.6.12}$$

Relative to the specific units, it becomes:

$$q'(t', w_{sat}) - q'(t, w) = (w_{sat} - w)h_{l\,water}(t') \tag{2.6.13}$$

This is an implicit equation in t′, involving the mixture formed partly by the dry gas and partly by water. It can be solved in a temperature range between a value close to the dew point as a minimum and the dry-bulb temperature as the maximum.

As stated above, we must ensure that the reference values of enthalpies are consistent. Moreover, when t or t′ falls below 0°C, the equations of liquid water must be replaced by those of solid water (ice).

2.6.2.5 Moist Mixture Diagrams (Psychrometric Diagrams)

The main thermodynamic relations that we have given show that moist mixture variables and state functions are connected by relatively complex equations, which justify seeking an easy-to-use graphical presentation.

Two broad families of diagrams exist: the diagrams derived from that proposed by Carrier, with the dry-bulb temperature as the abscissa and the specific humidity as the ordinate (Figures 2.6.20a and 2.6.21), and Mollier diagrams (Figures 2.6.20b and 2.6.22), with specific humidity as the abscissa and the specific enthalpy as the ordinate.

The gas–water saturation curve is the smooth concave curve upward in the Carrier diagram (Figure 2.6.20a) and downward in the Mollier diagram (Figure 2.6.20b).

In the Carrier diagram, the area above the saturation curve corresponds to cases where water is in excess and exists as condensed liquid or solid: the gas is supersaturated. In the Mollier diagram, the reverse is true: this zone is located below the saturation curve.

The diagrams in Figures 2.6.20a and 2.6.21 relate to air at atmospheric pressure. They are equipped with different iso-value curves:

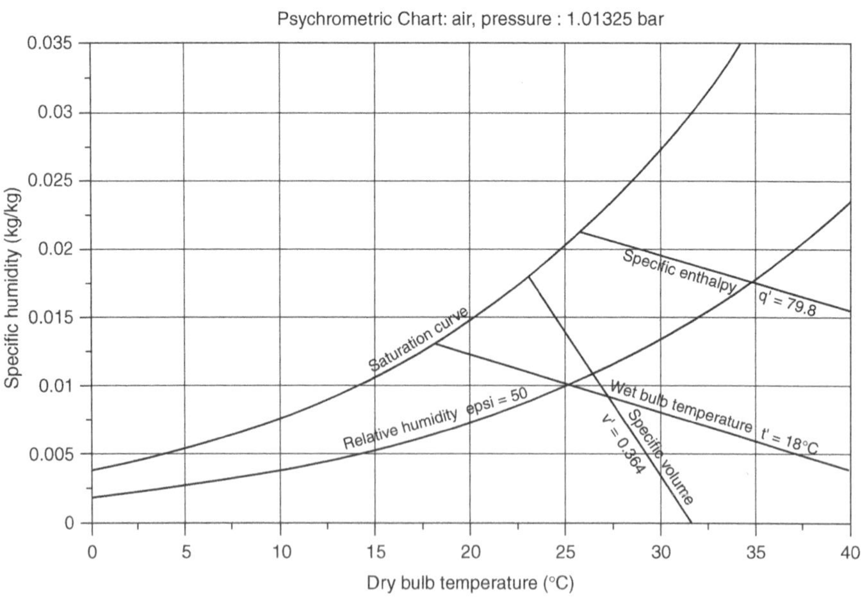

FIGURE 2.6.20A Carrier diagram.

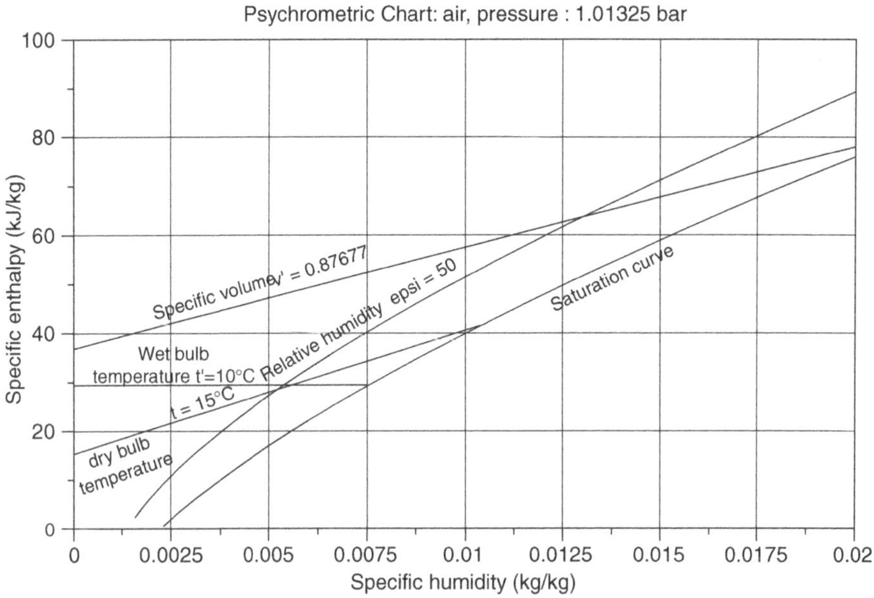

FIGURE 2.6.20B Mollier diagram.

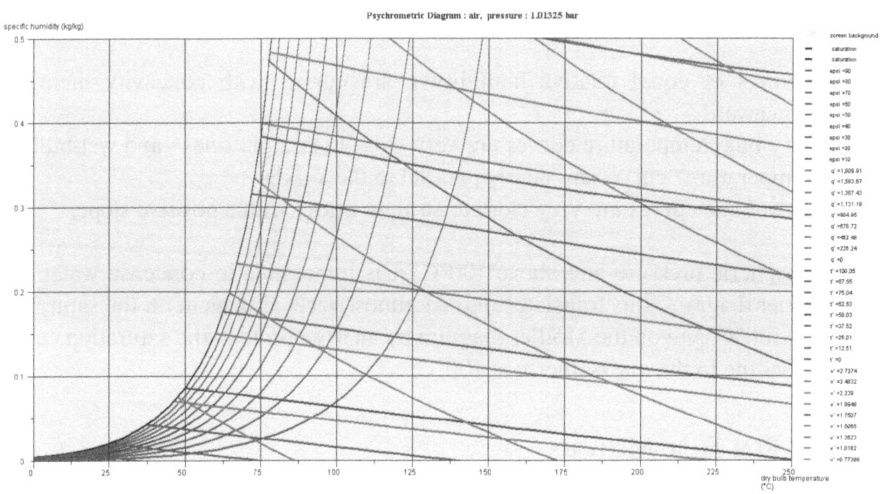

FIGURE 2.6.21 Carrier moist air psychrometric diagram.

- Curves of equal relative humidity ε, concave upward.
- Isenthalpic curves are very close to straight lines, and moist isotherm[1] curves are nearly parallel to them (usually not shown).
- Isovolume curves are very close to straight lines with a negative slope* stronger than the previous ones.

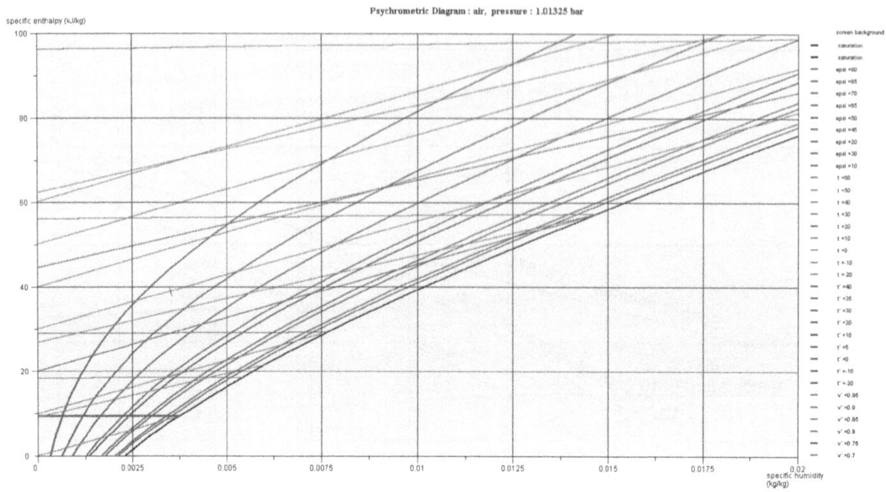

FIGURE 2.6.22 Moist air Mollier diagram.

The Mollier diagram in Figures 2.6.20b and 2.6.22 also relates to air at atmospheric pressure. It is equipped with different iso-value curves:

- Curves of equal relative humidity ε are curves with concavity facing downward.
- Dry-bulb temperature curves are very close to straight lines* and wet-bulb temperature* curves are nearly parallel to the x-axis.
- Isovolume curves are very close to straight lines* with a positive slope.

At atmospheric pressure and above 100°C, it is impossible to condense water. On the Carrier diagram, this translates into an almost vertical tangent on the saturation curve at 100°C, and on the Mollier diagram by an asymptote to the saturation curve, corresponding to the dry isotherm at 100°C.

2.6.3 Real Fluid Mixtures

In the previous section, we considered moist mixtures for which the dry gas could be assumed to behave as an ideal gas, allowing us to simplify the modeling. This section addresses real fluid mixtures that cannot be modeled as ideal gases.

Due to interactions between molecules of the various constituents, estimating thermodynamic properties of real fluid mixtures is considerably more difficult than for pure substances, whose diagrams are presented in Section 2.6.1. In the general case, this relates to chemical engineering and falls outside the scope of this book. However, we provide below some guidance on how to proceed, so that readers can understand the challenges involved.

2.6.3.1 Physical Phenomena Brought into Play

The behavior of mixtures in liquid–vapor equilibrium generally differs from that of pure substances. The molar or mass fraction of each component varies between limits that depend on pressure and temperature, due to the distillation that occurs. In the presence of several components, phase change is more complex than for a pure substance. Its graphical representation in a diagram is simple if the mixture is binary, the case to which we will restrict ourselves in this introduction.

To understand the phenomena that occur, consider the behavior of a binary mixture of propane and butane. Some properties of the pure components (critical pressure and temperature, boiling temperature at 1 bar and 2 bar) are shown in Table 2.6.2. They demonstrate that propane is much more volatile than butane, as at 1 bar they evaporate at −42.4°C and −0.7°C, respectively.

Consider a mixture of these two substances and examine how it behaves at constant pressure when its temperature changes. Figure 2.6.23, called an "equilibrium isobaric lens," shows the phase diagram of this mixture at a pressure of 2 bar.

TABLE 2.6.2
Pure Component Properties

	Pc (bar)	Tc (°C)	Ts (1 bar) (°C)	Ts (2 bar) (°C)
Propane	42.48	96.7	−42.4	−25.4
Butane	37.98	152.01	−0.7	18.8

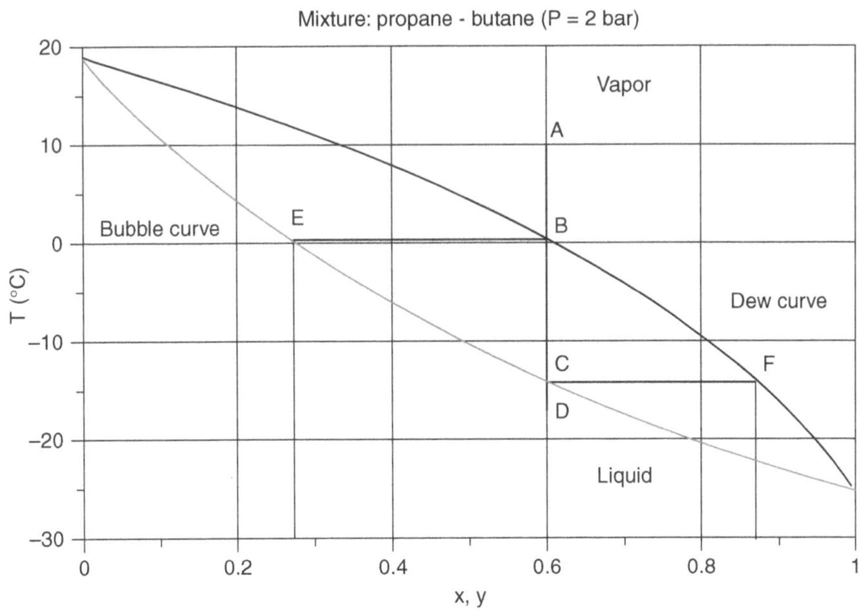

FIGURE 2.6.23 Isobaric equilibrium lens.

The vapor zone is at the upper right of the diagram, the liquid zone at the bottom left, and the liquid–vapor equilibrium in the lens delimited by the dew and bubble curves. The horizontal axis represents the mole fraction of propane, meaning that the value 0 corresponds to pure butane (T_s= 18.8°C) and the value 1 to pure propane (T_s=-25.4°C). As is customary when studying vapor mixtures, x represents the mole fractions in the liquid phase, y the mole fractions in the vapor phase, and z the overall mole fractions. The vertical axis is temperature in °C.

Consider a mixture of given composition (here $z_1=0.6$ for propane, $z_2=1-z_1=0.4$ for butane) in the vapor state, corresponding to point A at temperature 10°C. If the mixture is cooled at constant pressure (e.g., in a heat exchanger), the characteristic point moves along the vertical segment AB as it remains in the vapor state.

At B, at around 0°C, distillation begins on the dew curve (or drop curve), so named because the first drop of liquid appears. The two components liquefy together, but since butane is less volatile than propane, it condenses preferentially, so that the first drop composition differs from the average z. The mole fraction x in the liquid state of component 1 (here propane) is given by the abscissa of point E on the bubble curve, so called because in a reverse temperature change process it is on this curve that the first bubbles appear (here $x_1=0.28$), and the mole fraction of liquid butane is $x_2=1-x_1=0.72$.

Gradually, as cooling continues, the composition of the liquid phase evolves, with the mole fraction of propane given by the abscissa of the intersection of the bubble curve and the corresponding isotherm. The figurative point moves from E to C, while that of the gas phase moves from B to F. At point C, there is only liquid except for one last bubble of gas, whose composition is $y_1=0.85$ in propane and $y_2=1-y_1=0.15$ in butane. Beyond this point, the mixture remains in the liquid state with its original composition.

In other words, at a given pressure, the phase change of the mixture does not occur at constant temperature but with a temperature "glide" (distillation range) that can be more or less significant depending on the mixture. For example, for the above mixture at a pressure of 2 bar, the temperature glide is 14 K.

Now consider an evolution of the mixture at constant temperature and variable pressure. In the diagram of Figure 2.6.24, called an "equilibrium isotherm lens," the abscissa represents the propane quality and the ordinate represents the pressure in bar. Let us assume that the mixture has an overall composition ($z_1=0.6$ and $z_2=0.4$). If the pressure is lower than 2.78 bar, the mixture is entirely in the vapor state, and if it exceeds 4.3 bar, it is entirely liquid. Between these two pressures, it is in liquid–vapor equilibrium, and the mixture composition changes continuously in the liquid and vapor phases. At P=2.78 bar, the liquid phase has a composition ($x_1=0.28$, $x_2=0.72$). At P=3.7 bar, it is ($x_1=0.48$, $x_2=0.52$), and at P=4.3 bar it corresponds to the overall composition. Vapor phase compositions are as follows: ($y_1=0.6$, $y_2=0.4$) for P=2.78 bar, ($y_1=0.77$, $y_2=0.33$) for P=3.7 bar, and ($y_1=0.85$, $y_2=0.15$) for P=4.3 bar.

A special case is the so-called azeotropic mixtures, where the bubble and dew curves meet for a given overall composition. It can be shown that an azeotropic mixture represents a pressure extremum at constant temperature or a temperature extremum at constant pressure. Such a mixture behaves in practice as a pure substance.

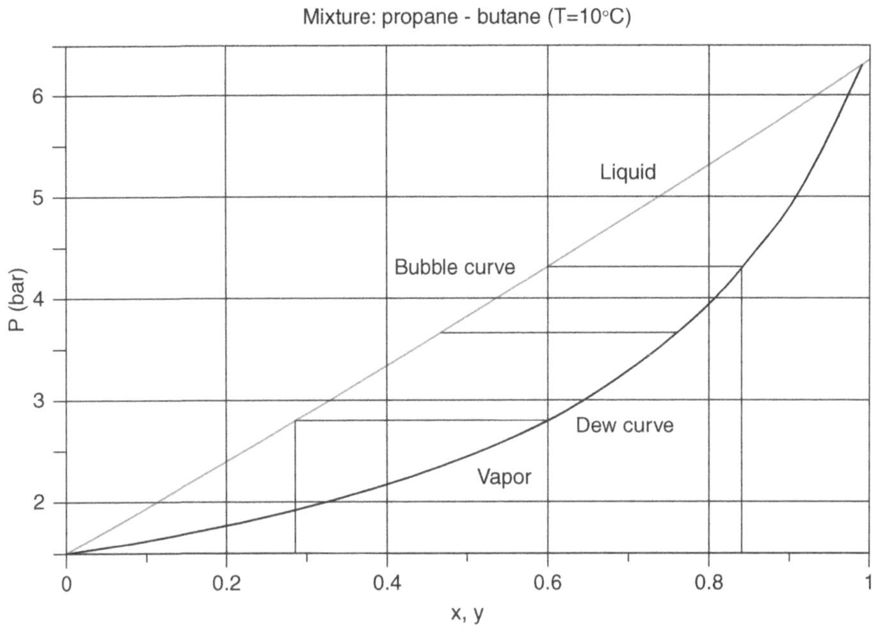

FIGURE 2.6.24 Isotherm equilibrium lens.

Finally, when the temperature glide is small (<1 K), the error committed by neglecting the distillation is very low. This is called a quasi-azeotropic mixture. We chose the propane/butane system here because the glide is significant and helps demonstrate the phenomena specific to mixtures.

We introduced above two diagrams—binary mixture equilibrium isothermal and isobaric lenses—which are very useful for qualitatively understanding the phenomena involved. However, quantitatively, they are only valid for a single pressure or temperature, and above all they contain no information on the enthalpies involved. They are of limited interest for applications in power engineering.

2.6.3.2 Diagrams for Refrigerants with Temperature Glide

For refrigeration applications, the composition of a refrigerant blend remains constant in the vapor and liquid phases (if we neglect the influence of lubricating oil). In the liquid–vapor equilibrium zone, this is not strictly true, but almost always one does not require a precise calculation of the exact composition of the gas phase and liquid phase at the outlet of the expansion valve and the evaporator. By analogy with pure substances, we generalize the concept of vapor quality to that of mean quality, equal to the mass of the vapor phase (all components combined) relative to the total mass of liquid and vapor phases. We can then use a fixed composition mixture model, which is much simpler. An equation numerically adjusted to the experimental values of a given mixture will thus be both more accurate and easier to calculate than the general formulation we have just described. Furthermore, equations of state have been developed specifically for these applications; the reader will find references in the literature.

We present below the diagrams of two blends, R404A and R407C (Figures 2.6.27 and 2.6.28). In what follows, we shall only point out the differences introduced by the temperature glide, which is lower for R404A than for R407C.

2.6.3.2.1 *Refrigerant Blend Diagrams*

Essentially, at a given pressure, the differences between the component saturation temperatures introduce a temperature glide in the liquid–vapor equilibrium zone: when boiling starts, the temperature (bubble temperature) is lower than when boiling ends (dew point), whereas they are equal for pure substances. Thus, pressure and temperature do not remain constant during boiling or condensation. In Clapeyron (P, v), entropy (T, s), and (h, ln(P)) diagrams, isobars or isotherms are not horizontal in the vapor–liquid equilibrium zone (Figures 2.6.27 and 2.6.28).

As shown in Figures 2.6.25–2.6.28, an isobar is now represented in this area by an ascending line segment in the entropy diagram, and an isotherm by a downward-sloping line in the (h, ln(P)) diagram. In the vapor–liquid equilibrium zone of Mollier (h, s) or exergy diagrams, the difference is much smaller: isobars are slightly more ascending than isotherms.

The existence of a temperature glide has a direct impact on the design of heat exchangers used in refrigeration cycles. In first approximation, these can be considered as isobaric, so that temperature differences between the coolant and other fluids are not the same as those observed in machinery using a pure fluid. Generally speaking, for blends with temperature glide, heat exchangers should be of the counter-flow type, which was not imperative with pure fluids.

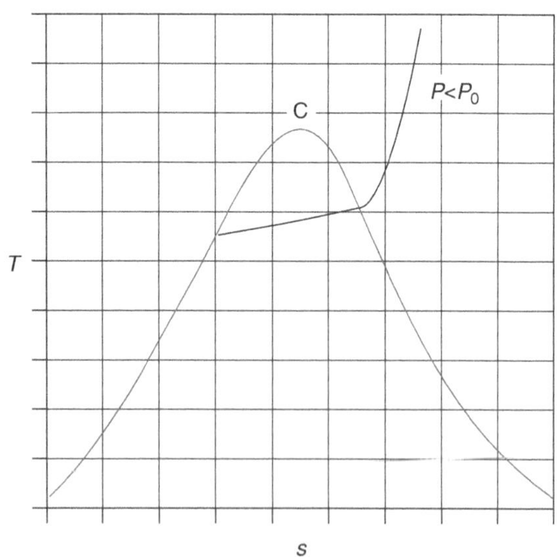

FIGURE 2.6.25 Entropy diagram with temperature glide.

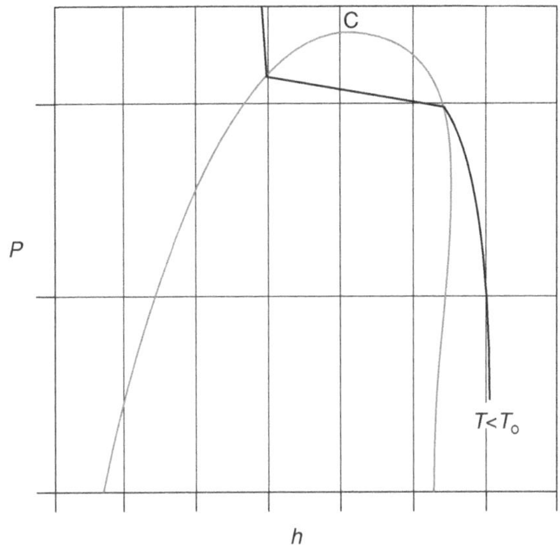

FIGURE 2.6.26 (h, ln(P)) diagram with temperature glide.

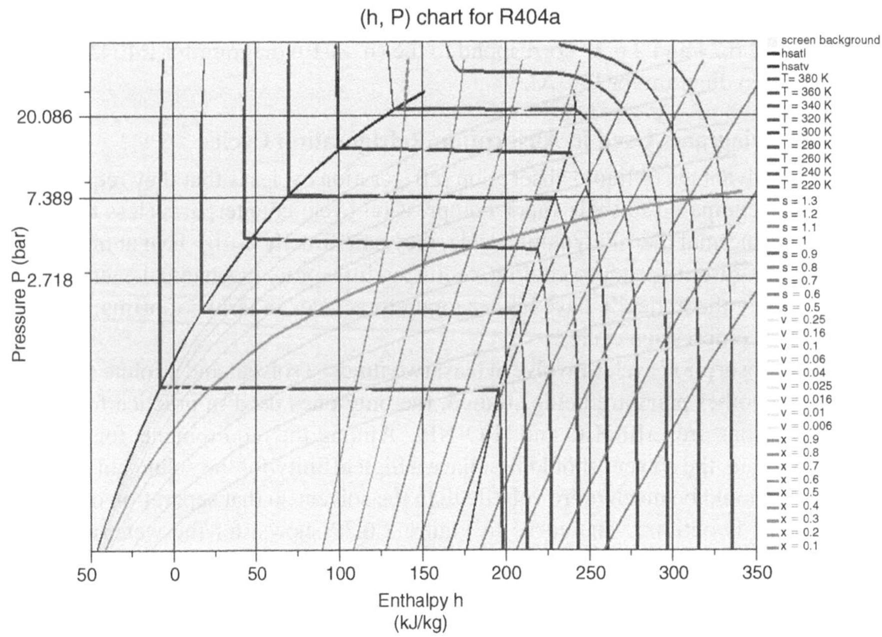

FIGURE 2.6.27 (h, ln(P)) diagram for R404a.

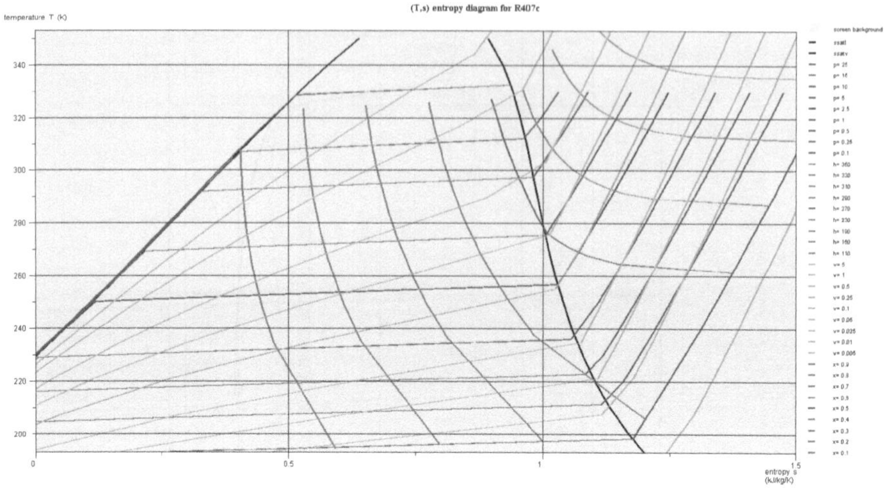

FIGURE 2.6.28 Entropy diagram for R407c.

Figures 2.6.27 and 2.6.28 correspond to the (h, ln(P)) diagram for R404A and the entropy (T, s) diagram for R407C.

2.6.3.3 Diagrams Used in Absorption Refrigeration Cycles

The main advantage of liquid absorption refrigeration cycles is that they require only low power compared to their vapor compression cycle counterparts (less than 1%). Using a tri-thermal thermodynamic cycle, they can directly utilize heat at medium or high temperatures to produce cooling, requiring little or no mechanical energy input. As such, they theoretically have higher total efficiencies in terms of primary energy than vapor compression cycles.

Liquid absorption cycles involve at least two fluids: a solvent and a solute (the coolant). While other pairs are being studied, the only ones used in practice for almost all applications are LiBr-H_2O and H_2O-NH_3. Among the requirements for a pair to be appropriate, the solvent should first have a high affinity for the solute, and second, the solute should be much more volatile than the solvent so that separation of the two components is optimal. The curve in Figure 2.6.29 shows, for the water-ammonia pair, the relationship between the mass fraction of ammonia in liquid and vapor at different pressures (1, 2, 5, 10, and 20 bar). It clearly shows that as soon as the liquid mass fraction of solute x exceeds 0.4, the ammonia content of the vapor exceeds 0.95. For the lithium bromide–water mixture, the separation is even stronger, so that one can legitimately assume that water vapor is pure.

With the LiBr-H_2O pair, water is the refrigerant, which imposes two constraints: first, working pressures are very low given the saturation pressure law of water, and second, the minimum cycle temperature must be greater than 0°C. Machines using the lithium bromide–water pair are only used for air conditioning.

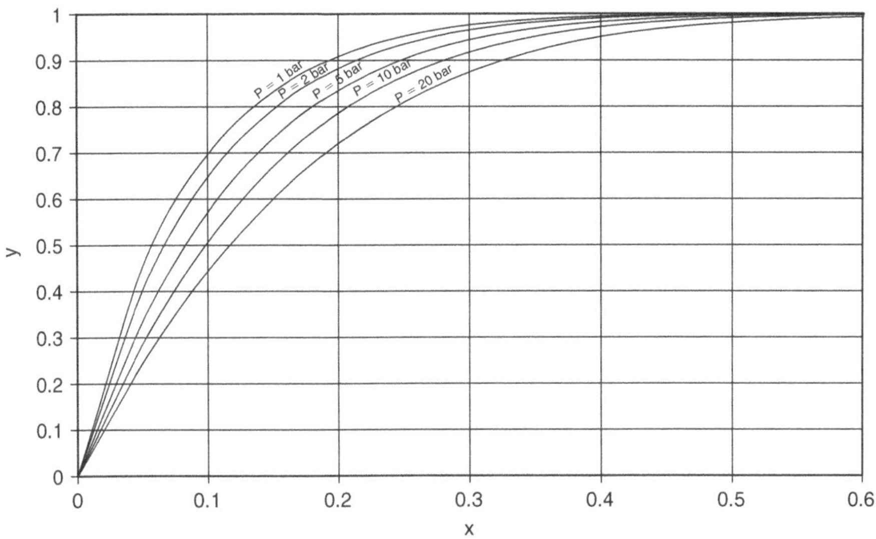

FIGURE 2.6.29 Vapor composition function of liquid NH_3-H_2O composition.

2.6.3.3.1 Oldham Diagram

The Oldham diagram displays the saturation pressure curves of the mixture in question for different values of its composition (mass fraction of solute or solvent, as appropriate). Pressure appears as the ordinate with a logarithmic scale, while the abscissa corresponds to the bubble temperature T_b. It is also possible to show the curves corresponding to the dew point, but this is of less interest.

If one chooses $(-1/T_b)$ as the temperature scale, the iso-quality curves are very close to straight lines. The extremes correspond to the vapor–liquid equilibrium of pure substances. An evolution of the rich or weak solution corresponds to a line segment tilted parallel to the nearest iso-quality curve. As the general shape of cycles in this diagram is close to a parallelogram, scientists customarily illustrate the architectures of the machines in the form of a diamond (Figure 2.6.32).

In machines using a LiBr-H_2O mixture, the difference in vapor pressure between the solvent (LiBr) and solute (H_2O) is such that we can neglect the mass fraction of solvent vapor, thereby simplifying calculations: the equilibrium diagram can then be directly calibrated as a function of the solution temperature. Incidentally, it is customary to present the LiBr-H_2O pair diagram as a function of solvent (LiBr) mass fraction rather than solute. Since water is likely to crystallize at low temperatures, the crystallization curve of the mixture often appears on the diagram, which corresponds to the lower operating limit of machines.

For this pair, ASHRAE (2001) provides equations (2.6.1) and (2.6.2), established by generalizing the mixture saturation law of the coolant (water) pressure, in which the water temperature t′ (°C) is replaced by a linear function of the solution temperature T (°C). P, expressed from the decimal logarithm, is the pressure in kPa, and X is the mass fraction of LiBr in the mixture (Table 2.6.3).

TABLE 2.6.3
Coefficients of Equations 2.6.14 and 2.6.15

A0	−2.00755	B0	124.937	C	7.05
A1	0.16976	B1	−7.71649	D	−1,596.49
A2	−3.13E-03	B2	0.152286	E	−104,095.5
A3	1.98E-05	B3	−7.95E-04		

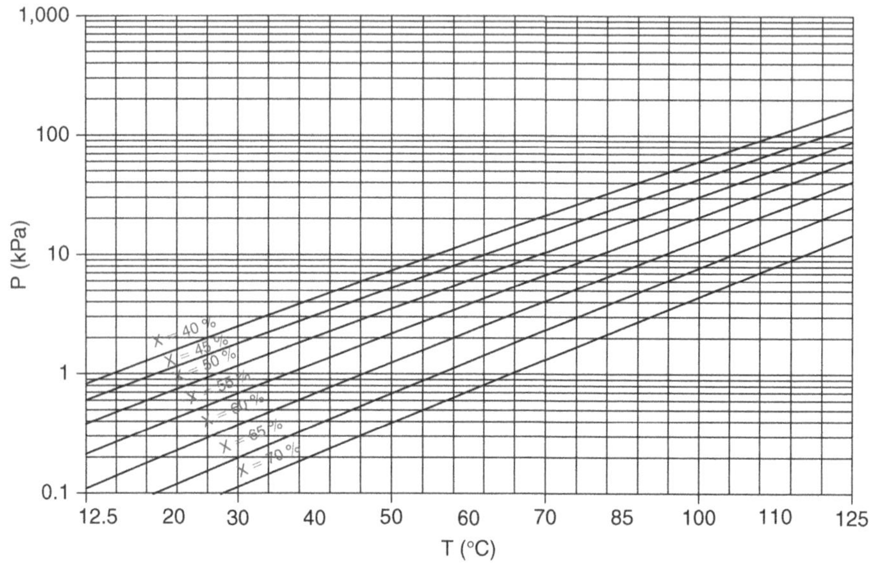

FIGURE 2.6.30 Oldham diagram for LiBr-H_2O.

These equations, which were used to prepare the diagram in Figure 2.6.30, are valid in the following ranges of values:

$$-15 < t' < 110°C, \quad 5 < t < 175°C, \quad 45 < X < 70\%$$

$$\log(P) = C + \frac{D}{t' + 273.15} + \frac{E}{(t' + 273.15)^2} \tag{2.6.14}$$

$$t' = \frac{t - \sum_{i=0}^{3} B_i X^i}{\sum_{i=0}^{3} A_i X^i} \tag{2.6.15}$$

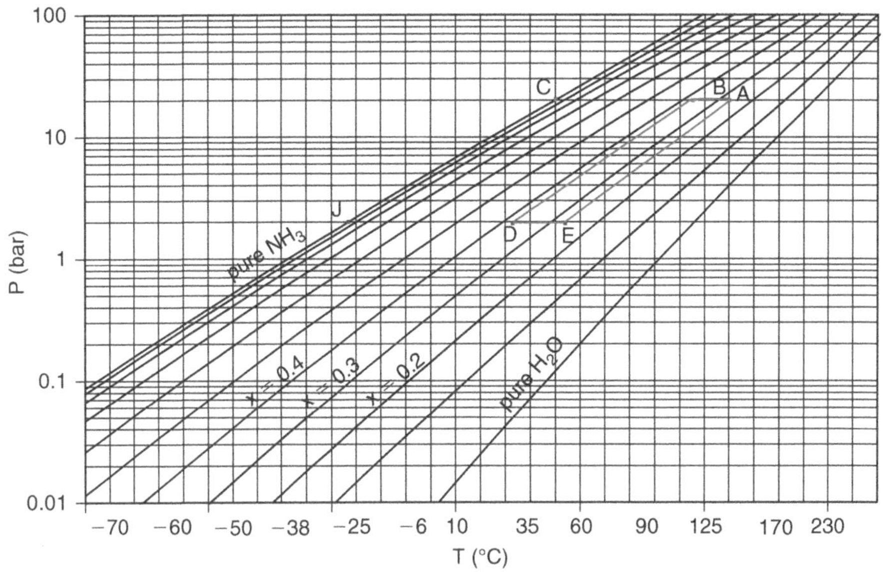

FIGURE 2.6.31 Cycle representation in the Oldham NH_3-H_2O diagram.

Figures 2.6.30 and 2.6.31 provide Oldham diagrams for LiBr-H_2O and H_2O-NH_3 pairs.

2.6.3.3.2 Merkel Diagram

The Oldham diagram is easy to understand and use as discussed above but, like isobar or isotherm equilibrium lenses, it has the disadvantage of not providing any information about the energies involved.

The Merkel diagram (or enthalpy-concentration diagram) is constructed with enthalpy as the ordinate and the mass fraction of solute (or often the solvent for the LiBr-H_2O pair) as the abscissa. It allows us to superimpose information from multiple isobar equilibrium lenses while showing enthalpies. Figure 2.6.32 presents the Merkel diagram for the ammonia–water mixture. There are five types of curves:

- A set of bubble curves at P = Const, continuous lines with concavity turned upward.
- A set of dew curves for the same pressures, solid lines with concavity turned downward.
- A set of liquid zone isotherms, dashed in the lower part of the diagram, which are valid regardless of pressure, since the influence of pressure on the enthalpy of the liquid is negligible in first approximation (for example, the difference in enthalpy of the liquid solution at 30% ammonia by weight is only 5 kJ/kg at 0°C when pressure increases from 0.5 to 50 bar, while the useful part of the diagram covers more than 1,000 kJ/kg in the liquid zone).

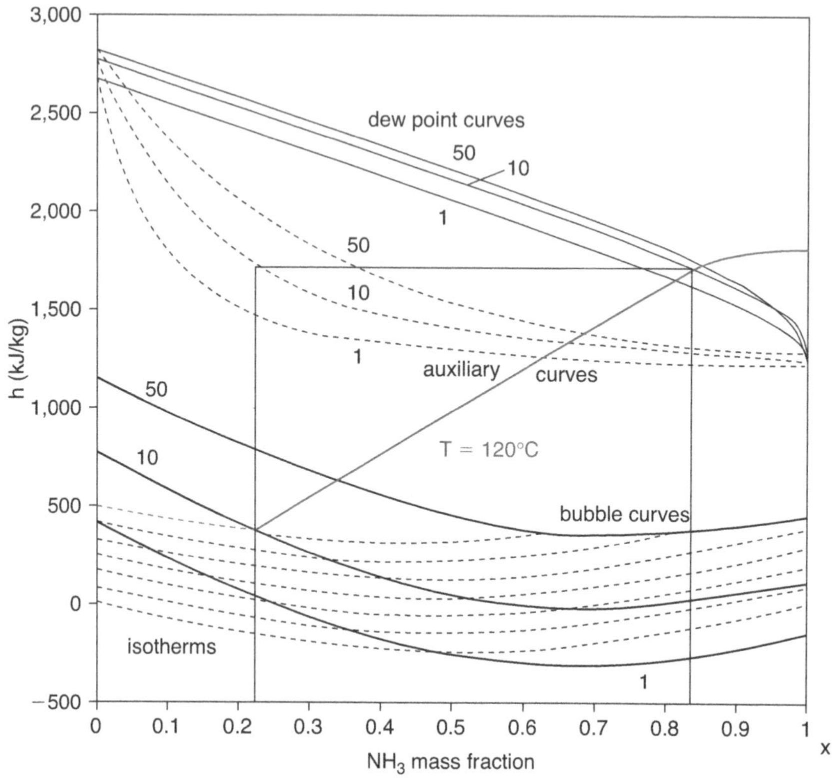

FIGURE 2.6.32 Isotherms in the Merkel diagram.

- A set of curves called auxiliary or construction curves, dashed in the upper part of the diagram, which, for a given pressure and knowing the solute liquid mass fraction x, determine the solute vapor mass fraction y.
- A set of refrigerant iso-quality curves in the vapor phase, which, for a given pressure, provide the mass fraction of liquid corresponding to the chosen vapor iso-quality (this last set of curves is sometimes omitted from the Merkel diagram, as in this example).

However, isotherms in the vapor zone do not appear in the Merkel diagram because they depend on pressure and the diagram would become unreadable.

Note that this diagram superimposes two kinds of curves: liquid–vapor equilibrium and isotherm curves in the liquid state. There may be difficulties in interpreting the cycle points if one does not specify whether they are in liquid–vapor equilibrium or not.

In the H_2O-NH_3 pair diagram of Figure 2.6.32, the saturation pressure for $x=0.22$ and $T=120°C$ is 10 bar. To determine the mass fraction value y of the vapor, we draw the vertical segment passing through $x=0.22$ to the auxiliary curve corresponding to 10 bar, then the horizontal segment from that point to the dew isobar for 10 bar, and then we draw the vertical to the horizontal axis, which gives $y=0.83$.

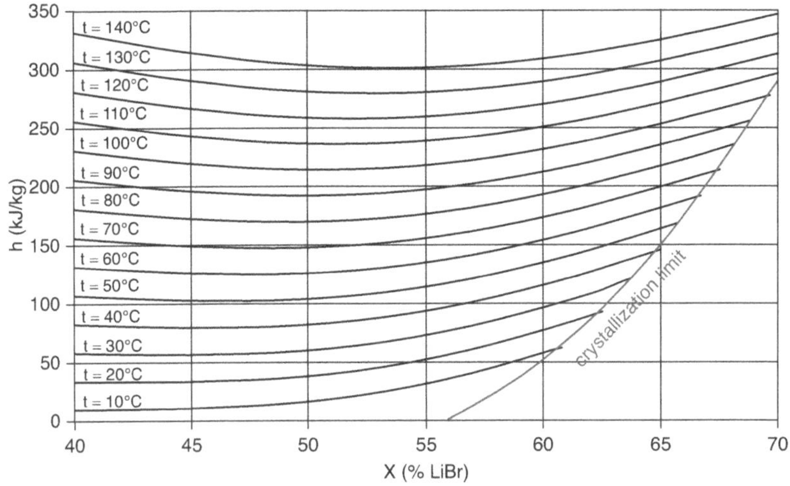

FIGURE 2.6.33 Merkel diagram of the LiBr-H_2O mixture.

TABLE 2.6.4
Coefficients of Equation 2.6.16

A0	−2024.33	B0	18.2829	C0	−0.037008214
A1	163.309	B1	−1.1691757	C1	2.89E-03
A2	−4.88161	B2	0.03248041	C2	−8.13E-05
A3	0.06302948	B3	−4.03E-04	C3	9.91E-07
A4	−2.91E-04	B4	1.85E-06	C4	−4.44E-09

We have already noted that for the LiBr-H_2O pair we can neglect the mass fraction of solvent vapor, that is, consider that the refrigerant vapor behaves like pure water. The Merkel diagram is thus simplified, and we generally limit ourselves to constructing it by showing only the liquid zone isotherms, bounded on the lower right by the crystallization curve (Figure 2.6.33).

Given this simplification, it is possible to determine the LiBr-H_2O mixture enthalpy through a simple equation, as proposed by ASHRAE (2.6.16) (Table 2.6.4, ASHRAE 2001), where t is the temperature of the solution in °C, h is its specific enthalpy in kJ/kg, and X is the LiBr mass fraction of the mixture. This equation, which was used to establish the diagram of Figure 2.6.33, is valid within the following ranges of values: $5 < T < 165°C$, $40 < X < 70\%$. If this equation is chosen, the origin of the water vapor enthalpy is chosen at 0°C in the liquid state (which corresponds to the usual conventions for pure water diagrams).

$$h = \sum_{i=0}^{4} A_i X^i + t \sum_{i=0}^{4} B_i X^i + t^2 \sum_{i=0}^{4} C_i X^i \tag{2.6.16}$$

These are the equations used in the external substance model of the LiBr-H_2O pair implemented in Thermoptim.

2.7 CONCLUSION

This chapter presented the fundamental thermodynamic knowledge necessary for the study of energy technologies, prioritizing simplicity and practical application.

Following the introduction of basic concepts (open/closed systems, state variables, reversible processes), the energy exchanges during thermodynamic processes were evaluated. The First and Second Laws of Thermodynamics provided the theoretical foundation for subsequent calculations. The concept of exergy was introduced to reconcile energy quantity and quality.

The study of substance properties showed that they can be modeled by a cascade of nested models: perfect gases, ideal gases, and real condensable fluids, with particular attention given to moist mixtures, refrigerants, and real mixtures. Several thermodynamic diagrams (Clapeyron, entropy, Mollier, enthalpy-pressure, exergy) were presented to visualize properties and plot cycles.

This simplified approach makes applied thermodynamics accessible while maintaining the rigor required for energy calculations.

Examples illustrating how the concepts presented in this chapter can be applied using Thermoptim are available on the book's website.[2]

NOTES

1 By equating the dry gas to an ideal gas and neglecting in some cases the contribution of water vapor, these expressions are simplified and correspond to those of straight lines.
2 Available at: https://server.s4e2.com/crc/amtc/chapters/chapter2/_resources/index.html.

BIBLIOGRAPHY

ASHRAE (2001). *2001 ASHRAE Handbook — Fundamentals Handbook (SI ed.).* American Society of Heating, Refrigerating and Air-Conditioning Engineers. Atlanta, GA.

Bejan, A., Tsatsaronis, G. & Moran, M. (1995). *Thermal Design and Optimization.* John Wiley & Sons, New York. ISBN 978 0 471 58467 4.

Borel, L. & Favrat, D. (1987). *Thermodynamique et énergétique.* Vol. 2: *Exercices corrigés.* Presses Polytechniques et Universitaires Romandes, Lausanne.

Borel, L. & Favrat, D. (2005). *Thermodynamique et énergétique.* Vol. 1: *De l'énergie à l'exergie.* Presses Polytechniques et Universitaires Romandes, Lausanne.

Çengel, Y.A. & Boles, M.A. (2006). *Thermodynamics: An Engineering Approach (SI Units).* 5th ed. McGraw-Hill Higher Education, Boston, MA. ISBN 978-0070606593.

Chase, M.W. Jr. et al. (1985). JANAF thermochemical tables. *Journal of Physical and Chemical Reference Data,* Vol. 14, Suppl. 1, p. 535.

Fer, F. (1970). *Thermodynamique macroscopique. Tome* 1*: Systèmes fermés.* Gordon & Breach Science Publishers, Paris.

Fer, F. (1971). *Thermodynamique macroscopique.* Tome 2*: Systèmes ouverts.* Gordon & Breach Science Publishers, Paris.

Holman, J.P. (1988). *Thermodynamics.* 4th ed. McGraw-Hill, New York.

Kotas, T.J. (1995). *The Exergy Method of Thermal Plant Analysis.* 2nd ed. Krieger Publishing Company, Malabar, FL.

Moran, M.J. (1989). *Availability Analysis: A Guide to Efficient Energy Use.* Prentice Hall, New York.

Moran, M.J. (2000). Exergy analysis. In Kreith, F. (ed.), *The CRC Handbook of Thermal Engineering*. CRC Press, Boca Raton, FL. ISBN 0-8493-9581-X.

Moran, M.J., Shapiro, H.N., Boettner, D.D. & Bailey, M.B. (2018). *Fundamentals of Engineering Thermodynamics*. 9th ed. John Wiley & Sons, Boca Raton, FL. ISBN 978 1 119 39138 8.

Sandler, S. (1999). *Chemical and Engineering Thermodynamics*. 3rd ed. John Wiley & Sons, New York.

Singh, O. (2009). *Applied Thermodynamics*. 3rd ed. New Age International Publishers, New Delhi. ISBN 978-8122426029.

Valero, A., Serra, L. & Uche, J. (2000). *Fundamentals of Thermoeconomics*. Lectures 1–3, Euro Summer Course on Sustainable Assessment of Clean Air Technologies, Lisbon, Centro Superior Técnico.

Wagner, W. & Kruse, A. (1998). *Properties of Water and Steam*. Springer, New York.

Wall, G. (2003). *Exergy: A Guide Book*. Available at: https://exergy.se/ftp/exergetics.pdf

Wang, S.K., Lavan, Z. & Norton, P. (2000). *Air Conditioning and Refrigeration Engineering*. CRC Press, Boca Raton, FL. ISBN 0-8493-0057-6.

3 Basic Components and Processes

INTRODUCTION

This chapter presents the physical phenomena governing the operation of the main types of components encountered in energy conversion technologies and how they can be calculated in the Thermoptim modeling environment. Energy technologies can be considered as assemblies of components through which thermodynamic fluids pass and undergo changes of varying complexity. We limit ourselves to steady-state representation of phenomena without attempting to model the detailed behavior of particular components, which can be extremely complex. For the technologies of interest, a component can be represented with sufficient accuracy when given the variables defining coupling with other system components and a small number of characteristics whose physical meaning must be well understood. We examine components available in Thermoptim core: compressions, expansions, combustions, throttling or flash, and various processes that moist mixtures can undergo. Supplements on technological sizing of these components are presented in Chapters 6–9, and heat exchangers are studied in Chapters 4 and 7.

3.1 COMPRESSIONS

In this section, we will mainly address compressors, which can be grouped into two broad classes: displacement compressors and dynamic compressors. The most common are displacement compressors, in which fluid is trapped in a closed volume that is gradually reduced to achieve compression. Dynamic compressors use a different principle: compression is achieved by converting into pressure the kinetic energy communicated to the fluid by moving blades.

Compressors are used for many industrial applications: refrigeration, air conditioning, natural gas transportation, etc. Special mention must be made of air compressors, used as a power source for public works and construction as well as in factories, with pneumatic tools offering many benefits.

In this section, having established the theoretical basis of compression valid for all compressor types, we will consider initially displacement compressors (reciprocating and rotary), then dynamic compressors (centrifugal and axial) and pumps and fans.

3.1.1 Thermodynamics of Compression

Compressing a fluid means raising it from suction pressure P_a to discharge pressure P_r (above P_a). The process involves work, called "compression work."

DOI: 10.1201/9781003606017-4

The fluid initial state "a" before compression is known: pressure P_a, temperature T_a, and specific volume v_a. However, of state "r" after compression, only discharge pressure P_r is determined. The final temperature depends on heat exchanges with the surroundings. The same applies to all functions referred to the "r" thermodynamic state, including compression work.

Assuming the process (a–r) is known, compression work τ is given by (2.3.6) written here as:

$$h_r - h_a + \Delta K = \tau + Q_a^r \tag{3.1.1}$$

Assuming that fluid velocities are low, which is legitimate if we consider state in the discharge tank, we obtain:

$$\tau = h_r - h_a - Q_a^r \tag{3.1.2}$$

3.1.2 Reference Compression

Compressors are compact machines through which gaseous fluid passes very briefly. Exchange surfaces are reduced and heat exchange coefficients are low. Consequently, heat transfer between working fluid and surroundings is generally negligible compared to compression work: the reference compression is therefore adiabatic compression. If it is reversible, it is isentropic.

A second compression type is of interest, at least theoretically: isothermal compression, which would be realized if we could cool the working fluid, which receives heat due to compression, so that its temperature remains constant.

The advantage of such compression is that it would minimize required work. Although generally not feasible in practice, it may be advantageous to cool fluid between two compression stages when multistage compression is needed. The process then performed approaches an isotherm and is called para-isothermal.

3.1.2.1 Reversible Isothermal Compression

Knowing the equation of state, it is easy to calculate τ_i by integration:

$$\delta\tau_i = v dP$$

For example, for an ideal gas:

$$Pv = rT, \quad \text{and} \quad \delta\tau_i = rT\frac{dP}{P}$$

As $T = T_a$ is constant:

$$\tau_i = r T_a \ln\left(\frac{P_r}{P_a}\right) = P_a v_a \ln\left(\frac{P_r}{P_a}\right) \tag{3.1.3}$$

3.1.2.2 Reversible Adiabatic Compression

The Second Law gives s = Const., and, in the entropy diagram, the process is represented by a vertical segment (a–s) (Figure 3.1.1).

The corresponding work is called isentropic or sometimes adiabatic work. We denote it by τ_s.

Relationship (3.1.2) gives here:

$$\tau_s = h_s - h_a \tag{3.1.4}$$

For perfect gases whose heat capacity is independent of temperature, the isentropic process law reads:

$$Pv^{\gamma} = \text{Const} \quad \text{or} \quad v_r = v_a \left(\frac{P_r}{P_a}\right)^{-1/\gamma}$$

The integration of the vdP term gives the isentropic work expression:

$$\tau_s = \frac{\gamma}{\gamma - 1} P_a v_a \left[\left(\frac{P_r}{P_a}\right)^{(\gamma-1)/\gamma} - 1\right] \tag{3.1.5}$$

Given the equation of state and relations between r and heat capacity, the above equality can be written as:

$$\tau_s = c_p T_a \left[\left(\frac{P_r}{P_a}\right)^{(\gamma-1)/\gamma} - 1\right] \tag{3.1.6}$$

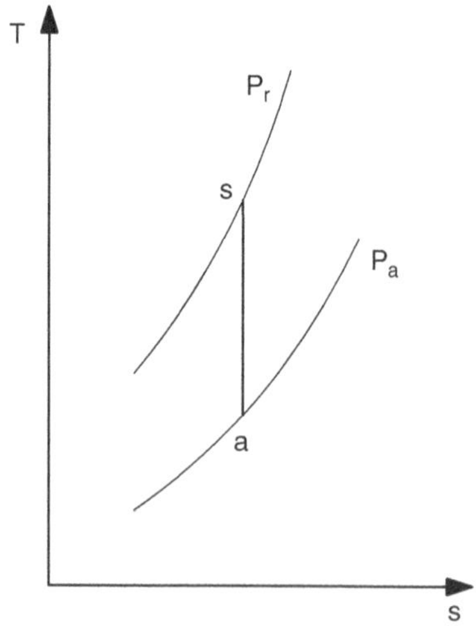

FIGURE 3.1.1 Isentropic compression.

For ideal or real gases, integration of the vdP term is not so simple. The calculation must be done by solving the equations of state. In all cases, we know pressure and suction temperature, which determine entropy and enthalpy at suction. We know discharge pressure and entropy, thereby determining temperature and corresponding enthalpy. Work τ is deduced by (3.1.4).

3.1.3 Actual Compressions

3.1.3.1 Actual Adiabatic

When compression, still adiabatic, is no longer reversible, irreversibilities (friction, impact, etc.) cause entropy generation and, in the entropy diagram (Figure 3.1.2), adiabatic compression is represented by a real curve (a–r) located to the right of the theoretical vertical (a–s). The temperature at compression end is greater than that which would result from reversible adiabatic compression.

Formula (3.1.2) follows directly from the First Law, which does not involve any assumptions about process reversibility. The one that interests us being adiabatic Q=0, we obtain:

$$\tau = h_r - h_a$$

It is clearly greater than isentropic work τ_s.

If the gas can be regarded as perfect:

$$\tau = c_p (T_r - T_a)$$

Isentropic efficiency, or adiabatic efficiency, is defined as the ratio of isentropic work to real work:

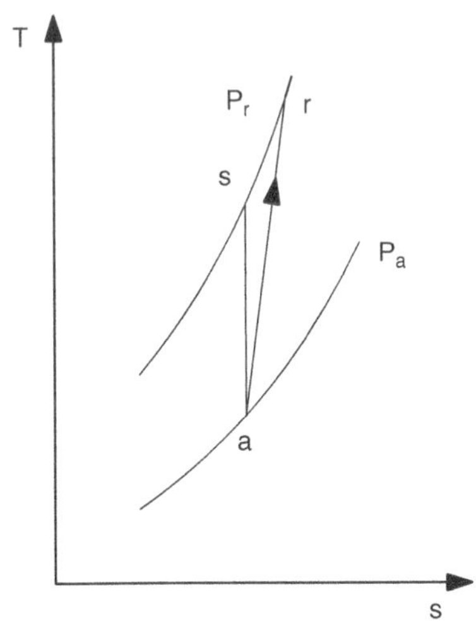

FIGURE 3.1.2 Real compression.

$$\eta_s = \frac{\tau_s}{\tau} = \frac{h_s - h_a}{h_r - h_a} \quad (3.1.7)$$

This efficiency measures the process imperfection from the reversible adiabatic. It is important to note that adiabatic efficiency η_S is thus the main characteristic of an irreversible adiabatic process, but its determination also presents the greatest difficulties.

When the machine considered is conventional, that is, has no major innovations compared to manufactured and tested machines, we take as provisional value the one resulting from testing these machines. Where no such data are available, or if they are not satisfactory, the preliminary evaluation of η_S is always very difficult. It results from detailed loss study and is only possible if one has numerous basic experimental data. The problem thus posed (calculation of losses, leaks, wall friction, etc.) falls within experimental fluid mechanics, not thermodynamics.

3.1.3.2 Polytropic Adiabatic Compression

Note that knowing isentropic efficiency η_S provides no guidance on the law followed by fluid during irreversible compression, so that we cannot integrate the vdP term. To do so, we must make additional assumptions. One of the most common assumptions leads to the widely used concept of a polytropic process, which may cover slightly different definitions according to authors. For some it is an irreversible process, while for others it is inherently reversible. We begin by presenting the first way of seeing things, which is justified for adiabatic compression, then we will show how the second is well suited to studying non-adiabatic compressions. Additionally, we will discuss this concept with two approaches often given as equivalent, whereas they are strictly so for perfect gas only, that is, for gases following law $Pv = rT$, and whose specific heat capacity is constant.

3.1.3.2.1 Differential Approach of the Polytropic

The assumption we make here is to consider that irreversibilities are uniformly distributed throughout compression, which is to assume that during any infinitesimal process stage, isentropic efficiency keeps constant value, equal by definition to polytropic efficiency η_p (Figure 3.1.3).

For multistage machines, this efficiency has clear physical meaning: it is the basic stage efficiency.

An infinitely small compression leads to state function variations dP, dT, dh, and ds.

The process being adiabatic, actual work $\delta\tau$ is equal to dh, while reversible work is $\delta\tau_s = vdP$.

$$\text{By definition: } \eta_p = \frac{\delta\tau_s}{\delta\tau}$$

The assumption η_p constant thus results in the differential equation:

$$\eta_p = \frac{vdP}{vdP + Tds} = \frac{dh - Tds}{dh} = \text{Const}$$

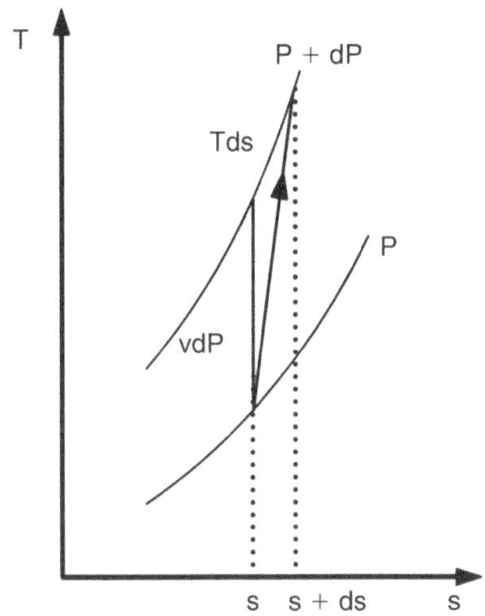

FIGURE 3.1.3 Elementary compression.

which can be written in different equivalent forms, such as:

$$\frac{Tds}{dh} = 1 - \eta_p = Const \quad \text{or} \quad \frac{1-\eta_p}{\eta_p} vdP - Tds = 0 \tag{3.1.8}$$

Note that all these equations express the existence of proportionality between fluid enthalpy change and work or heat dissipated by irreversibilities.

3.1.3.2.2 Integral Approach of the Polytropic

The assumption we make here is to apply the law followed by fluid between the compressor inlet and outlet, which is of type:

$$Pv^k = Const \tag{3.1.9}$$

k is called the polytropic coefficient of the process.

One of the great interests of the polytropic concept lies in equation (3.1.9) simplicity, which generalizes that of isentropic $Pv^\gamma = Const$. By varying k, it generates numerous elementary thermodynamic processes:

$k = 0$	$P = Const$	isobaric process
$k = 1$	$Pv = Const$	isotherm for an ideal gas
$k = \gamma$	$Pv^\gamma = Const$	isentrope for a perfect gas
$k = \infty$	$v = Const$	isovolume (isochoric process)

3.1.3.2.3 Equivalence between the Two Approaches for Perfect Gas

Although different in the general case, these two approaches are equivalent for perfect gas.

For an ideal gas, the first form of equation (3.1.8) gives:

$$Tds = (1-\eta_p)c_p dT \quad \text{or} \quad ds = (1-\eta_p)c_p \frac{dT}{T}$$

Moreover, as $ds = c_p \frac{dT}{T} - r\frac{dP}{P}$ we have $\eta_p\, c_p \frac{dT}{T} - r\frac{dP}{P} = 0$

whose integration is straightforward and gives, if c_p = Const :

$$\frac{T}{P^{\beta}} = \text{Const} \quad \text{with} \quad \beta = \frac{r}{c_p\,\eta_p} = \frac{\gamma - 1}{\gamma\eta_p}$$

Posing $\frac{k-1}{k} = \beta$, we find all calculations done: $\mathbf{Pv}^k$ = Const or

$$Tv^{k-1} = \text{Const} \quad \text{or} \quad TP^{(1-k)/k} = \text{Const} \tag{3.1.10}$$

We thus find the form postulated by the integral approach, the polytropic coefficient k and efficiency η_p being linked by relationship:

$$\frac{k-1}{k} = \frac{\gamma - 1}{\gamma\eta_p} \tag{3.1.11}$$

For adiabatic compression, polytropic coefficient k is always greater than γ, especially since the process is irreversible.

Keep in mind that the equivalence between these two approaches is only valid for perfect gases. If a gas's heat capacity depends on temperature, and even more on pressure, the differential form and integral form are not equivalent: if one assumes that polytropic law is defined by equation Pv^k = Const, it is no longer true that irreversibilities are uniformly distributed. Conversely, relation (3.1.11) involving $c_p(T)$, k is no longer constant.

Differences generally exist at this level between presentations made by authors. However, it is true that in most cases they do use in practice the polytropic under the assumption of perfect gas, so it loses some of its importance.

Knowing polytropic coefficient k, it is easy to calculate the final state of adiabatic compression, and therefore the enthalpy increase and compression work.

For open systems, and, remember, for perfect gas, we obtain:

$$\tau = \Delta h = c_p\left(T_r - T_a\right)$$

$$\tau = c_p T_a\left[\left(\frac{P_r}{P_a}\right)^{(k-1)/k} - 1\right] = \frac{\gamma}{\gamma - 1} P_a v_a\left[\left(\frac{P_r}{P_a}\right)^{(k-1)/k} - 1\right] \tag{3.1.12}$$

Note that:

$$\int_{P_a}^{P_r} vdP = \frac{k}{k-1} P_a v_a\left[\left(\frac{P_r}{P_a}\right)^{(k-1)/k} - 1\right] = \eta_p \frac{\gamma}{\gamma - 1} P_a v_a\left[\left(\frac{P_r}{P_a}\right)^{(k-1)/k} - 1\right]$$

$$\int_{P_a}^{P_r} vdP = \eta_p \tau \tag{3.1.13}$$

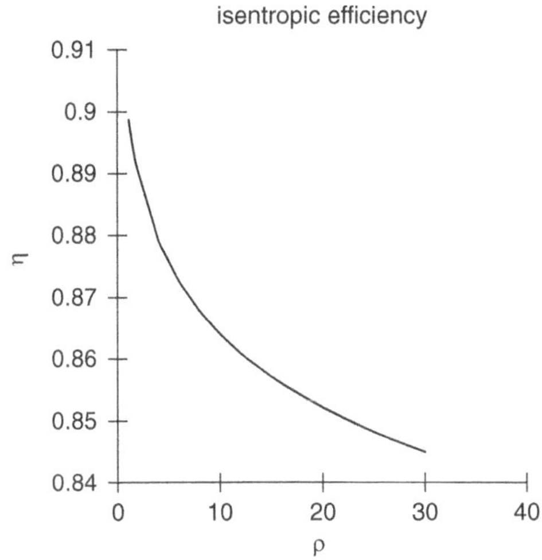

FIGURE 3.1.4 Isentropic efficiency function of compression ratio.

The integration of vdP along the polytropic provides what might be called the "reversible polytropic work" $(\tau_p = \eta_p \tau)$ of the reversible non-adiabatic process in which heat $Q = T\Delta s$ would be provided to the fluid.

Knowing polytropic compression work, one can calculate the process isentropic efficiency, since we know isentropic work τ_S. For perfect gas:

$$\eta_s = \frac{\left(\frac{P_r}{P_a}\right)^{(\gamma-1)/\gamma} - 1}{\left(\frac{P_r}{P_a}\right)^{(k-1)/k} - 1}$$

Figure 3.1.4 shows that for constant polytropic efficiency (here equal to 0.9), compression isentropic efficiency decreases as compression ratio increases.

3.1.3.2.4 Formulations for Ideal Gases

For an ideal gas, if one starts from differential equation (3.1.8), the calculation can be done as follows:

$$Tds = (1 - \eta_p)dh \quad \text{or, for closed systems,} \quad Tds = (1 - \eta_p)du$$

$$dh = c_p dT \qquad ds = (1 - \eta_p)c_p \frac{dT}{T}$$

But as $ds = c_p \frac{dT}{T} - r\frac{dP}{P}$

$$\eta_p c_p \frac{dT}{T} - r\frac{dP}{p} = 0$$

$$ds = (1-\eta_p)\frac{r}{\eta_p}\frac{dP}{P} = \frac{r(1-\eta_p)}{\eta_p}\frac{dP}{P}$$

$$\int_{T_a}^{T_r} ds(t)\,dt = \frac{r(1-\eta_p)}{\eta_p}\int_{P_a}^{P_r}\frac{dP}{P} \tag{3.1.14}$$

The entropy difference is independent of c_p and T for polytropic compression and can be expressed as:

$$\text{open systems: } \Delta s = \frac{1-\eta_p}{\eta_p} r \ln\left(\frac{P_r}{P_a}\right) \tag{3.1.15}$$

or

$$\text{closed systems: } \Delta s = \frac{1-\eta_p}{\eta_p} r \ln\left(\frac{v_a}{v_r}\right) \tag{3.1.16}$$

The calculation of outlet compression temperature T_r can be done by reversing the entropy function at discharge pressure for value $s = s_a + \Delta s$. Knowing inlet and outlet point states, k can then be identified by reversing $Pv^k = \text{Const}$.

If one starts from polytropic law $Pv^k = \text{Const}$, T_r is calculated by:

$$T_r = T_a\left(\frac{P_r}{P_a}\right)^{(k-1)/k}$$

η_p is then given by:

$$\eta_p = \frac{r \ln\left(\frac{P_r}{P_a}\right)}{s_r - s_a + r \ln\left(\frac{P_r}{P_a}\right)} \tag{3.1.17}$$

These relations, which generalize equation (3.1.11) for ideal gases, are valid for any polytropic, and allow us to obtain k from η_p, and vice versa. They are even more interesting as, since they do not explicitly call for the $c_p(T)$ law, they can be used as first approximation to study real gases, even if they are not strictly applicable.

To calculate the complete process, they must be completed by the equation providing compression work. In the case of adiabatic compression (Q=0), it is very simple:

$$\tau = \Delta h$$

We will see in the next section what they become for non-adiabatic compressions.

For closed systems, we write:

$$\delta W_s = -Pdv = du - Tds = \eta_p\,du \quad W_s = \eta_p \Delta u$$

3.1.3.3 Non-Adiabatic Polytropic Compression

As mentioned above, it is also possible to present a polytropic process as (non-adiabatic) reversible by definition, verifying law $Pv^k = \text{Const}$, which passes through points corresponding to inlet and outlet.

The compression reversible work τ_r involved in this evolution can be directly calculated from the polytropic differential equation, noting that, η_p being constant:

$$\delta\tau_p = vdP = dh - Tds = \eta_p\, dh \quad \text{which integrates directly to give} \quad \tau_r = \eta_p \Delta h$$

Since actual compression work $\tau = \Delta h$ is equal to $\frac{\tau_r}{\eta_p}$, it is reasonable to assume that polytropic law $Pv^k = \text{Const}$ is a reversible evolution. The real process is derived by taking into account polytropic efficiency η_p. This way of presenting it thus leads, for adiabatic compression, to formalism similar to that obtained by considering the polytropic as irreversible.

If now we consider non-adiabatic compression, it appears clearly in differential equations (3.1.8) defining the polytropic, that parameter η_p can be considered an elementary isentropic efficiency.

Using the term efficiency to describe it is confusing, and it seems better to rename it and consider it simply as parameter α characterizing the polytropic, representative of enthalpy distribution between compression work on one hand, and irreversibilities and exchanged heat on the other hand.

$$\alpha = \frac{vdP}{vdP + Tds} \tag{3.1.18}$$

For ideal gas, non-adiabatic reversible polytropic law $Pv^k = \text{Const}$ implies the following equality:

$$\alpha = \frac{\tau_r}{\tau_r + Q_r} \quad \text{with} \quad \tau_r + Q_r = \Delta h \quad \text{and, as we have seen above,} \quad \tau_r = \alpha \Delta h$$

By introducing efficiency η to characterize real compression irreversibilities compared to reversible polytropic:

$$\eta = \frac{\tau_r}{\tau_r + \pi} = \frac{\tau_r}{\tau} \tag{3.1.19}$$

with $\tau + Q = \Delta h$ and π uncompensated work.

Knowing α and η, it is possible to calculate compression work and deduce heat Q exchanged with surroundings.

We easily obtain: $\tau = \frac{\alpha}{\eta} \Delta h$ and

$$Q = \left(1 - \frac{\alpha}{\eta}\right) \Delta h \tag{3.1.20}$$

The relevance of considering a polytropic process as reversible by definition and characterizing actual compression in relation to it by its polytropic efficiency η is clear.

One might criticize this approach by noting that the physical meaning of polytropic efficiency η is not proven and in fact that it is perhaps just an arbitrary coefficient. To answer this, note first that one verifies that if the process is adiabatic $\alpha = \eta$, and we find the previous result. We will show later that in the case of isothermal compression, which is the limit of cooled compression, η still retains the sense of efficiency.

If the fluid is a non-perfect ideal gas, the differential approach and integral approach are not exactly the same here, but the above relations can still be used as first approximation, provided that α varies slightly if k is constant, and vice versa.

3.1.3.3.1 Practical Calculation

3.1.3.3.1.1 Direct Problem If we know polytropic coefficient k of the process, T_r is calculated by:

$$T_r = T_a\left(\frac{P_r}{P_a}\right)^{(k-1)/k}$$

We obtain characteristic α of the reference reversible polytropic passing by inlet and outlet points by:

$$\alpha = \frac{r\ln\left(\frac{P_r}{P_a}\right)}{s_r - s_a + r\ln\left(\frac{P_r}{P_a}\right)} \tag{3.1.21}$$

The work of reversible polytropic is then:

$$\tau_r = \alpha(h_r - h_a) \tag{3.1.22}$$

Knowing polytropic efficiency η of the real process from reversible polytropic, we can deduce actual compression work:

$$\tau = \frac{\alpha}{\eta}(h_r - h_a) \tag{3.1.23}$$

The heat exchanged with surroundings is:

$$Q = h_r - h_a - \tau = \left(1 - \frac{\alpha}{\eta}\right)(h_r - h_a) \tag{3.1.24}$$

3.1.3.3.1.2 Inverse Problem Knowing T_r, we calculate polytropic coefficient k of the process by:

$$T_r = T_a\left(\frac{P_r}{P_a}\right)^{(k-1)/k}$$

We obtain characteristic α of reversible polytropic reference through inlet and outlet relationship Equation 3.1.21.

Knowing heat exchanged with surroundings Q, η is determined by:

$$\eta = \frac{\alpha(h_r - h_a)}{h_r - h_a + Q} \tag{3.1.25}$$

We deduce $\tau = \frac{\alpha}{\eta}(h_r - h_a)$.

These results generalize easily to closed systems.

Since they do not explicitly involve the $c_p(T)$ law, they can be used as first approximation to study real gases, even if they do not strictly apply.

3.1.3.3.2 Limiting Case of Isothermal Compression

Isothermal compression is the limiting case of cooled compression. The gas temperature at compressor outlet being the same as at input, the polytropic law reduces to ideal gas law Pv = const, and k = 1.

By taking equations written for non-adiabatic open system compressions, we have, for ideal gas:

$$\tau_r = \alpha(h_r - h_a) = \alpha c_p T_a \left[\left(\frac{P_r}{P_a} \right)^{(k-1)/k} - 1 \right]$$

$$\text{If } k \to 1, \frac{k-1}{k} \to 0 \quad \text{and} \quad \left[\left(\frac{P_r}{P_a} \right)^{(k-1)/k} - 1 \right] \to \frac{k-1}{k} \ln\left(\frac{P_r}{P_a} \right)$$

$$\tau_r = \alpha c_p T_a \frac{k-1}{k} \ln\left(\frac{P_r}{P_a} \right)$$

We get: $\tau_r = r T_a \ln\left(\frac{P_r}{P_a} \right)$

What we called non-adiabatic polytropic reversible compression work does correspond to work obtained by direct integration of vdP with Pv = Const. Polytropic efficiency η is then also efficiency in the usual sense.

3.1.4 Staged Compression

In turbomachines, one often makes use of dimensionless parameters, whose values must be within relatively narrow ranges. One in particular defines the power factor

$$\mu = \frac{\tau}{U^2} \tag{3.1.26}$$

where U is the peripheral velocity of the impeller. This factor cannot exceed certain values if we want proper functioning (in the range of 0.4–0.5 for a centrifugal impeller, much less for an axial ring).

To obtain high discharge pressure, it would theoretically be sufficient to increase peripheral velocity U. But the stresses induced in metal by centrifugal force being proportional to U^2, we are led in actual devices to limit peripheral velocity to around 200–300 m/s. As the elementary compression ratio cannot exceed a certain value, achieving high compression ratios requires putting several stages in series.

By analogy with hydraulic pumps, another quantity is also sometimes used to characterize a compressor: its pressure head, $H = \tau/g$, g being gravity acceleration.

In positive displacement or piston compressors, this problem does not arise, and one can theoretically achieve very important compression ratios in one section. In fact, other technological problems (dead space, leaks) arise, that become predominant at high compression ratios, as will be discussed later.

Relationship (3.1.6) shows that isentropic compression work is proportional to suction absolute temperature. When compression is to be staged, it is interesting in terms of thermodynamics (but not necessarily economically) to cool the working fluid in a heat exchanger between two stages, by reducing its temperature to a value close to that of suction. In this case, compression is called para-isothermal. As the number of compression stages increases, the overall process approaches the isotherm. Assuming that gas is perfect, that fluid is cooled to its initial temperature and that isentropic efficiencies of n compression stages are all equal, it is possible to show that total compression work is minimal if intermediate pressures are a geometrical progression of ratio equal to the overall compression nth root:

$$\frac{P_{i+1}}{P_i} = \sqrt[n]{\frac{P_r}{P_a}} \quad (3.1.27)$$

When the above assumptions are not verified, the calculation of intermediate pressures must be done iteratively, but they often differ slightly from values given by equation (3.1.27).

3.1.5 Calculation of a Compression in Thermoptim

The definition and calculation of processes involve a series of screens that differ by process type in question, but which nevertheless share certain characteristics (Figure 3.1.5):

- The process name is set at the top left.
- At the top right are options and fields used to configure the process: "open" or "closed" system, "set flow" process, "observed" process (the latter two concepts are used to drive automatic recalculation of a project).
- The left part is used to define inlet and outlet points. To connect a point from this screen (if not already done from the diagram editor), double-click in the corresponding name field, and select it from the list.
- Buttons such as "Calculate," "Save," "Close," and "Delete" are also common.
- Energy type can be changed by double-clicking in the corresponding field.
- The mass flow value is by default initialized to 1. Its unit is deliberately not specified, so the user can choose in the simulator screen to work in g/s, kg/s,

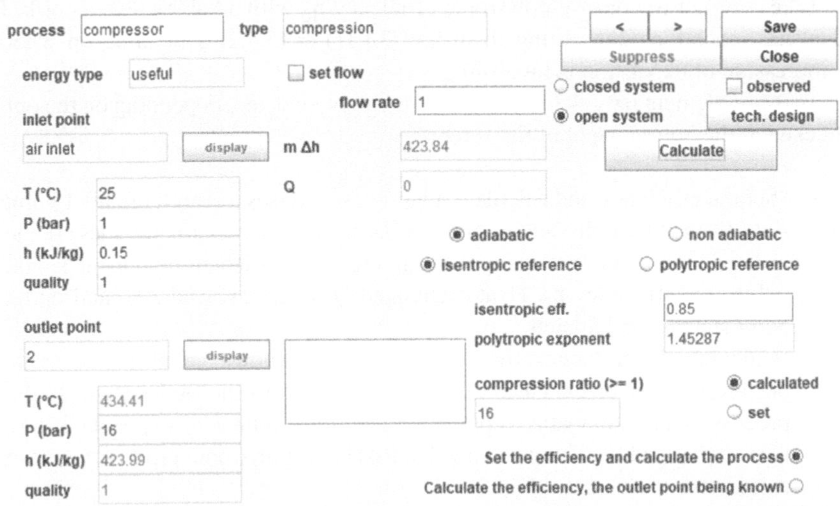

FIGURE 3.1.5 Compression screen.

or t/s depending on the problem studied, while keeping high readability. Obviously, the results of calculations on energies or enthalpies (respectively on power outputs) involved depend on the unit selected for flow, and can be expressed in J, kJ, or MJ (respectively, W, kW, or MW).

- Finally, a text field can document the process.

For compressions and expansions (Figure 3.1.5), the calculation options available are:

- Calculations can be performed either in an open system, in which case input variables are temperature, pressure, and enthalpy, or in a closed system, in which case it is temperature, volume, and internal energy.
- The compression ratio ρ can be set, in which case outlet point pressure (or volume) is determined from that of the inlet point and the value of ρ chosen, or calculated, in which case the value of ρ is evaluated from pressures or volumes of inlet and outlet points, considered as given. Whether for compression or expansion, ρ is greater than unity.
- The process can be adiabatic or not.

When adiabatic, it can be calculated using an isentropic or polytropic reference, and is characterized by isentropic or polytropic efficiency η_s or η_p, as defined above, between 0 and 1. In this case, polytropic exponent k appearing in law $Pv^k = Const$ is calculated by Thermoptim.

When it is non-adiabatic, the process is necessarily of polytropic type and is based on principles outlined in Section 3.1.3.3. It is characterized by polytropic exponent k and polytropic efficiency η_p (for compression, it is the ratio of reversible work to real work, and for expansion it is the reverse). The heat exchanged with surroundings Q is displayed in the field under compression or expansion work Delta H or Delta U.

Take care not to confuse polytropic efficiency η_p with adiabatic exponent k. For ideal gas, equations connecting them are (3.1.11) or (3.4.2) depending on whether compression or expansion is involved.

Two complementary calculation methods can be selected, depending on the option checked at the bottom right of the screen:

- "Set the efficiency and calculate the process" causes a search of outlet point state from inlet point and compression or expansion characteristics. If the process is non-adiabatic, input parameters are adiabatic exponent k and polytropic efficiency η_p. Heat exchanged Q, work involved ΔH, and outlet point state are calculated.
- "Calculate the efficiency, the outlet point being known" identifies the value of compression or expansion efficiency leading to the outlet state. If the process is non-adiabatic, input parameters are adiabatic exponent k and heat exchanged Q. The work involved ΔH and polytropic efficiency η_p are calculated.

3.2 DISPLACEMENT COMPRESSOR

A displacement compressor is characterized by encapsulation of flow passing through it in a closed volume that is gradually reduced. Return of this fluid in the direction of decreasing pressure is prevented by the presence of one or more movable walls. In this machine type, kinetic energy imparted to fluid usually does not play any useful role, unlike what happens in turbomachinery.

By design, displacement compressors are particularly suitable for processing relatively low fluid flow rates, possibly highly variable, and under relatively high pressure ratios.

Their operating principle is: a fixed mass of gas at suction pressure P_1 is trapped in a chamber of variable volume (Figure 3.2.1). To increase pressure, volume is gradually reduced in a manner that differs according to the technique used. Generally, the process follows a law close to polytropic $Pv^k = Const.$

By compression end, the chamber is in communication with the discharge circuit so that compressed gas at pressure P_2 can emerge.

A new mass of gas at pressure P_1 is then sucked into the inlet pipe, and so on, machine operation is cyclical. The cycle is represented on the "Watt chart," which gives the fluid pressure as a function of chamber volume (Figure 3.2.2).

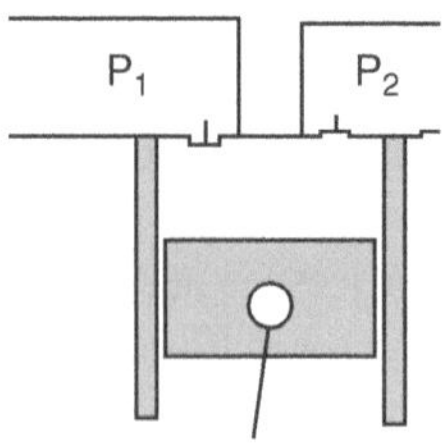

FIGURE 3.2.1 Sketch of a piston compressor.

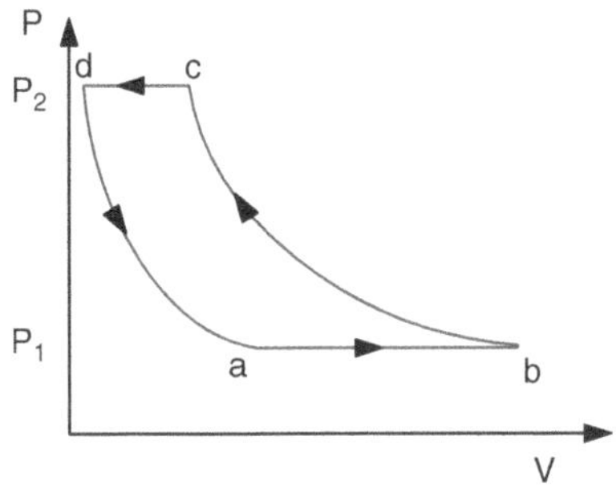

FIGURE 3.2.2 Theoretical Watt diagram.

3.2.1 Piston Compressors

The chamber is the volume defined by a cylinder, one of its bases being fixed and the other being a movable piston in the cylinder bore, driven by a connecting rod system (Figure 3.2.1).

The organs that control discharge or admission are, in piston compressors, valves operated automatically by pressure differences between the chamber and discharge or admission manifold.

One denotes Vs the volume swept by the piston between its two extreme positions, and "dead space" εVs the minimum volume of the compression chamber. In current achievements, ε is on the order of 3%–5%.

3.2.1.1 Theoretical Energy Balance

On its stroke of length L, piston displacement sweeps Vs. If A is cylinder cross-sectional area, one clearly has: $Vs = AL$.

The chart in Figure 3.2.2 is used to describe the machine cycle by four processes:

- a–b: intake of fluid mass (Δm), which mixes with that (Δm_0) already contained in dead volume.
- b–c: isentropic compression of fluid mass $(\Delta m + \Delta m_0)$ to state 2.
- c–d: expulsion of mass Δm.
- d–a: isentropic expansion of mass Δm_0 to state 1.

Note that fluid mass in the chamber is not the same in the four cycle phases and that this will have to be taken into account when setting balances.

In the theoretical cycle, we assume that fluid undergoes no state change during intake (a–b) and discharge (c–d) phases. Another major simplification is to admit that the two evolutions (b–c) and (d–a) are isentropic.

Under these conditions, the thermodynamic cycle undergone by mass Δm_0 trapped in dead volume is limited to one round trip between isentropic states 1 and 2, which does not involve any net work.

The assimilation of fluid to perfect gas leads to law $Pv^{\gamma} = \text{Const}$ and allows integration of cycle work on evolution 1–2. Expression (3.1.5) becomes:

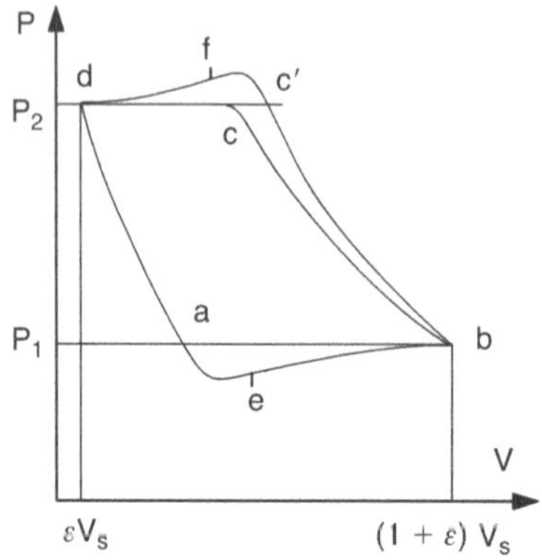

FIGURE 3.2.3 Real Watt diagram.

$$\tau_s = P_1 v_1 \frac{\gamma}{\gamma - 1}\left(\left(\frac{P_2}{P_1}\right)^{(\gamma-1)/\gamma} - 1\right) \tag{3.2.1}$$

Note that it is this expression that must be used, even if compression is done in a closed system, because of the existence of intake and discharge phases, occurring in open systems, and involving transfer work $\Delta(Pv)$.

3.2.1.2 Actual Piston Compressor Cycle

Compared to the theoretical cycle just described, the actual Watt chart has several differences (Figure 3.2.3).

- Valves do not operate instantly and only at points marked e and f are they at their maximum aperture, and even then they create throttling. It follows that pressure is less than P_1 during intake stroke, and greater than P_2 for discharge stroke.
- The compression curve (b–c′) does not follow a perfect adiabatic process. Irreversibilities increase gas entropy, and, for a given pressure, its specific volume. Point c′ is to the right of theoretical c. It is characterized either by the polytropic law of exponent k, or by isentropic efficiency η.

For all these reasons, actual work is greater than theoretical work. There are also other losses not mentioned above: they are due to wall thermal actions, where temperature stabilizes at a value intermediate between that of suction and discharge T_1 and T_2, creating a "thermal shunt" that allows heat to pass from heated fluid (on discharge) to cold fluid (on admission), which is additional irreversibility.

3.2.1.3 Determination of Work per Cycle

Relation (3.2.1) is not sufficient to calculate real machine efficiency, and it is necessary to rely on experimental results. Lacking theory capable of determining the quantitative impact of all irreversibilities, one can generally define real internal compressor work τ compared to isentropic work τ_s calculated from the technological reference magnitude: swept volume Vs (displacement).

If phenomena were reversible, a compressor of swept volume Vs would contain, in the absence of any dead volume, useful mass Vs/v_1. But mass $\Delta m_0 = \varepsilon Vs / v_2$ is recirculated due to dead space presence, and useful mass becomes:

$$\Delta m_{th} = (1+\varepsilon)\frac{Vs}{v_1} - \Delta m_0 = \frac{Vs}{v_1}\left(1 - \varepsilon\left(\frac{v_1}{v_2} - 1\right)\right)$$

$$\Delta m_{th} = \frac{Vs}{v_1}\left(1 - \varepsilon\left(\left(\frac{P_2}{P_1}\right)^{1/\gamma} - 1\right)\right) \tag{3.2.2}$$

This relation shows that the more compression ratio increases, the more important is recirculating mass and the more useful mass decreases. There is therefore a maximum compression ratio beyond which the compressor runs without effect. Piston compressors have the advantage of working with variable compression ratios and even automatically adjust over wide operating ranges, but at the cost of decreased efficiency when the compression ratio exceeds a certain limit.

In fact, useful mass is less than the value given by equation (3.2.2), due to many technological imperfections. To characterize it, we introduce the concept of volumetric efficiency λ, defined as the ratio of mass flow actually discharged to theoretical mass flow, the latter being referred either to swept volume, or to Δm_{th} depending on authors. We retain here the first definition, which leads to:

$$\Delta m = \lambda \frac{Vs}{v_1} \tag{3.2.3}$$

Introducing isentropic efficiency η_c (often called compression efficiency or indicated efficiency) and mechanical efficiency η_m, characteristic of friction in the compressor, specific compression work is usually written:

$$\tau = \frac{\tau_s}{\eta_c \eta_m}$$

and compression work per cycle is:

$$\Omega = \Delta m \tau = \frac{Vs}{v_1}\frac{\lambda}{\eta_c \eta_m}\tau_s \tag{3.2.4}$$

If N is the rotation speed (expressed in revolutions per minute), compression power (assuming one cycle per revolution) is given by:

$$Pc = \frac{N\Omega}{60} = \frac{N}{60}\frac{Vs}{v_1}\frac{\lambda}{\eta_c \eta_m}\tau_s \tag{3.2.5}$$

The compression power is expressed as a function of volumetric efficiency λ, representing compressor filling losses, compression efficiency η_c, characteristic of actual process deviation compared to isentropic, and mechanical efficiency η_m.

3.2.1.4 Volumetric Efficiency λ

Note that λ differs from other efficiencies, since it appears in the numerator of relations (3.2.4) and (3.2.5). Although one is used to calling it volumetric efficiency, its meaning is very different: it represents the fraction of displacement that produces useful effect. Low volumetric efficiency is not inherently disadvantageous in terms of energy: it simply means that mass flow passing through the compressor is less than that theoretically corresponding to displacement. However, it is economically disadvantageous, since for a given specification, it leads to cylinder oversizing, and thus to higher investment.

Analysis and experimentation have shown that λ corresponds to losses due to various origins: dead space, pressure drop in intake and discharge manifolds, wall effects (thermal shunt), sealing defects (at piston rings), and finally pressure drop in inlet and outlet valves.

Overall, these losses combine to make volumetric efficiency decrease substantially linearly with the compression ratio:

$$\lambda = 1 - \alpha \frac{P_2}{P_1} \tag{3.2.6}$$

The α value is between 0.02 and 0.07.

3.2.1.5 Compression Efficiency η

The norm in the profession is to call compression efficiency the usual isentropic efficiency.

In a piston compressor, three sources of losses, connected to previous ones, mainly affect compression efficiency. Wall effects are very disadvantageous because, as illustrated by the Watt chart in Figure 3.2.4, they seriously increase compression work by twisting to the right (b–c′) the isentropic (b–c). Pressure drop in valves is next, followed by losses due to sealing leakages.

At a given speed, the shape of compression efficiency as a function of compression ratio shows the existence of a maximum around a rather low compression ratio (3–5). A relation of the Dehausse type can be used when no constructor data are available:

$$\eta = 0.9 - 0.004\left(\frac{P_2}{P_1} - 5\right)^2 - \frac{0.5}{\frac{P_2}{P_1} - 0.3} \tag{3.2.7}$$

Volumetric and compression efficiencies intervening only by their ratios in the expression of work per cycle Ω, we can represent η/λ according to the compression ratio. We obtain for a given range of compressors a set of relatively grouped curves, each curve corresponding to a suction temperature. Figure 3.2.4 shows curves giving

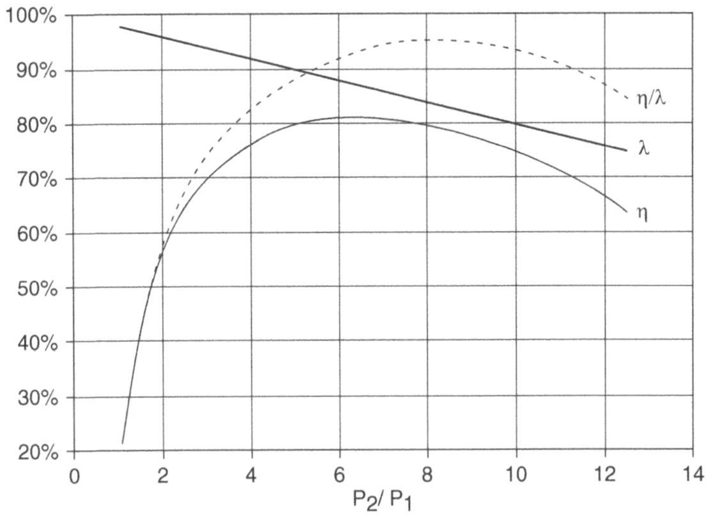

FIGURE 3.2.4 Volumetric and compression efficiencies.

volumetric efficiency λ (thin lines), compression efficiency η (bold line), and their ratio η/λ (dashed).

3.2.1.6 Practical Calculation of a Piston Compressor

We first determine inlet conditions (P_1, v_1) and fluid mass flow $\dot{m}$. We deduce theoretical volume flow $\dot{V}_{th} = \dot{m}v$, which must enter the compressor.

The pressure ratio r being known, manufacturer's charts provide values of η_c and λ. We deduce the actual volume flow rate to be transferred:

$$\dot{v} = \frac{\dot{V}_{th}}{\lambda} \tag{3.2.8}$$

Knowing swept volume Vs of each compression unit and compressor rotation speed, we can select the most appropriate machine in the range.

Compression power is equal to the product of specific work $\tau = \tau_s/\eta_c$ and mass flow $\dot{m}, \tau_s$ being determined by application of the appropriate equation of state, or by reading on the thermodynamic diagram, or from Thermoptim.

Mechanical losses are then added to compression power, to provide total power consumption. These mechanical losses are taken into account either by mechanical efficiency, or by fixed value equal to unloaded drive power.

3.2.2 Rotary Positive Displacement Compressors (RPDCs)

RPDCs are a class of machines that compress gas by trapping a fixed volume and mechanically reducing its space using rotating elements. Unlike reciprocating compressors, which rely on pistons moving back and forth, rotary compressors use continuous rotary motion to achieve compression, resulting in smoother operation and often more compact designs.

One of the most common types is the rotary screw compressor, which uses two intermeshing helical screws to compress air as it moves through the threads. These compressors are highly efficient, reliable, and well suited for industrial applications requiring continuous airflow. Another widely used variant is the rotary vane compressor, where a rotor with sliding vanes rotates inside an eccentric cavity. The vanes maintain contact with the housing, creating chambers that decrease in volume and compress the gas. These are often found in refrigeration and low-pressure systems.

Scroll compressors operate with two spiral-shaped elements—one fixed and one orbiting—that progressively compress the gas toward the center. Known for their quiet operation and high efficiency, scroll compressors are popular in HVAC systems and heat pumps. Lobe compressors, such as Roots blowers, use two rotating lobes that trap and move air without internal compression; pressure builds externally, making them ideal for high-volume, low-pressure applications. Lastly, rolling piston compressors feature a rotor that rolls within a cylinder, with a vane separating intake and compression zones. These are compact and commonly used in small air conditioning units.

Each type offers distinct advantages depending on the application, from industrial manufacturing to residential climate control. Their rotary nature allows for continuous operation, reduced vibration, and often lower maintenance compared to piston-based systems.

3.2.2.1 Compression Efficiency

In RPDCs, the compression ratio is determined by machine geometry, which determines the internal volume ratio $V_i = V_1/V_{2c}$, a construction volumetric ratio that can only be modified by acting on exhaust opening shape (the index c refers to constructional characteristics).

It follows that their compression ratio is fixed, and the machine does not automatically adjust to different compression ratios, as do reciprocating compressors with valves. However, valve absence is an advantage in terms of maintenance, the latter being mechanically highly stressed parts.

If discharge pressure P_r is less than constructive pressure P_{2c}, rapid expansion from P_{2c} to P_r occurs at exhaust port opening. Otherwise, there is sudden compression of gas cell from P_r to P_{2c}, and then discharge at this pressure.

In both cases, this induces loss, which can be evaluated as follows.

Compression work is equal to the sum of transfer work $\Delta(Pv)$ and external forces $W_A = -\int_1^2 P dv$:

$$\tau = -\int_{v_1}^{v_{2c}} P dv + P_r v_{2c} - P_1 v_1$$

Introducing the internal pressure ratio π_i, given by:

$$\pi_i = V_i^k \tag{3.2.9}$$

k being the compression polytropic exponent, and with $\rho = \frac{P_r}{P_1}$, we obtain:

$$\tau = \frac{P_1 v_1}{k-1}\left[\pi_i^{(k-1)/k} + \frac{\rho}{V_i}(k-1) - k\right]$$

The compression efficiency is then: $\eta_s = \dfrac{\frac{\gamma}{\gamma-1} P_1 v_1 \left[\rho^{(\gamma-1)/\gamma} - 1\right]}{\frac{P_1 v_1}{k-1}\left[\pi_i^{(k-1)/k} + \frac{\rho}{V_i}(k-1) - k\right]}$

$$\eta_s = \frac{\frac{(k-1)\gamma}{\gamma-1}\left[\rho^{(\gamma-1)/\gamma} - 1\right]}{\left[\pi_i^{(k-1)/k} + \frac{\rho}{V_i}(k-1) - k\right]} \tag{3.2.10}$$

However, some RPDCs such as screw compressors have the advantage of being able to vary internal volume ratio Vi, through a sliding drawer, which alters discharge port. The "V_i" can thus vary continuously within a range of values around 2.2–5, depending on the power and pressure ratio required.

Figure 3.2.5 gives compression efficiency curves for fixed and variable V_i compressors. It shows that for a given internal pressure ratio π_i, the compression efficiency varies slightly with the operating compression ratio, with maximum at $P_2/P_1 = \pi_i$.

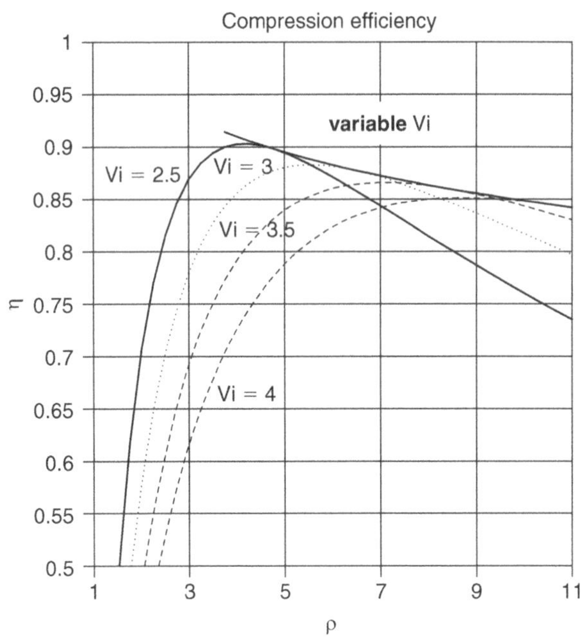

FIGURE 3.2.5 Compression efficiency.

However, the curve is fairly flat, so that a screw compressor maintains proper efficiency even for compression ratios well above the nominal value. Given a curve shape, much steeper on the left than on the right, the operating point is chosen slightly right of optimum, to maintain good efficiency even in the case of operating regime disturbance.

Figure 3.2.5 also shows efficiency gain that can be expected if one adopts variable V_i. The counterpart is increased compressor cost, due to sliding drawer presence. For refrigeration installations where power control drawers already exist, the addition of V_i variation drawer is generally justified.

3.2.2.2 Volumetric Efficiency

Volumetric efficiency λ of screw compressors is much better than that of piston engines, for the following reasons:

- There is almost no expansion loss due to dead space.
- Pressure drop at suction and discharge is low due to valve absence.
- There is no thermal bridge: in the steady state, machine temperatures remain stable.
- Conversely, losses due to sealing defects may be more important because of section leakage length and difficulties of sealing the ends.

Volumetric efficiency depends on the compression ratio and internal pressure ratio π_i. It varies almost linearly with the compression ratio, from a value of 0.98 or 0.99 at a compression ratio of 1 to values of 0.75 or 0.8 for compression ratios up to 20. With variable V_i, excellent volumetric efficiencies are obtained over a very wide range.

Practical calculation of an RPDC is done similarly to that of a piston compressor. Once design V_i is determined, a comparison is made between the performance of variable V_i and machine in the manufacturer's range with the nearest V_i. Choice is then made according to extra cost of the sliding drawer and its control.

3.3 DYNAMIC COMPRESSORS

3.3.1 General

Unlike positive displacement machines where fluid is enclosed in a closed volume, in a dynamic compressor, continuous fluid flow takes place to which energy is communicated through moving blades driven by a rotor.

In this chapter, we limit our ambition to establishing the main results needed to understand dynamic compressor operation, and, given that expansion turbine operation is quite similar to that of dynamic compressors, we will directly present results valid for all turbomachinery.

There are two main types of fluid flow relative to turbomachine rotor: axial flow, almost always performed in turbojet dynamic compressors, and radial flow which is widely used for centrifugal dynamic compressors, including for refrigeration or automobile engine supercharging.

We will always consider the working fluid to be compressible, either an ideal gas or a condensable real fluid. Consequently, phenomena are governed by fundamental laws of steady compressible flows which we begin by establishing in Section 3.3.2. We will show that by varying gas velocity, it is possible to change its pressure.

In general, a turbomachine consists of four elements in series (Figure 3.3.1).

- An inlet convergent C, or distributor, which is a fixed part whose function is to properly orient fluid streams entering the impeller, and accelerate them slightly.
- The impeller I, or rotor, driven by rotational movement around an axis. This wheel has vanes defining channels between which fluid flow is distributed. It communicates to the fluid the mechanical energy of blades in the form of kinetic, thermal, and pressure energy.
- The diffuser D is a fixed component that serves to transform into the pressure part of the kinetic energy gained by fluid when passing through the wheel. Depending on circumstances, the diffuser may or may not involve blades. It is said to be partitioned or smooth.
- A volute V, also fixed, corrects fluid lines on the wheel periphery, and guides them to the device outlet.

For reasons discussed in Section 3.1.4, turbomachinery frequently has several stages, each having a limited pressure head. This is called multistage turbomachinery.

In axial turbomachinery, successive stages are juxtaposed by compact groups in which the volute is no longer necessary, and where the convergent and diffuser situated between two moving wheels can be combined to form a single crown whose role is to straighten fluid lines so they appear correctly on the second impeller inlet. Under such conditions, middle stages have only two components.

It also happens that, for simplicity reasons, a turbomachine even single staged may be composed of only two components, one fixed and one mobile.

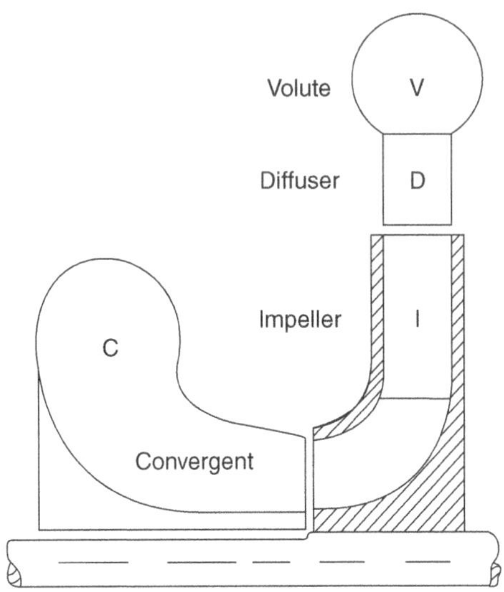

FIGURE 3.3.1 Sketch of a centrifugal compressor.

In a turbine, it is essential to have expansion and guide nozzles at the impeller inlet, while the diffuser plays a secondary role and may eventually disappear.

In a dynamic compressor, conversely, the wheel inlet guide plays a secondary role in outlet kinetic energy recovery. We can possibly suppress the inlet distributor.

3.3.2 Thermodynamics of Permanent Flow

In previous sections, we assumed kinetic energy variations of fluids undergoing processes were negligible, which allowed us to eliminate dK in our calculations. In dynamic compressors, this assumption is no longer valid, the practical effect being obtained by converting into pressure the fluid's kinetic energy.

3.3.2.1 Compressible Perfect Fluid in Steady State

We restrict ourselves initially to steady flows where absolute pressure and the three velocity components are assumed constant over time. Flow occurs in fixed pipes, which implies the absence of moving walls. These conditions are obtained in turbomachine convergent and diffuser.

We assume initially the fluid is perfect from a hydrodynamic viewpoint, that is, without viscosity (note that an ideal gas is not necessarily a perfect fluid from this viewpoint).

We obtain compressible fluid flow law applying the First Law of thermodynamics to a stream tube, taking into account gravity forces in enthalpy calculation.

According to (2.4.7), we have: $\delta\tau = vdP + dK + gdz + \delta\pi$

The walls being fixed, $\delta\tau = 0$, and the fluid being perfect, $\delta\pi = 0$.

We thus have: $dK + vdP + gdz = 0$

which, in integral form, reads:

$$\frac{C^2}{2} + \int vdP + gz = \text{Const} \tag{3.3.1}$$

This relationship is a generalization of Bernoulli's equation for incompressible flows, which corresponds to the particular case when $v = \text{Const}$.

$$\frac{C^2}{2} + \frac{P}{\rho} + gz = \text{Const}$$

In form (3.3.1), it shows that even in the case of reversible process, flow law is complex because it depends on many factors (fluid initial state, final pressure, external heat exchange).

Relationship (3.3.1) has been established in the case of absolute steady flow. It can be generalized in the case of relative steady flow, where pressure and velocity values are fixed at every point of a system driven by uniform movement. These are conditions encountered in turbomachine impeller.

This requires us to express that useful work equals 0 (fixed walls), plus work of inertial forces corresponding to driving acceleration. These are the following laws:

- System driven by uniform translation motion (axial turbomachinery case): driving acceleration is zero, and equation (3.3.1) remains valid.

- System driven by uniform rotation (centrifugal turbomachinery case): drive acceleration is radial, and work of corresponding centrifugal inertia force is $\Delta\omega^2 R^2/2$. Posing $U = \omega R$, drive velocity, equation (3.3.1) becomes:

$$\frac{C^2 - U^2}{2} + \int vdP + gz = \text{Const} \tag{3.3.2}$$

3.3.2.2 Viscous Steady Compressible Fluid

Viscosity changes the above relations only slightly. Useful work remains zero, walls being stationary, and work of viscous forces is zero, if one applies the First Law to a stream extending to the wall, because the velocity of viscous fluid is always zero at the wall.

According to (2.4.7), we have: $\delta\tau = 0 = vdP + dK + gdz + \delta\pi$

We deduce:

$$\frac{C^2}{2} + \int vdP + gz + \int \delta\pi = \text{Const} \tag{3.3.3}$$

Friction effect is ultimately a decrease in head. We thus find the concept of "pressure drop" π.

$$\pi = \int \delta\pi$$

In the present state of our knowledge, as we do not a priori know how to compute π, it is experimentally determined.

3.3.2.3 Adiabatic Flows

3.3.2.3.1 Basic Law

In the foregoing, no assumption was made on flow heat exchange with surroundings, which may vary from perfect isolation (adiabatic process) to infinitely developed contact with a source (isothermal process).

Real changes are always very close to adiabatic since gases are very poor heat conductors (see Section 2.3.6). Therefore, it is assumed that processes are always adiabatic, the actual process being irreversible, while theoretical reference is isentropic.

In what follows, we neglect potential energy gz, which is perfectly legitimate, given relatively low altitude variations of fluid, compared to kinetic energy and enthalpy variations (see Section 2.2.1).

The fundamental law of adiabatic flow in fixed reference follows directly from (3.3.3). It reflects conservation of total enthalpy $h + K$, and is written along a stream tube:

$$h + \frac{C^2}{2} = \text{Const} \tag{3.3.4}$$

Assuming further that flow is reversible adiabatic (ds = 0), then:

$$dh = vdP, \quad \text{and}$$
$$vdP + CdC = 0 \tag{3.3.5}$$

It is possible to change fluid pressure acting on its velocity and vice versa.

In the case where reference is no longer fixed, but driven by relative permanent rotation movement of drive velocity $U = \omega R$, the relation is written:

$$\frac{C^2 - U^2}{2} + h = \text{Const} \tag{3.3.6}$$

or

$$vdP + CdC - UdU = 0 \tag{3.3.7}$$

This relation shows that in a dynamic compressor rotor, compression can be achieved by reducing fluid velocity C or increasing driving velocity U. In an axial compressor, wheel diameter is approximately constant, and the first effect is dominant, while a centrifugal compressor can combine both, which allows it to provide much higher compression ratios in one stage.

3.3.2.3.2 Stagnation Properties

We have already introduced in the previous section total enthalpy h + K, and in Section 2.3.1 total energy u + K. We call stagnation pressure the pressure P_i indicated by a Pitot tube placed in perfect gas flow, and stagnation temperature the isentropic temperature T_i measured by a thermometer placed in flow so that the process bringing flow to rest in front of the thermometer would be isentropic.

Given (3.3.4)

$\Delta h = \frac{C^2}{2}$ the gas being perfect, we have: $\Delta h = c_p \Delta T$

Total or stagnation temperature is given by:

$$T_i = T + \frac{C^2}{2c_p} \tag{3.3.8}$$

The isentropic equation gives stagnation or total pressure:

$$PT^{\gamma/(\gamma-1)} = \text{Const} \quad \text{or} \quad P_i = P\left(\frac{T_i}{T}\right)^{\gamma/(\gamma-1)} = \text{Const}$$

Introducing the flow Mach number:

$Ma = \frac{C}{\sqrt{\gamma r T}}$ and noticing that: $c_p = \frac{\gamma r}{\gamma - 1}$

we obtain:

$$P_i = P\left(1 + \frac{\gamma - 1}{2} Ma^2\right)^{\gamma/(\gamma-1)} = \text{Const} \tag{3.3.9}$$

These two relationships are interpreted as follows: in any isentropic flow of perfect gas in a tube with fixed walls, stagnation temperature and pressure are conservative.

Additionally, the value of stagnation properties is that they allow us to write equations governing flow by formally removing velocity C, which greatly simplifies the expressions.

3.3.2.3.3 Varying the Section of a Stream

Let us study what should be the section variation of a fluid stream undergoing adiabatic expansion or compression in a reversible manner. In what follows, we assume that fluid is perfect gas.

For flow to be reversible and adiabatic, three equations must simultaneously be verified:

- The continuity equation, which indicates that flow is constant in any section of the flow. If S is the section, $\dot{m}$ being fluid flow, we obtain $SC/v = \dot{m}$;

  It follows: $\frac{dS}{S} + \frac{dC}{C} = \frac{dv}{v}$

- The isentropic equation $Pv^{\gamma} = \text{Const}$

  Combined, and introducing Mach number Ma, the first two equations lead to:

$$\frac{dS}{S} = \frac{vdP}{C^2}\left(1 - M_a^2\right) \tag{3.3.10}$$

- The kinetic equation (3.3.5) $vdP + CdC = 0$

These last two equations, for various sonic operation modes, allow one to simply analyze pressure and velocity changes in convergent and divergent, as shown in Figure 3.3.2.

This shows that, for subsonic flow, compression corresponds to fluid deceleration, and expansion corresponds to acceleration, while the reverse is true for supersonic regimes.

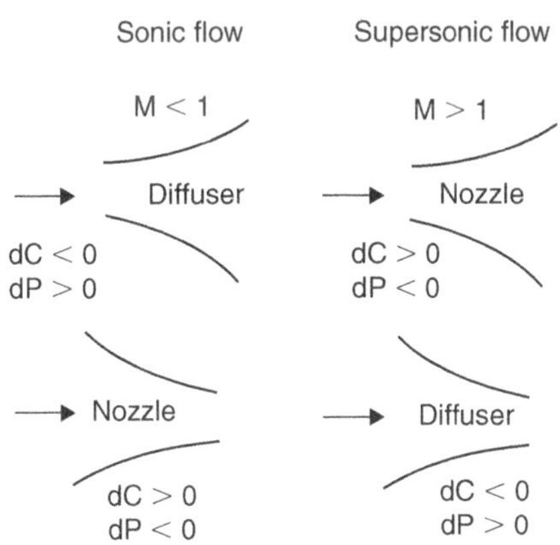

FIGURE 3.3.2 Adiabatic flow geometries.

We find a well-known result of fluid mechanics: a sonic nozzle must be comprised, at the throat inlet, of a convergent of decreasing section, and at the throat outlet, of a diverging increasing section. We call such a converging–diverging nozzle configuration a Laval nozzle.

3.3.2.4 Changes in Pressure and Fluid Velocity in a Compressor

In a dynamic compressor, fluid evolution is increased pressure, which, for the subsonic regime, requires that stream section be increasing, while velocity decreases. Evolution is a two-time process (Figure 3.3.3): in the impeller, relative velocity drops sharply, while absolute velocity increases. The stator (diffuser) then slows absolute velocity down.

To analyze in detail dynamic compressor operation, it would be necessary to make a detailed kinematic study of fluid evolution through various organs.

Developments that would require such analysis do not seem justified given our approach, oriented rather on systems than components, so we content ourselves at this stage to present the principle of these calculations, before considering compressor overall performance, based on dimensionless analysis. Further analysis will be provided in Chapter 8. We refer interested readers to books of Vivier (1965) and de Vlaminck and Wauters (1988) cited in the bibliography.

We denote with index a the fluid suction, r the discharge, 1 the rotor outlet, and R the relative velocity.

In the impeller, application of (3.3.6) gives:

$$2(h_1 - h_a) = \left(U_1^2 - U_a^2\right) + \left(R_a^2 - R_1^2\right)$$

In the diffuser:

$$2(h_r - h_1) = C_1^2 - C_a^2$$

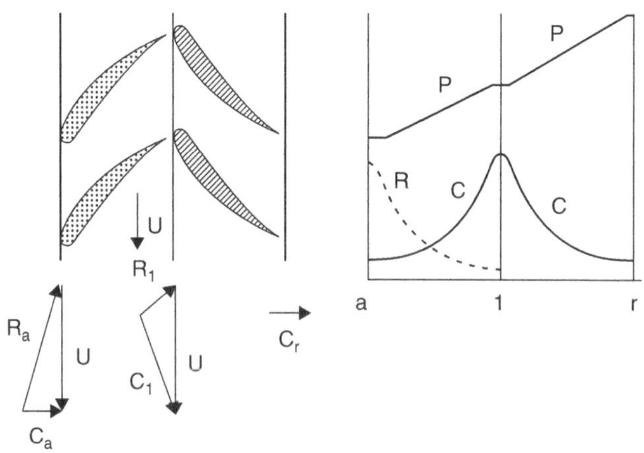

FIGURE 3.3.3 Velocity profile in an axial compressor.

Summing, we **obtain:**

$$2(h_r - h_a) = (U_l^2 - U_a^2) + (R_a^2 - R_l^2) + (C_l^2 - C_r^2)$$

Since C_r is somewhat equal to C_a, one can write:

$$2(h_r - h_a) = U_l^2 + C_l^2 - R_l^2 - (U_a^2 + C_a^2 - R_a^2)$$

However, as shown in the velocity triangle (Figure 3.3.3):

$$\vec{R} = \vec{C} - \vec{U} \quad \text{and} \quad R^2 = C^2 + U^2 - 2\vec{U} \cdot \vec{C}$$

$$\tau = h_r - h_a = \overrightarrow{U_l} \cdot \overrightarrow{C_l} - \overrightarrow{U_a} \cdot \overrightarrow{C_a} \tag{3.3.11}$$

This relationship is Euler's theorem, which connects enthalpy change and thus useful work to the velocity profile in the compressor.

In an axial compressor, drive velocity being substantially the same at the wheel inlet and outlet, this relationship is simplified, and yields:

$$\tau = h_r - h_a = \vec{U} \cdot (\overrightarrow{C_l} - \overrightarrow{C_a}) \tag{3.3.12}$$

Generally, one ensures that $\overrightarrow{C_a}$ be perpendicular to $\vec{U}$ at the dynamic compressor inlet, so that these relations become simpler yet:

$$\tau = \overrightarrow{U_l} \cdot \overrightarrow{C_l} \tag{3.3.13}$$

3.3.2.5 Velocity Triangle and Shape of the Blades

Euler's relation is used to define the blade profile: in order for fluid to flow in rotor with least possible irreversibilities, shocks must be avoided, whether from discontinuities in fluid streams or from shock waves which can propagate if velocity is too high. Flow thermodynamics finds here one of its limits: exact blade profile shape calls upon fluid mechanics and is thus outside this presentation scope. We will just give some indication of results, only for axial compressors.

Given the Euler relationship, we seek to maximize scalar product $\overrightarrow{U_l} \cdot \overrightarrow{C_l} = UC_l \cos \alpha_l$. Therefore, angle α_l must be as small as possible.

Suppose driving velocity U is known. To avoid shock in blades, relative velocity vector $\vec{R}$ must at all times be parallel to blades, which sets its angle β with drive velocity $\vec{U}$.

Entry speed $\overrightarrow{C_a}$ being perpendicular to $\vec{U}$, the inlet velocity triangle immediately gives angle β_a and defines vector $\overrightarrow{R_a}$.

At outlet we obtain similarly β_l and $\overrightarrow{R_l}$, knowing that absolute velocity $\overrightarrow{C_l}$ must generally remain subsonic to avoid shock in diffuser vanes.

By continuity the blade profile can be plotted (Figure 3.3.4).

However as discussed later, technological issues lead manufacturers to choose wheel type and to set some reduced design parameters. We do not, therefore,

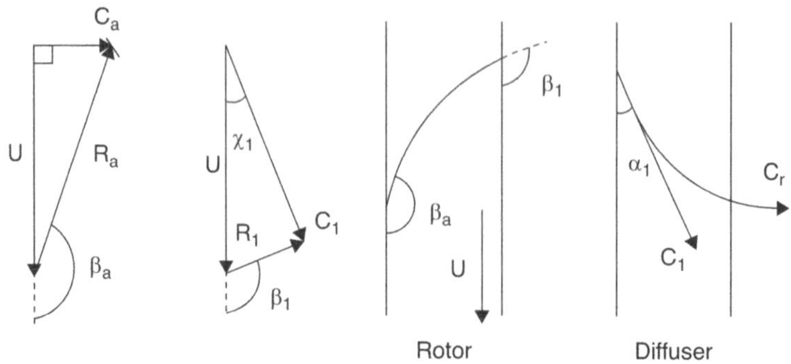

FIGURE 3.3.4 Velocity triangle.

generally know U as has been assumed here, but relationships exist between velocity triangle quantities, so that it is determined by reasoning similar to that just presented.

To determine blade thickness, whose distance between them defines the cross-section available for fluid, we must give, in addition to the Euler relation, the continuity equation. Moreover, even more than for the nozzle profile, it is necessary to reduce curvature to prevent detachment of fluid streams along compressor blades, because the pressure gradient exerted here is unfavorable, both in rotor and in diffuser. This constraint is much stronger for dynamic compressors than for turbines.

3.3.3 Similarity and Performance of Turbomachines

When the dynamic compressor is not working under its design conditions, performance generally suffers in larger proportion than in the case of positive displacement compressors. The reason is that dynamic irreversibilities (shock, detachment of fluid streams, etc.) are rising quite rapidly once flow no longer corresponds to blade geometry.

We will now further analyze turbomachinery characteristics, and machine adaptation at different operating regimes. These regimes depend not only on basic variables considered in design ($N, \dot{V}, \Delta h$), but also on characteristic parameters of fluid passing through the machine.

In most common conditions, seven independent physical variables may affect turbomachine efficiency: a characteristic dimension (e.g., wheel diameter D), rotation speed N, fluid mass flow $\dot{m}$, the stagnation pressures at inlet and outlet P_a and P_r, and the total enthalpy at the entrance h_a and exit h_r.

As mentioned in Section 3.3.2.3, we choose stagnation properties to eliminate from equations fluid velocity at the machine entrance and exit.

Among these seven variables are involved three basic units (M, L, T). Application of the Vaschy-Buckingham theorem reduces to $7-3=4$ the number of dimensionless variables characteristic of machine operation.

The variables most commonly chosen are:

- A flow velocity C_f Mach number $(M_a)_c = \dfrac{C_f}{C_s}$

$$\text{with } C_f = \frac{\dot{m} v_a}{A} = \frac{4 \dot{m} v_a}{\pi D^2} = \frac{4 \dot{m} r T_a}{\pi D^2 P_a}$$

and C_S speed of sound in fluid ($C_s = \sqrt{\gamma r T_a}$ by equating fluid to perfect gas), thus:

$$(M_a)_c = \frac{4\dot{m}\sqrt{rT_a}}{\pi D^2 P_a \sqrt{\gamma}} \quad (3.3.14)$$

- a wheel Mach number $(M_a)_u = \frac{U}{C_s}$ with $U = \frac{\pi DN}{60}$

$$(M_a)_u = \frac{\pi DN}{60\sqrt{\gamma r T_a}} \quad (3.3.15)$$

- the ratio of inlet and outlet stagnation pressures: $\frac{P_r}{P_a}$
- the isentropic efficiency of the stage:

$$\eta_s = \frac{(h_r)_s - h_a}{h_r - h_a} \quad \text{or} \quad \eta_s = \frac{h_r - h_a}{(h_r)_s - h_a}$$

depending on whether a compressor or a turbine.

When choosing a given machine and particular fluid, dimensionless numbers $(Ma)_c$ and $(Ma)_u$ become proportional to reduced variables of simpler expressions:

$$(M_a)_c \div \frac{\dot{m}\sqrt{T_a}}{P_a} = (\text{corrected mass flow } \dot{m}_c) \quad (3.3.16)$$

$$(M_a)_u \div \frac{N}{\sqrt{T_a}} = (\text{ corrected rotation speed } N_c) \quad (3.3.17)$$

Other dimensionless quantities are also commonly used by manufacturers: flow factor φ and enthalpy factor ψ, or Rateau coefficients μ and δ.

The flow factor φ

It is natural to look at both Mach numbers representative of flows in machines, $(Ma)_c$ and $(Ma)_u$. The ratio of these two quantities, independent of fluid properties, determines velocity triangle shape, and corresponds to first dimensionless property: flow factor φ which ensures kinematic similarity over entire flow boundaries.

$$\varphi = \frac{(M_a)_c}{(M_a)_u} = \frac{4\dot{m}\sqrt{rT_a}}{\pi D^2 P_a \sqrt{\gamma}} \frac{60\sqrt{\gamma r T_a}}{\pi DN}$$

$$\varphi = \frac{C_f}{U} = \frac{240\dot{V}}{\pi^2 ND^3} = \frac{240}{\pi}\delta \quad (3.3.18)$$

This factor φ is proportional to another widely used dimensionless quantity: second Rateau coefficient δ.

3.3.3.1 The Enthalpy Factor ψ

Turbomachinery is used either to expand fluid to produce energy or to provide energy to fluid. Flow being, as we noted earlier, close to adiabatic, because of low exchange surfaces and velocities, it is logical to take as reference energy the expansion or compression isentropic work $|\Delta h_s|$.

$$\text{We have}: \ |\Delta h_s| = \frac{\gamma}{\gamma - 1} P_a v_a \left| \left(\left(\frac{P_r}{P_a} \right)^{(\gamma-1)/\gamma} - 1 \right) \right| \tag{3.3.19}$$

The ratio of reference energy to well-chosen kinetic energy is a dimensionless quantity of interest. A first idea is to use sound speed C_s. We obtain:

$$\frac{|\Delta h_s|}{1/2C_s^2} = \frac{2}{\gamma - 1} \left| \left(\left(\frac{P_r}{P_a} \right)^{(\gamma-1)/\gamma} - 1 \right) \right| = \Omega \tag{3.3.20}$$

Usually this kinetic energy is taken equal to $1/2U^2$, corresponding to the maximum kinetic energy in rotor. This way of doing is equivalent to multiplying Ω by $(M_a)_u^2$. This defines enthalpy factor ψ, equal to twice first Rateau coefficient μ (itself equal to the power factor introduced in Section 3.1.4).

In practice, machine performance is usually given in the form of characteristic curves for constant values of corrected rotation speed N_c:

$$\frac{P_r}{P_a} = f(\dot{m}_c) \quad \eta_s = f\left(\frac{P_r}{P_a}\right) \quad \text{or} \quad f(\dot{m}_c)$$

It should be noted that curves of equal efficiency can be directly plotted on charts.

Sometimes, curves are presented in relation to reference values of corrected mass flow rate or corrected speed.

Note that performance maps of turbomachinery (as well as displacement machines) play a unique role in that they constrain intensive variables with extensive variables, whereas usually they are independent: flow affects the compression ratio.

Examples of experimental results are presented in the following pages, corrected flows and corrected speeds being reported to their nominal values.

3.3.3.2 Performance Maps of Dynamic Compressors

We content ourselves in this section to present some general results concerning dynamic compressor performance maps, which will be completed in Chapter 9 on the off-design behavior of these machines.

For dynamic compressors (Figures 3.3.5–3.3.7), the corrected mass flow is used as the abscissa. Ordinate is the pressure ratio or isentropic efficiency. Corrected rotation speed is still used as a parameter.

They have the same shape as diffuser characteristics, for different rotation speeds.

Corrected rotation speed strongly influences dynamic compressor performance. This is easily explained considering that in such machine, energy is transmitted to fluid by rotor in the form of kinetic energy. At most, this energy is $1/2\,U^2$, that is, it

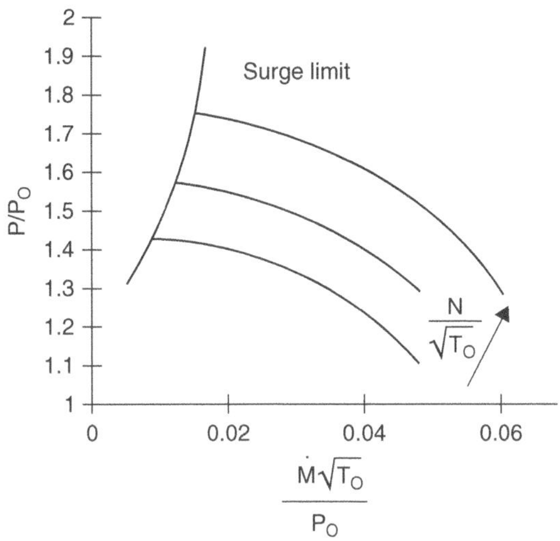

FIGURE 3.3.5 Dynamic compressor performance map.

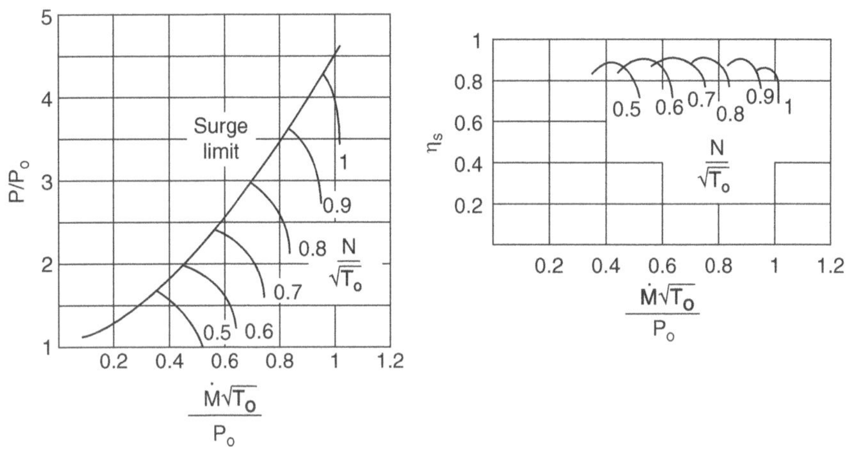

FIGURE 3.3.6 Axial compressor performance map.

is proportional to N^2. It is therefore natural that the sensitivity of these machines to regime change is dramatic.

In addition, fluid mechanics tells us that fluid flow in blades is destabilized here by the pressure gradient. As soon as one deviates too much from design operating conditions, there is a significant risk of stream separation along blades. Besides the strong sensitivity of isentropic efficiency, this results in a double limitation of machine use range: risk of surge at low flow (which depends on the network in which compressor discharges, cf. next section), and stalling on the side of higher flow rates. Some flexibility is indeed possible in this machine type only if you can

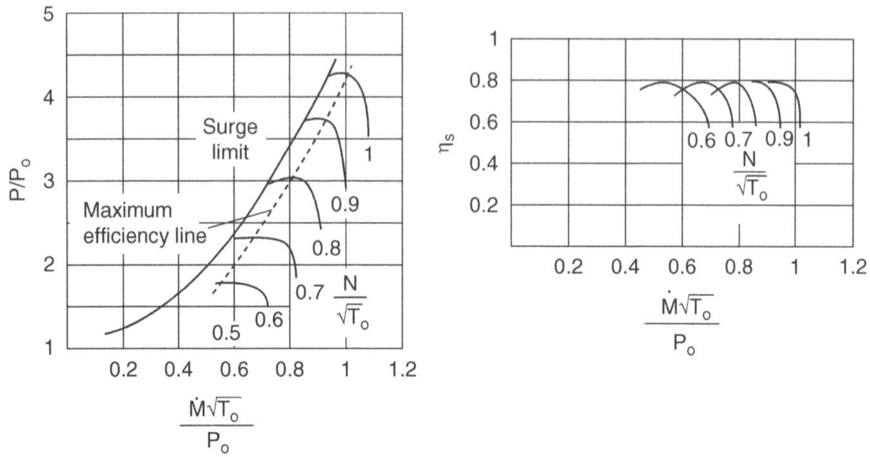

FIGURE 3.3.7 Centrifugal compressor performance map.

freely adjust the rotation speed and possibly angle of incidence of certain blade rows (guide vanes).

3.3.3.2.1 Axial Compressors

Compression ratios per stage that axial compressors can provide are relatively low, generally between 1.2 and 2.

3.3.3.2.2 Centrifugal Compressors

Characteristics of centrifugal compressors are steeper than those of axial compressors, making them more stable. Due to the acceleration that the fluid receives during its passage through the rotor, the compression ratio per stage that they can provide is much more important, between 2.5 and 9.

It will be noted on the characteristic of Figure 3.3.11 that we can equip the (pressure - flow) performance map by including iso-isentropic efficiency η lines. The locus of these values corresponds substantially to portions of ellipses. They are commonly called efficiency islands.

3.3.3.3 Surge in Dynamic Compressors

When a dynamic compressor delivers in a receiver circuit, its operating point is determined by the intersection of its characteristic curve and that of the circuit.

Circuit losses being in the first approximation proportional to square of flow rate, circuit characteristic shape is given in Figure 3.3.8. The ordinate intercept, sometimes called static load, corresponds to terms independent of flow in the generalized Bernoulli equation (3.3.3).

The compressor operating point evolves thus according to its rotation speed, so that at any time, power dissipated in the circuit equals machine shaft power, and passes from A to C when velocity increases (Figure 3.3.9).

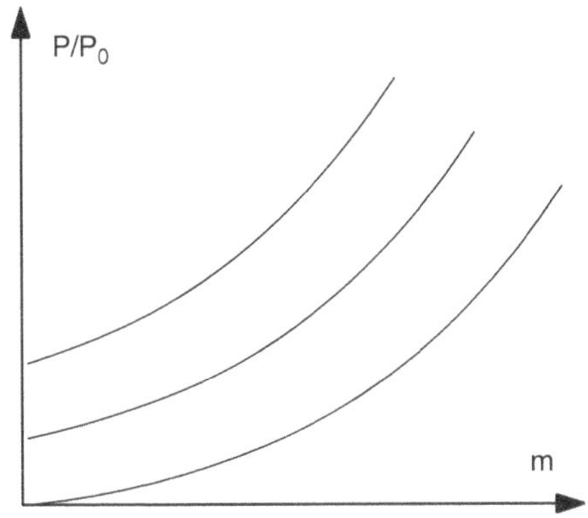

FIGURE 3.3.8 Circuit losses.

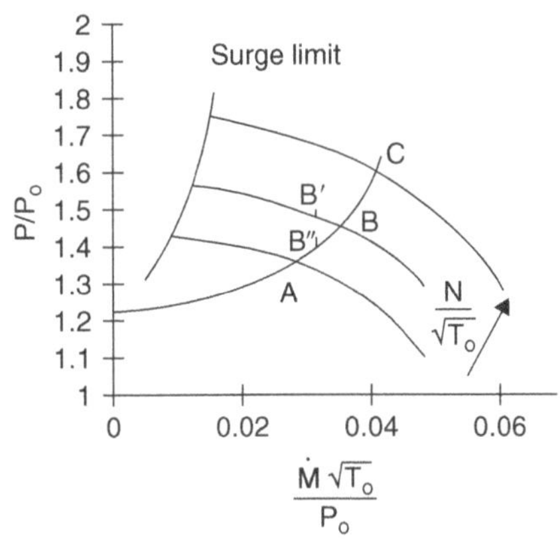

FIGURE 3.3.9 Stable operation point.

Things do not always happen as simply, as a zone of unstable operation may appear in the left part of dynamic compressor performance map maximum for reasons explained below.

Let us first check that a point such as B is stable (Figure 3.3.9). Suppose the machine is operating nominally at point B and at constant speed when the flow suddenly drops. Given the shape of the machine characteristic, its operating point tends to move to B′, while the circuit resistance shifts to B″.

As a result, the compressor delivers more power (B′ – B″) than that required by the circuit, which increases the flow rate and brings the operating point back to B. The system is therefore naturally stable.

If the operating point lies to thc lcft of the compressor characteristic maximum, two cases may arise:

- If the circuit characteristic slope is greater than that of the machine, reasoning similar to the previous shows that the system is stable.

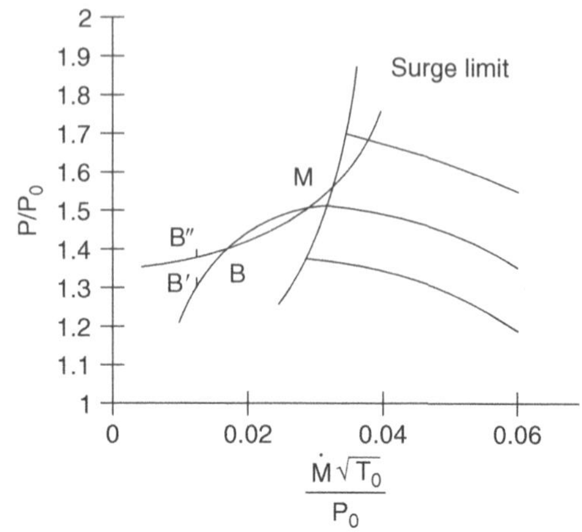

FIGURE 3.3.10 Unstable operation point.

- If, as shown in the chart of Figure 3.3.10, the compressor characteristic is the steepest, any reduction in the flow rate results in a faster decrease of compressor power than that of the circuit resistance. The initial disturbance is therefore amplified, and the system becomes unstable. Complex phenomena then occur, depending on the mechanical inertia of the entire system. The flow rate drops very rapidly (within 20–80 ms) and may even reverse.

Circuit resistance falls accordingly, discharge manifold empties, and compressor available power becomes surplus, which leads to a very rapid increase in flow. Very quickly, the functioning point returns to the right of B′. The discharge pipe is filled, pressure rises, and flow lowers again: cyclic conditions are met. We call this instability a surge cycle, whose period varies between 0.5 and 5 seconds. Surge is generally very detrimental to mechanical compressor, which is subjected to intense vibration that can put rotor and stator in contact.

Phenomena involved are very complex because of inertial effects associated with the filling and emptying collector network. They are still poorly known. Manufacturers therefore recommend users avoid operating in the left of the performance map maximum of their machines, bounded by the surge limit called the surge line.

As it is not always possible to meet this constraint, it is often necessary to provide check valves at the machine outlet, or more complex devices when it is not enough. Instrumentation and modern control provide opportunities to detect surge onset, which is accompanied by changes in pressure and flow as well as in suction temperature for axial compressors. Anti-surge control must be able to react very quickly given phenomenon speed, increasing flow through the opening of the bypass valve.

3.3.4 Practical Calculation of Dynamic Compressors

To be able to size their compressors quickly with sufficient accuracy, manufacturers take full advantage of similarity rules. Main difficulty, however, is that situation is not as simple as in the case of positive displacement compressors.

In the latter case, in fact, an entire range of machines can be represented with sufficient accuracy by the same set of charts giving volumetric efficiency and compression ratio as a function of the compression ratio.

In the case of turbomachinery, this approximation is generally valid, and manufacturers usually prefer to use characteristics defined per wheel when high accuracy is required, since the shapes of efficiency islands may vary significantly from one machine to another, even within the same range. However, similarity laws are applied to the flow, and performance maps are often plotted using a specific system of coordinates that combines enthalpy and flow coefficients (or Rateau coefficients) with the wheel Mach number.

We find generally the head coefficient in ordinate and flow coefficient in abscissa. York in France, a leading manufacturer of compressors for industrial refrigeration, uses as ordinate the Ω factor introduced earlier, and as abscissa factor θ product of δ and $(M_a)_u$ and therefore proportional to $(M_a)_c$.

$$\Omega = \frac{\Delta h_s}{C_s^2} = \mu (M_a)_u^2$$

$$\theta = \frac{\dot{V}}{C_s D^2} = \delta (M_a)_u$$

$C_s = \sqrt{\gamma r T_a}$ being sound speed at the inlet and μ and δ the Rateau coefficients.

The characteristic obtained has form given in Figure 3.3.11.

The practical calculation of a compressor runs as follows.

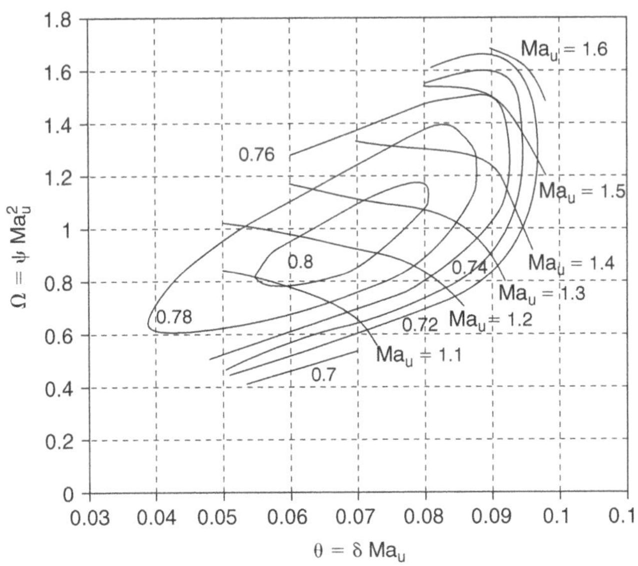

FIGURE 3.3.11 Centrifugal compressor performance map.

We are given the volume flow at the inlet and desired compression ratio, as well as suction conditions, which enables us to calculate sound speed C_s.

On the thermodynamic diagram, using either tables or equations of state or Thermoptim, we determine isentropic compression work τ_s, which gives the Ω value.

We choose a priori a machine, i.e. wheel diameter, which determines the θ value.

On the performance map, it is then possible to check if the chosen wheel is well suited, that is, if the operation point is to the right of the efficiency maximum to avoid any risk of surge.

Knowledge of wheel coefficient μ allows us to determine the Mach number and hence tangential velocity U.

Since $U = \frac{\pi DN}{60}$, the rotation speed can be deduced.

It is then possible to study behavior at variable flow and constant speed $\left((M_a)_u = \text{Const}\right)$, and possible operation at part load and variable speed, to determine if the proposed operating ranges are compatible with machine possibilities.

3.3.5 Pumps and Fans

Mechanical and thermodynamic study of flows in pumps and fans is to very large extent similar to that of dynamic compressors, but with important simplification: for them it is quite legitimate to assume that fluid is incompressible.

Expression for isentropic work is considerably simplified, since $\Delta h_s = v\Delta P$ or $\Delta h_s = \Delta P/\rho$. For pump, the data provided is generally pressure head H, and g being acceleration due to gravity, $\Delta h_s = gH$.

Under these conditions, pump or fan will be represented in Thermoptim by the same component as compressor.

We will not detail here various technological solutions that exist for pumps or fans.

Note, however, that a critical phenomenon can occur at the pump inlet: cavitation, which occurs when suction pressure falls below the saturated vapor pressure of the fluid at the temperature considered. No precaution is taken in Thermoptim to ensure that cavitation in the pump is prevented.

Let us recall that non-cavitation conditions are given by manufacturers as the hydraulic minimum load that must exist at the pump inlet. Called NPSHR (Net Positive Suction Head Required), it is determined experimentally and depends on wheel rotation speed. NPSHR equals net positive suction height available (NPSHA) minus the absolute saturation vapor pressure of the fluid. It is expressed in meters.

NPSHA, i.e., that which exists at the pump inlet, must exceed NPSHR, whose value is generally determined as the threshold below which cavitation appears in the pump and degrades its efficiency by 3%.

To avoid cavitation in the pump, reduce pressure drop in suction to prevent tank pressure becoming too low and, in the case of the suction pump, prevent suction height being too great.

3.4 EXPANSION

Nowadays, most expanders are turbines.

Turbine theory is very close to that of dynamic compressors, expansion thermodynamics being distinct from that of compression essentially by some definition changes.

3.4.1 Thermodynamics of Expansion

In preceding pages we explained compression laws, indicating that those of expansion differ only by a different definition of isentropic efficiency. This short section provides relations for expansion.

Isentropic work is still given by (3.1.5) or (3.1.6).

3.4.1.1 Irreversible Expansion

In irreversible expansion, shaft work is less than isentropic work, because irreversibilities have the effect of increasing fluid entropy and its temperature, and are therefore reflected by shift (a–r) to the right of isentropic (a–s) (Figure 3.4.1).

We define in this case isentropic efficiency by:

$$\eta_s = \frac{\tau}{\tau_s} = \frac{h_r - h_a}{h_s - h_a} \tag{3.4.1}$$

This definition, reverse of that of isentropic compression efficiency, leads to value $\eta_s < 1$.

Again, it must be stressed that given the current state of our knowledge, we are not able to assess a priori the value of expansion isentropic efficiency, which is measured experimentally.

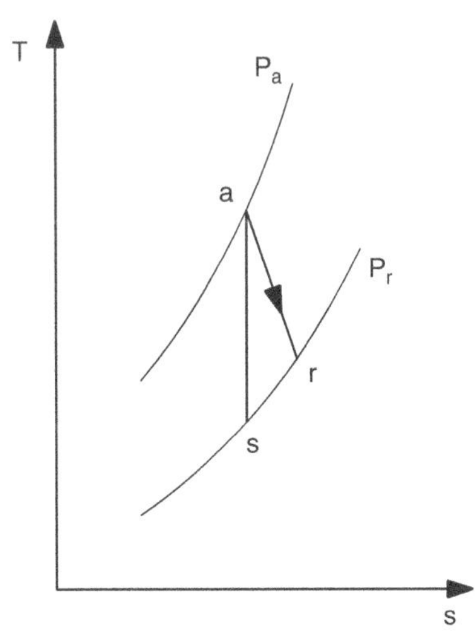

FIGURE 3.4.1 Irreversible expansion process.

3.4.1.2 Polytropic Expansion

Similarly to what we did for compression, we define polytropic expansion as the process in which, for any infinitely small stage, isentropic efficiency remains constant and equal to polytropic efficiency η_p.

$$\eta_p = \frac{vdP + Tds}{vdP} = \frac{dh}{dh - Tds} \qquad \delta\tau_s = \frac{1}{\eta_p} dh$$

Posing:

$$\frac{k-1}{k} = \eta_p \frac{\gamma - 1}{\gamma} \tag{3.4.2}$$

We find polytropic law (3.1.9): $Pv^k = \text{Const}$.

With this convention, shaft work in polytropic expansion can also be written according to equation (3.1.6):

$$\tau = \frac{\gamma}{\gamma - 1} P_a v_a \left[\left(\frac{P_r}{P_a} \right)^{(k-1)/k} - 1 \right]$$

Entropy difference is independent of c_p and T for a polytropic expansion and can be expressed as:

$$\Delta s = (1 - \eta_p) R \ln\left(\frac{P_a}{P_r} \right) \quad \text{(open systems) or}$$

$$\Delta s = (1 - \eta_p) R \ln\left(\frac{v_r}{v_a} \right) \quad \text{(closed systems)}$$

3.4.1.3 Calculation of Heated Polytropic for Open System

If you know polytropic coefficient k of the process, T_r is calculated by:

$$T_r = T_a \left(\frac{P_r}{P_a} \right)^{(k-1)/k}$$

If one assumes that reversible polytropic reference passes through inlet and outlet points, we must first identify polytropic efficiency η_{pref} by applying between these two points $\Delta s = (1 - \eta_p) R \ln\left(\frac{P_r}{P_a} \right)$.

Shaft work for reversible polytropic is then: $\tau_{ref} = \frac{1}{\eta_p}(h_r - h_a)$.

Knowing process efficiency η (isentropic or polytropic), we can deduce actual expansion: $\tau = \frac{\eta}{\eta_{pref}}(h - h_a)$ work.

Heat to provide is then:

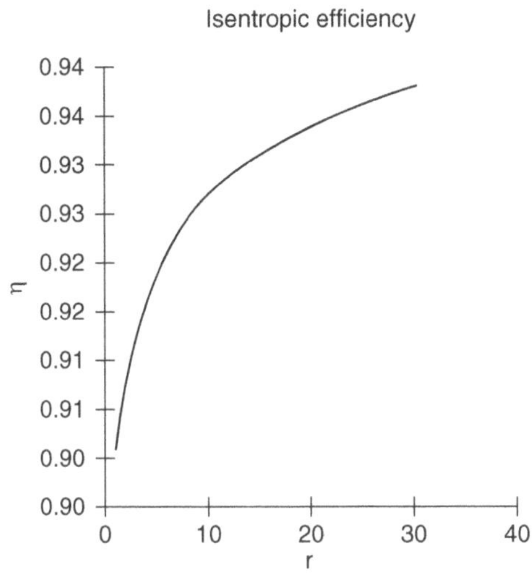

FIGURE 3.4.2 Isentropic efficiency function of expansion ratio.

$$Q = h_r - h_a - \tau = \left(1 - \frac{\eta}{\eta_{pref}}\right)(h_r - h_a) \tag{3.4.3}$$

While polytropic compression leads to $k > \gamma$, we obtain in the case of polytropic expansion: $k < \gamma$.

Unlike what happens in compression, given constant polytropic efficiency (0.9 in Figure 3.4.2), the expansion isentropic efficiency increases when the expansion ratio increases.

This arises because irreversibilities taking place at high temperature (and pressure) are partially recovered in subsequent expansion stages, as they have effect of warming fluid.

3.4.2 Calculation of an Expansion in Thermoptim

In Thermoptim, the calculation of expansion is done in a manner analogous to that of compression, with the same screen. You should therefore refer to what was presented in Section 3.1.5.

3.4.3 Turbines

3.4.3.1 Geometrical Arrangement

The general layout of turbine is given in Figure 3.4.3. The stator is made of guide nozzles, which accelerate fluid, while the wheel (rotor) converts into mechanical energy at least part of available enthalpy.

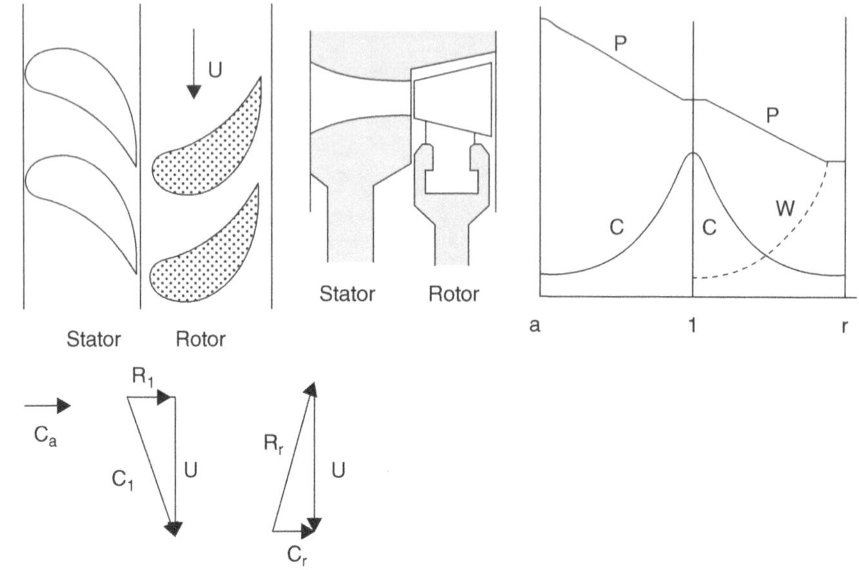

FIGURE 3.4.3 Velocity and pressure profile in an axial turbine stage.

3.4.3.2 Changes in Pressure and Fluid Velocity in a Turbine

In a turbine, fluid evolution is expansion. Compressible flow equations indicate that for the subsonic regime, the stream section should decrease and velocity increase. This evolution takes place in two steps (Figure 3.4.3): in the stator, absolute velocity increases, while in the impeller, relative velocity increases, and absolute velocity decreases.

3.4.4 Turbine Performance Maps

Just as we did in Section 3.3.3.1, we will only present in this section some general results concerning turbine performance maps, which will be completed in Chapter 9 dealing with the off-design behavior of machines.

In the case of turbine (Figure 3.4.4), it is conventionally the pressure ratio which is used as abscissa. As ordinate, we find the corrected mass flow or isentropic efficiency of machine. The curve parameter is still corrected rotation speed, which plays a secondary role.

We can see turbine flexibility to adapt to various operating conditions: efficiency tends to deteriorate only when we dramatically reduce the pressure ratio or speed. This flexibility is particularly due to flow stability in blades due to the pressure gradient therein. But what is most remarkable is flow stability for high pressure ratios, which comes from the supersonic regime, which takes place in at least part of machine (flow is choked at place where sound speed is reached).

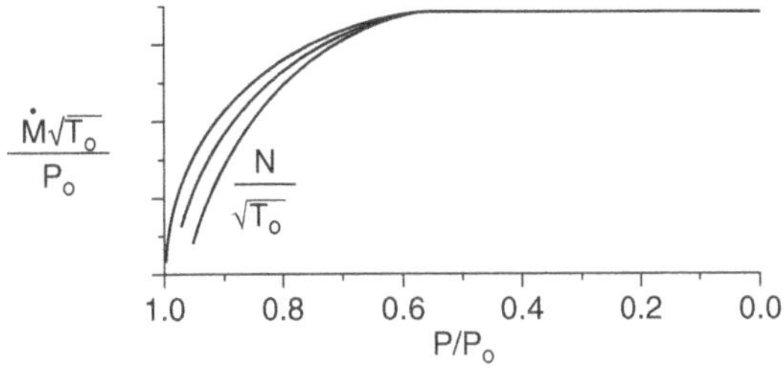

FIGURE 3.4.4 Turbine performance map.

Limiting value reached by flow, when the pressure ratio exceeds the critical ratio, is called critical flow. It is proportional to the throat section, which is obviously independent of rotation speed, which explains the weak influence of this parameter.

In supersonic nozzle, the corrected mass flow is proportional to throat section and independent of inlet conditions. When nozzle is caused to operate outside rated conditions, fluid flow somehow adapts to duct shape. Efficiency decrease resulting from this adaptation is especially important when one wants to exceed the design pressure ratio.

As the expansion ratio remains above its critical value, flow stays below the critical flow. When its value decreases beyond this value, flow maintains a constant value.

If you want to go beyond a critical expansion ratio in a converging–diverging nozzle, flow cannot increase sufficiently, and expansion continues beyond the exit section with large oscillations at the nozzle outlet.

Indeed, if the flow is supersonic and the pressure in the outlet chamber differs from the design value, the following phenomenon occurs. When the outlet pressure is higher than the nominal value, overexpansion takes place in the nozzle, followed by a sharp contraction at the exit. Conversely, when the outlet pressure is lower than nominal, underexpansion occurs, leading to a sudden expansion at the exit.

The explanation for this phenomenon is straightforward: beyond the throat, the flow velocity exceeds the speed of sound. Consequently, the conditions imposed on the outlet stream can no longer influence the flow upstream of the throat. The flow therefore continues as if the downstream pressure were equal to the design outlet pressure, and if the actual pressure differs, a discontinuity inevitably appears.

3.4.5 Degree of Reaction of a Stage

Study of stage energy balance showed that enthalpy change takes place partly in the rotor, and partly in the diffuser. We call the degree of reaction ε the fraction of enthalpy change that takes place in the rotor. With notation of Figure 3.3.3, we have:

$$\varepsilon = \frac{h_l - h_a}{h_r - h_a} = \frac{h_l - h_a}{\Delta h} \tag{3.4.4}$$

By definition, ε is between 0 and 1.

In the dynamic compressor stage $0.5 \le \varepsilon \le 1$ is commonly obtained. The lower limit ($\varepsilon = 0.5$) is approximated in axial dynamic compressors. As for the upper limit (pure reaction: $\varepsilon = 1$), it is reached in the case of single-stage machine whose diffuser is nonexistent or of negligible effectiveness.

In the turbine stage, pure reaction is impossible. Two limiting cases frequently arise:

- Impulse turbines, in which $\varepsilon = 0$: any fluid expansion is then carried out in fixed blades or nozzles, at the wheel inlet, and inlet and outlet pressures of the rotor are equal.
- Reaction turbines, where $\varepsilon = 0.5$: expansion is then evenly distributed between nozzle and wheel.

Each of these two turbine types has advantages and disadvantages of its own. As we have said, it is not our intention to develop here in detail turbine theory. Simply note that impulse turbines are generally used for head stages of multistage sections or for small capacity units, while reaction turbines are proving well suited for low-pressure sections.

Indeed, the first advantage of using impulse turbines in high pressure sections is that entire expansion being made in the stator, the rotor is not subjected to a high pressure difference, which limits mechanical constraints. The second advantage is that flow in these turbines can be reduced by using partial injection, which is to only supply fraction of fluid in stator vanes. This operation type is made possible in this case because pressure is the same on both sides of the rotor, and no parasitic flow is expected in non-injected parts.

However, efficiencies of these turbines are slightly (2%–3%) worse than those of reaction turbines, which are subjected to significant axial thrust, and do not use partial injection.

Because irreversibilities taking place in top stages are partially recovered in subsequent stages, we can tolerate that their efficiency is slightly lower, which allows us to use impulse turbines in this case.

We will return to this topic in Section 9.1.2.

3.5 COMBUSTION

Combustion phenomena are of particular importance in energy technology studies, because they are the source of most heat and power production in the world: more than 90% of global primary commercial energy consumption comes from burning coal, oil, or natural gas.

Combustion study determines the state and composition of combustion products, and consequently, their thermochemical properties including the amount of energy involved in reactions. Moreover, combustion conditions largely determine pollutant quantities emitted by energy technologies.

This chapter's purpose is to determine as closely as possible combustion characteristics that are important in energy terms: product thermodynamic properties (c_p, M, h, s), energy released by combustion (ΔH_r), the concentration of unburned hydrocarbons, and pollutants.

From these elements, it becomes possible to optimize combustion, i.e., to obtain the best combustion efficiency by adjusting the air/fuel ratio or excess air, which determines flame temperatures and losses through flue gases.

To achieve this goal, it is necessary to know how to characterize fuels and write complete reactions, stoichiometric and with excess or lack of air, taking into account dissociation where it exists, which requires knowledge of general principles governing equilibrium reactions (law of mass action), and finally calculate energies released.

In the remainder of this chapter, we will always assume that combustion reaction involves gaseous species, and that various components are treated as ideal gases. Dalton's law then allows us to consider that combustion products follow ideal gas law.

3.5.1 Combustion Phenomena, Basic Mechanisms

Combustion reactions are rather unusual chemical reactions, typically violent, where non-equilibrium is the rule, equilibrium the exception. In these circumstances, it is still impossible to determine exactly during reaction the evolution of thermodynamic functions characterizing involved systems. They can only be calculated assuming that equilibrium conditions are met before and after combustion.

3.5.1.1 Complete Reactions, Dissociation

Generally, a chemical reaction involving two reactants and two products can be written as:

$$\nu_1\, A_1 + \nu_2\, A_2 \leftrightarrow \nu_3\, A_3 + \nu_4\, A_4 \quad (3.5.1)$$

Experience shows that combustion reaction unfolding is highly dependent on temperature.

At low temperature, reactants generally do not react spontaneously. For reaction to start, it is necessary to provide activation energy, usually in the form of electric spark. They say that the mixture is in false equilibrium. If reaction is triggered artificially, it is violent, and total if temperature remains low enough. Combustion is said to be complete, or without dissociation.

At high temperature, reaction is usually incomplete. Reaction medium composition depends on the reaction's degree of progress, the ratio between the number of reactant moles that have been converted and the total number of reactant moles. This degree of progress depends on kinetics and mass action law at equilibrium. Reactants are not completely oxidized, and we speak of dissociation. If, starting from such a situation, we gradually decrease reactant temperature, composition stabilizes and remains unchanged only above a certain threshold. It is said there is freezing of reaction.

In energy systems we study, combustion will be either complete or incomplete, depending on cases. We show in this section how it can be calculated with some simplifying assumptions.

General thermodynamic study of combustion reactions is part of chemical process thermodynamics. In this chapter, we limit ourselves to only those aspects necessary for our purposes, i.e., mainly study of reactive mixture thermodynamic function variations (internal energy, enthalpy, entropy) caused by chemical composition variations.

Study of the reaction kinetics mechanism (false equilibrium limits, means of triggering reaction mixture in false equilibrium, reaction propagation methods, kinetics, etc.) is much more complex. We will discuss it here only very briefly and qualitatively.

In what follows, as stated in introduction, we limit ourselves mainly to studying combustion reactions in the gaseous phase. Even when fuel is introduced as liquid (oil, diesel), products are then present in gaseous form.

3.5.1.2 Combustion of Gaseous Fuel Mixtures

A gaseous fuel mixture, in practice generally composed of air and fuel gas, is defined by its molar or mass composition.

It is customary to relate any mixture composition to that of a stoichiometric mixture, i.e., with exactly the amount of oxygen required for complete combustion. Richness R is defined as the ratio of the number of moles (or mass) of fuel in the predetermined amount of mixture, to the number of moles (or mass) of fuel in the stoichiometric mixture.

$R=1$ corresponds to the stoichiometric mixture, $R<1$ to excess air, and $R>1$ to excess fuel (lack of air). In Section 3.5.2.4, we will introduce two more quantities directly related to richness, which are also widely used for the non-stoichiometric combustion study: air factor λ and excess air e.

3.5.1.2.1 Conditions of Self-Ignition and Combustion Propagation

We have seen that the fuel mixture based on the given fuel is generally stable at room temperature. By heating, it is possible to trigger combustion reaction from certain temperature, called self-ignition. H_2 mixtures burn spontaneously at 550°C, those based on CO and CH_4 around 650°C. It sometimes happens that self-ignition delay appears, which can reach 200 K for CH_4.

Auto-ignition temperature depends of course on the fuel and mixture composition, but it is also decreasing function of pressure, with some peculiarities (cool flames, long self-ignition delays), for which we refer to the literature.

The self-ignition temperature value defines the upper limit, which must not be reached, whether for safety reasons when handling fuel, or for more specific applications, such as burning in gasoline engines.

When the fuel mixture is heated to temperature above its self-ignition temperature, reaction is triggered after some delay, the said auto-ignition delay, which becomes shorter at higher temperatures. This delay is related to basic combustion mechanisms presented in Section 3.5.1.5.

Moreover, in the given mixture of air and fuel, combustion, initiated locally by spark, can only spread if richness is between two limits: lower inflammable Ri and higher inflammable Rs. If richness is outside this interval, the mixture is either too lean or too rich, and combustion cannot be maintained. These limits depend on the temperature and fuel type. At room temperature, limits Ri and Rs for most hydrocarbons are respectively close to 0.5 and 4. They diverge as temperature increases.

3.5.1.2.2 Flame Propagation

To avoid self-ignition of gaseous fuels, it is necessary that the mixture is ignited locally before reaching self-ignition temperature. In practice, electric spark is generally used for this purpose.

Once the mixture is ignited, combustion spreads from the ignition point, in the reaction zone called the flame front, corresponding to the surface separating the burned gases from the mixture still intact. The case where the fuel and oxidizer form a homogeneous mixture before the reaction starts is referred to as the premixed flame. Experience shows that propagation can take place in two very different modes: deflagration and detonation. Difference between them is relative speed at which flame front moves in the mixture.

In deflagration, this speed is low, order of several cm/s to few m/s, whereas in detonation, it far exceeds sound speed in the mixture (several thousands of m/s).

In deflagration, combustion is mainly propagated by conduction: the combustion gas layer heats the next layer of the still intact mixture. When it is in turn brought to auto-ignition temperature, it ignites, and so on. Gases being poor conductors, propagation speed is low.

In detonation, combustion is propagated by shockwave that moves at speed exceeding sound speed in the mixture, causing almost instantaneous reaction at its passage, with sudden changes in velocity, pressure, and temperature. Given this shockwave, detonation is very dangerous combustion mode, which causes engines to deteriorate rapidly. Very abrupt pressure changes induce, in any parts submitted to gas thrust, very destructive steep front elastic waves. Joints are subjected to shocks, valves, and seats leak, deformed by waves. When it occurs in the engine, it is usually called knocking combustion mode. Additionally, when fuel is hydrocarbon, detonating combustion is always initiated by molecule decomposition, releasing black smoke and engine gets dirty.

3.5.1.2.3 Study of Deflagration

A mixture whose richness is between flammability limits burns with certain deflagration speed. This speed, often called laminar velocity, is flame velocity relative to the reaction mixture that feeds it, in laminar flow. Its value, which also depends strongly on fuel nature, is a function of richness, temperature, and pressure.

In general, for given temperature, maximum deflagration speed corresponds to richness slightly above 1, often around 1.2. Deflagration speeds are also low, few cm/s to few m/s. For given richness, they are increasing function of temperature and decreasing function of pressure.

Turbulence exerts complex effect on deflagration speed: it increases it significantly if it is already noticeable at rest; it decreases near flammability limit otherwise, and under these conditions can prevent combustion spread (blowing on fire stirs it up if it is intense, blows it out otherwise).

3.5.1.3 Combustion of Liquid Fuels

Phenomena described above for gaseous fuels are similar in the case of liquid fuels, with some differences.

A droplet of liquid fuel injected into air mass ignites spontaneously when the temperature exceeds a certain value, called fuel self-ignition temperature T_{ai}. This temperature depends mainly on fuel, and incidentally on air pressure, diameter, and drop speed.

Self-ignition temperatures thus measured range from 350°C to 500°C depending on fuel. Note that they are relatively low, much lower than those of gaseous fuel mixtures.

3.5.1.3.1 *Ignition and Combustion Delays*

Experience shows that, when injecting liquid fuel droplet into air mass whose temperature T exceeds self-ignition temperature T_{ai}, it does not catch fire instantly, but after some delay, the said inflammation delay, which depends mainly on difference $(T - T_{ai})$ and fuel type, secondarily on diameter and drop velocity. Mechanisms involved are complex, part of liquid fuel generally evaporating to form, mixing with oxidant, reactive medium that self-ignites depending on its composition and pressure and temperature conditions.

Additionally, complete combustion requires additional delay, the said combustion delay, which, apart from difference $(T - T_{ai})$, depends essentially on diameter and drop velocity as well as air turbulence.

When T exceeds T_{ai}, ignition delay is never very large, but it rarely falls below 2 ms. Combustion delay is similar, provided that fuel is finely pulverized (average diameter of drops of few microns) and injected at high speed, at pressure exceeding 100 bar. Nevertheless, overall ignition and combustion delay is of order of several milliseconds.

3.5.1.3.2 *Diffusion Flames*

Unlike premixed flames, where the fuel and oxidizer are mixed before the reaction starts, combustion of fuel introduced in the liquid state in the oxidizer is governed by speed at which reactive chemical species inter-diffuse. This is called diffusion flame. The reaction zone of such flame is thicker than that of the premixed flame. Richness varies from an infinite value in fuel to zero in the oxidizer. There are laminar diffusion flames (candles, oil lamps) and turbulent confined diffusion flames whose applications are numerous, either in industry or in conventional diesel engines.

3.5.1.4 Thermodynamics of Combustion at Equilibrium

Mass action law is fundamental to studying incomplete combustion reactions at equilibrium. It can be easily derived from Second Law of thermodynamics. Let us first express Gibbs energy of gas involved.

We can calculate molar Gibbs energy of ideal gas from $G = H - TS$:

$$H - H_0 = \int_{T_0}^{T} C_p(t)\,dt \qquad S = S_0 + \int_{T_0}^{T} \frac{C_p(t)}{t}\,dt - R\ln\frac{P}{P_0}$$

If index 0 values represent the reference state, the Gibbs energy value at temperature T and any pressure is given by:

$$G = H - TS = G_0^*(T) + RT\ln\frac{P}{P_0}$$

$G_0^*(T)$ representing the term independent of P in Gibbs energy.

Note that $G_0^*(T)$ is not reference state Gibbs energy G_0. This way of decomposing G has a sole purpose of separating variables P and T.

If, as is customary, $P_0 = 1$ bar, we obtain:

$$G - G_0^* = RT\ln P \qquad (3.5.2)$$

Let us consider combustion equation (3.5.1). We will investigate how one writes equilibrium condition (2.4.22). By definition, the reaction degree is ratio ε of the number of moles of the reactant (usually first of equation) that reacted to the total number of moles of this reactant capable of reacting.

At equilibrium, $dT = 0$, $dP = 0$, and equation (2.4.22) $\left(\Sigma\mu_i\, dn_i\right)_{T,P} = 0$ becomes:

$$\left(-\mu_1 \nu_1 - \mu_2 \nu_2 + \mu_3 \nu_3 + \mu_4 \nu_4\right) d\varepsilon = 0$$

This equation must be verified regardless of the ε value. We have:

$$-\mu_1 \nu_1 - \mu_2 \nu_2 + \mu_3 \nu_3 + \mu_4 \nu_4 = 0$$

However, for the ideal mixture, chemical potential equals partial Gibbs energy (2.4.23):

$$\mu_i = G(T, P_i)$$

Using equation (3.5.2) and separating variables, we find:

$$-rT\left(-\nu_1 \ln P_1 - \nu_2 \ln P_2 + \nu_3 \ln P_3 + \nu_4 \ln P_4\right) = -\nu_1 G_{01}^* - \nu_2 G_{02}^* + \nu_3 G_{03}^* + \nu_4 G_{04}^*$$

The second part of the equation represents the difference between pressure-independent terms of products and reactants free energies, and is sometimes called the free enthalpy change of reaction, denoted ΔG_0, expressed in molar variables.

ΔG_0 is defined, recall, for the given reference state, which we choose here as mixture temperature and $P_0 = 1$ bar: $\Delta G_0 = f(T, P_0)$. Logarithmic terms can be grouped in form:

$$\Delta G_0 = -RT\ln\left(\frac{P_3^{\nu 3} P_4^{\nu 4}}{P_1^{\nu 1} P_2^{\nu 2}}\right)$$

ΔG_0 being function only of temperature, the term under logarithm is itself a function of temperature only, and is independent of system pressure (but of course depends on reference pressure P_0).

Kp is defined as the equilibrium constant, with:

$$Kp = \frac{P_3^{\nu 3} P_4^{\nu 4}}{P_1^{\nu 1} P_2^{\nu 2}} \quad \text{and} \quad \Delta G_0 = -RT\ln Kp$$

At equilibrium, component partial pressures are proportional to their mole fractions: $P_i = x_i P$. Therefore:

$$Kp = \frac{x_3^{v_3} x_4^{v_4}}{x_1^{v_1} x_2^{v_2}} P^{v_3+v_4-v_1-v_2} \tag{3.5.3}$$

P is the system pressure, v_i are combustion reaction stoichiometric coefficients, and x_i are mole fractions of various constituents in products.

Equation (3.5.3) is called Law of Mass Action. Combined with balance equations of involved chemical species, it allows us, if we know Kp variation with temperature, to determine equilibrium concentrations of components for the given reaction.

The presence of non-reactant gases does not question mass action law. However, it can strongly influence reaction progress, since partial pressures are changed, because of the increased total number of moles in the mixture.

When several reactions can take place in the same mixture, mass action laws for different reactions must be checked simultaneously, knowing they are coupled together because, of course, partial pressure calculation cannot be done independently.

Equilibrium constant values for main reactions that may occur in combustion can easily be found in the literature.

One can obtain Kp estimate from the approximate formula valid in limited temperature ranges:

$$\ln(Kp) = A - \frac{B}{T}$$

For the reaction known as water gas $CO_2 + H_2 \leftrightarrow CO + H_2O$, in the range of 1,000–2,200 K, covering much of usual combustions, we take as values $A = 3.387$ and $B = 3{,}753$, P being expressed in bar and T in Kelvin.

3.5.1.5 Elements of Chemical Kinetics

In many cases, and as soon as we are to take into account pollutant emissions, complete combustion assumptions are too restrictive. It is then necessary to refine calculations to determine as precisely as possible the reaction degree. This is possible by using thermodynamics, if time is sufficient for equilibrium to be reached, and otherwise by using chemical kinetics. In this section, we give some elements to address this particularly difficult issue.

As indicated above, the fuel mixture is generally stable at room temperature. Statistical thermodynamics can explain this: reaction depends on the number of shocks that may arise between various reactants, and hence temperature.

Specifically, chemical kinetics reveals that the actual combustion mechanism can be decomposed into series of simple reactions, called basic steps, which involve reactive species called free radicals (atoms or groups of atoms with one or more unpaired electrons).

For example, the following reaction as such has negligible probability of occurring:

$$H_2 + \frac{1}{2}O_2 \rightarrow H_2O$$

For it to take place, where M is any collision (O_2, H_2, N_2, …), a chain like this must be put into place:

- initiation $O_2 + M \rightarrow 2O^{\bullet} + M$
- branching $O^{\bullet} + H_2 \rightarrow OH + H^{\bullet}$
- branching $O_2 + H \rightarrow {}^{\bullet}OH + O^{\bullet}$
- propagation ${}^{\bullet}OH + H_2 \rightarrow H_2O + H^{\bullet}$

Initiation reaction produces first free radicals. In reaction involving hydrocarbon, it usually corresponds to thermal decomposition of molecules, releasing $H^{\bullet}$. Branching is an essential step that multiplies the number of available radicals in the reaction, while propagation preserves their number, while participating in intermediate or final product production. Breaking reactions can finally complete the chain by transforming free radicals into stable chemical species, either in the homogeneous phase or in contact with walls.

The reaction path can be explained in light of what was just seen: when temperature is low (less than self-ignition temperature), the free radical number is very low, and breaking reactions consume them all, so that combustion cannot take place. Above certain temperature, this situation is reversed, and the free radical number increases. Conditions are then ripe for burning to develop.

All these phenomena explain why combustion reaction actually has a very high degree of complexity. If you want to be precise, it is necessary to follow in detail various chemical species that may appear, and consider all relevant equations.

For example, for methane, you can find in literature mechanisms that consider 46 species reacting as the function of 234 elementary coupled equations: 129 decomposition reactions of CH_4, 21 reactions of hydrogen (H, H_2) and oxygen (O, O_2), and 84 reactions characteristic of nitrogen oxide (NO) formation during combustion.

It is certainly possible to build simplified models of overall reaction, according to various methods. To understand main kinetic phenomena, experience proves that it is necessary to distinguish at least three phases in hydrocarbon combustion:

1. Fuel molecule decomposition by pyrolysis, producing olefins C_nH_2, also called intermediate hydrocarbons and free radicals $H^{\bullet}$, $O^{\bullet}$, and ${}^{\bullet}OH$.
2. Hydrocarbon radical and intermediate oxidation to CO and H_2.
3. CO and H_2 oxidation to give final combustion products CO_2 and H_2O.

3.5.1.6 Freezing Temperature

Combustion thermodynamics at equilibrium allows one, from mass action law, to determine the reaction progress degree depending on temperature and pressure. For this law to be applicable, it is necessary first that reaction speeds are larger than reactant residence time in the combustion chamber, and second that cooling conditions for flue gases are known.

For most combustion reactions, assumption of medium at equilibrium can be chosen without introducing too much error. Exception should however be made as we will see in Section 3.5.5 for nitrogen oxidation reaction, which leads to NO_x formation. This reaction has very slow kinetics, so that the reaction degree is generally much less advanced than at equilibrium.

Combustion kinetics study is also necessary if one wants to explain freezing phenomena that appear when dissociation takes place (at temperatures above 1,800 K), followed by sudden cooling of flue gases.

Due to dissociation, some combustion reactions are not complete. Slow reaction speed may then not be sufficient in the case of rapid temperature decrease, so that molecular recombinations that would have taken place normally (that is, at equilibrium, reactions being reversible) cannot be attained.

In these conditions, unburned hydrocarbons may remain. It is said that there is freezing of reaction. This explains the CO presence in flue gas, even with excess oxygen and for the homogeneous mixture. We call freezing temperature the temperature to be used in mass action law to get actual exhaust gas composition.

You should refer to specialized literature for further developments on this subject. In what follows, unless otherwise stated, we always assume that chemical reactions could develop and achieve equilibrium without kinetic considerations affecting these phenomena.

3.5.2 Study of Complete Combustion

In this section, we will deal with complete combustion, that is, such that all possible oxidations are carried out. After presenting different ways of representing oxidizer (usually air) and fuel, we will study successively stoichiometric and non-stoichiometric combustion. In the next section, we will see how to take into account the existence of unburned hydrocarbons.

3.5.2.1 Air Composition

A combustion reaction involves oxidation of fuel with oxygen or substance containing oxygen, such as air. In most cases in practice, it is air, and it is accepted as first approximation that all components other than oxygen are inert during reaction.

Normal air composition is given in Table 3.5.1. To simplify combustion reaction study, other inert gases are combined with nitrogen, by defining atmospheric nitrogen which allows us to consider dry air as the mixture of two ideal gases, whose composition is indicated below. From these values, it follows that the ratio (atmospheric nitrogen)/oxygen is 3.76 in volume, and 3.31 in mass.

In Thermoptim, calculations are made with a greater precision, the component number being up to the user. Among protected ideal gases, i.e., whose composition cannot be changed by the user, we find the above defined atmospheric air.

TABLE 3.5.1
Composition of Normal Air

Component	Molar Fraction %	Molar Mass kg/kmol	Mass Fraction %
N_2	78.1	28	75.6
$Ar+CO_2$	0.9	(40)	1.2
O_2	21.0	32	23.2
N_2 atmosph.	79.0	(28.15)	76.8

3.5.2.2 Fuel Composition

To study combustion quality, it is convenient to represent fuel by the standardized formula reflecting its actual composition. For ash-free fuel, chemical elements to be retained are: carbon, hydrogen, oxygen, nitrogen, and sulfur.

Fuel can therefore be described by the formula $CH_yO_xN_zS_u$.

Stoichiometric coefficients y, x, z, and u can be easily determined:

- From mass fractions (C), (H), (O), (N), and (S), which gives:

$$y = \frac{12.01\ (H)}{1.008\ (C)},\quad x = \frac{12.01\ (O)}{16\quad (C)},\quad z = \frac{12.01\ \ (N)}{14.008\ (C)},\quad u = \frac{12.01\ (S)}{32.07\ (C)} \tag{3.5.4}$$

- From the exact formula if it is a pure substance or mixture of pure chemically defined substances, relating the chemical formula to the unit of carbon. For compound $C_nH_mO_p$, we obtain:

$$y = \frac{m}{n}, x = \frac{p}{n} \tag{3.5.5}$$

- From volume or mole fractions if it is gaseous fuel.

For the mixture of fractions $[H_2]$, $[CO]$, $\Sigma[C_nH_m]$, $[CO_2]$, $[N_2]$, $[H_2S]$, and $[SO_2]$, we obtain:

$$y = \frac{2[H_2]+\sum m[C_nH_m]+2[H_2S]}{[CO]+\sum n[C_nH_m]+[CO_2]} \qquad x = \frac{[CO]+2[CO_2]+2[SO_2]}{[CO]+\sum n[C_nH_m]+[CO_2]}$$

$$z = \frac{2[N_2]}{[CO]+\sum n[C_nH_m]+[CO_2]} \qquad u = \frac{[H_2S]+[SO_2]}{[CO]+\sum n[C_nH_m]+[CO_2]} \tag{3.5.6}$$

3.5.2.3 Stoichiometric Combustion

We call stoichiometric combustion a combustion without excess or lack of air, where all available oxygen is completely consumed.

Assuming that nitrogen reacts only in negligible proportions and is found after combustion in molecular form, the general equation of stoichiometric combustion is:

$$CH_yO_xN_zS_u + \left(1+\frac{y}{4}+u-\frac{x}{2}\right)O_2 \rightarrow CO_2+\frac{y}{2}H_2O+uSO_2+\frac{z}{2}N_2 \tag{3.5.7}$$

Depending on circumstances, the coefficient $\left(1+\frac{y}{4}+u-\frac{x}{2}\right)$ may be greater or less than unity, depending on the existing oxygen availability in fuel. In most cases, $x \leq y/2+2u$, but for some lean gases, a reverse situation may exist.

For most conventional fuels, values of z and u are very low and often negligible, while the condition $x \leq y/2$ is always verified. We can then simplify symbolic formulation above by writing:

$$1+\frac{y}{4}-\frac{x}{2}=1+\frac{a}{4} \tag{3.5.8}$$

which leads to the conclusion that the formula CH_yO_x becomes:

$$CH_a + x\,H_2O$$

a represents what is called hydrogen available for combustion referred to complete oxidation of the carbon unit, and $a = y - 2x$.

With these conventions, complete oxidation of fuels of the formula $(CH_a + xH_2O)$ can be described by the following equation:

$$CH_a+\left(1+\frac{a}{4}\right)(O_2+3.76N_2)\rightarrow CO_2+\frac{a}{2}H_2O+3.76\left(1+\frac{a}{4}\right)N_2 \tag{3.5.9}$$

while there is conservation of the constitutive water xH_2O.

This reaction is fuel combustion without excess or lack of air.

If there is lack of air, some carbon is oxidized into carbon monoxide CO. To completely determine reaction, we have seen above that it is necessary to study chemical equilibrium conditions of carbon dioxide dissociation reaction, which depend primarily on temperature and correspond to the following equation:

$$CO_2 \leftrightarrow CO+\frac{1}{2}O_2$$

3.5.2.3.1 Combustion Quality

Combustion quality is primarily determined by two factors:

- Combustion efficiency, which characterizes the degree of chemical reaction fulfillment, or if one prefers, the absence of unburned hydrocarbons or incompletely oxidized compounds
- Level reached by combustion temperature, determined by the ratio of heat amount released by burning to the amount of reactants and products

In practice, we can optimize combustion by adjusting excess air. Large excess air ensures that at any point, enough oxygen is available for combustion and therefore reduces risk of unburned hydrocarbons. However, it dilutes flue gases, whose temperature drops.

3.5.2.4 Non-Stoichiometric Combustion

When combustion is not stoichiometric, it can be defined in several ways, usually by its excess air e, or its lack of air (−e), or richness R already introduced in Section 3.5.1.2, or its inverse the air factor λ. Formulations vary according to existing professional practice. For example, it is customary to speak of richness for gasoline engines, or excess air for boilers. In Thermoptim, we have selected air factor, very useful in practice.

Excess air is the percentage of air not consumed by stoichiometric reaction.

Richness R is defined as the ratio of the number of moles of fuel in the given quantity of mixture, to the number of moles of fuel in the stoichiometric mixture.

From these definitions, it follows the equivalence:

$$R = \frac{1}{1+e} \quad \text{and} \quad \lambda = 1 + e = \frac{1}{R} \tag{3.5.10}$$

Let us consider how non-stoichiometric reaction is written as the function of air factor λ.

In the case of excess air, λ is greater than 1, and there is too much oxygen. Reaction writes:

$$CH_a + \lambda\left(1+\frac{a}{4}\right)(O_2 + 3.76N_2) \rightarrow CO_2 + \frac{a}{2}H_2O + (\lambda-1)\left(1+\frac{a}{4}\right)O_2 + 3.76\lambda\left(1+\frac{a}{4}\right)N_2 \tag{3.5.11}$$

In the case of lack of air, λ is less than 1, and while there is enough oxygen to oxidize all carbon in carbon monoxide, and fraction in carbon dioxide, reaction becomes:

$$1 > \lambda > \frac{\left(1+\frac{a}{2}\right)}{2\left(1+\frac{a}{4}\right)} \quad \text{(neglecting dissociation of } CO_2\text{)}$$

$$\begin{aligned} CH_a + \lambda\left(1+\frac{a}{4}\right)(O_2 + 3.76N_2) &\rightarrow \left(2\lambda\left(1+\frac{a}{4}\right) - \left(1+\frac{a}{2}\right)\right)CO_2 \\ &+ \left(2(1-\lambda)\left(1+\frac{a}{4}\right)\right)CO + \frac{a}{2}H_2O + 3.76\lambda\left(1+\frac{a}{4}\right)N_2 \end{aligned} \tag{3.5.12}$$

If the lack of air is more important, there is not even enough oxygen to form all carbon monoxide.

3.5.3 Study of Incomplete Combustion

In most cases, combustion is not complete: for many reasons, various products including unwanted pollutants exist in flue gas. Efforts are generally made to minimize this phenomenon penalizing both energy production and environment, but it cannot be completely avoided.

In this section, we explain the principle of calculations that can be performed to characterize incomplete combustion reactions. Results we obtain are basis of combustion characterization methods from flue gas analysis, which are of great practical importance. Note that gas analysis can be made before or after condensation of water they contain. In principle, of course, one passes from one to another without difficulty, provided the humidity of air before combustion is known.

If these data are missing (which is most common), it is best to proceed on basis of dry gas, that is, after they have been sufficiently cooled so that water they contain is condensed and eliminated.

3.5.3.1 Mechanisms of Formation of Unburned Hydrocarbons

Unburned hydrocarbon formation can come from two main causes: either from oxygen lack, related to poor homogeneity of the oxidizer/fuel mixture or from lack of combustion reactivity, particularly due to too low temperature.

Mechanisms involved are very complex and cannot be described exhaustively in this book context. However, we can consider three limiting cases corresponding to previous causes.

In the first case, fuel CH_a can be locally heated at high temperature in a CO_2 and H_2O environment, complete combustion products having not diffused to make room for oxygen. Fuel may then, due to high temperature, be decomposed into carbon and hydrogen, strongly reducing, which will separate CO_2 and H_2O molecules.

This situation type occurs very frequently: in boilers or gas turbine combustion chambers, where fuel is fed into the burner, then mixed with the oxidizer, and burned. Whatever fuel nature (gas, liquid, or solid) and burner technology, the homogeneous mixture is not immediately realized, so that local areas may exist lacking air, even in the presence of overall excess air.

The reaction chain between these components is then established, as determined in first approximation by mass action law, and therefore pressure and ambient temperature conditions.

Global chemical equations involved are as follows with of course many more basic steps (see Section 3.5.1.5):

$$\begin{aligned}
&CH_a \rightarrow C + \frac{a}{2}H_2 \\
&CO_2 + C \rightarrow 2CO \\
&H_2O + C \rightarrow CO + H_2 \\
&CO_2 + H_2 \rightarrow CO + H_2O
\end{aligned} \tag{3.5.13}$$

In the second case, fuel is decomposed in a more complex manner, most volatile parts being easily vaporized in the gas phase and oxidized, while heavier undergo cracking which enriches them in carbon. The phenomenon leads to the formation of dry or oily soot depending on whether or not hydrogen is completely released. Tar or heavily polluting smoke can also be formed. This situation exists in some boilers in the presence of significant excess air and can occur in diesel engines incorrectly set.

In the third case, reaction is locally frozen on combustion chamber walls, due to the double phenomenon: on the one hand, surface temperature is lower than flame, and on the other hand, the wall inhibits branching and propagation of free radical reactions, which reduces dramatically combustion kinetics.

This occurs especially in reciprocating internal combustion engines. In gasoline engines, there is deposition on different surfaces (cylinder head, cylinder, piston, valves) of layers of partially processed fuel mixture. The result is presence in the exhaust of unburned hydrocarbons (often referred to as UHCs), which can to some extent be equated to methane CH_4 equivalent.

In diesel engines, the phenomenon is due to strong combustion heterogeneity: liquid fuel is injected into the mixture of hot air and combustion products, which burns by self-ignition, and whose nature changes during combustion between pure air and exhaust gases. At jet center, richness is infinite, whereas it is null in some areas. There can thus be simultaneously excess air in one chamber zone, and lack of air in another. Thus, wrongly set diesel engine can simultaneously produce soot, unburned hydrocarbons of CH_4 type and CO.

In what follows, we will show how calculations can be conducted to characterize combustion of three main categories of incomplete reactions. It is possible to study combustion phenomena in a finer way, but at cost of significant calculation complication, which must then be made with heavy computer calculation programs, which bring into play simultaneously up to several hundreds of dissociation equations.

3.5.3.2 Composition of Flue Gases without Soot or Unburned Hydrocarbons

In the first case where only unburned hydrocarbons are CO and H_2, we can write the combustion equation as follows:

$$CH_a + \lambda\left(1+\frac{a}{4}\right)(O_2 + 3.76\,N_2) \rightarrow (1-k_1)\,CO_2 + k_1\,CO + (1-k_2)\frac{a}{2}\,H_2O + k_2\frac{a}{2}\,H_2$$

$$+\left((\lambda-1)\left(1+\frac{a}{4}\right)+\frac{k_1}{2}+k_2\frac{a}{4}\right)O_2 + 3.76\lambda\left(1+\frac{a}{4}\right)N_2 \tag{3.5.14}$$

The equilibrium between species giving unburned hydrocarbons is governed by the following equation:

$$CO_2 + H_2 \leftrightarrow CO + H_2O \tag{3.5.15}$$

As can be seen, because the reaction involves a constant number of moles, this equilibrium is independent of pressure, and is function of temperature alone. According to mass action law, we have here:

$$Kp = \frac{[CO][H_2O]}{[CO_2][H_2]} = \frac{k_1(1-k_2)\frac{a}{2}}{(1-k_1)k_2\frac{a}{2}} = \frac{k_1(1-k_2)}{(1-k_1)k_2} \tag{3.5.16}$$

We get thus: $k_2(1-k_1)Kp = k_1(1-k_2)$:

$$k_2 = \frac{k_1}{k_1 + Kp(1-k_1)} \tag{3.5.17}$$

Let $k = k_1$, to simplify the notations.

For the given fuel, combustion can then be described by only three parameters: λ, air factor, k, CO_2 dissociation rate, and K_p.

Analysis of dry flue gas (after condensation and removal of H_2O) then provides volume or partial molar fractions. Letting Σ be sum of various constituent fractions, we obtain:

$$\Sigma = 4.76\lambda\left(1+\frac{a}{4}\right)+\frac{a}{2}+\frac{k}{2}-\frac{3Kp(1-k)}{k+Kp(1-k)}\frac{a}{4} \tag{3.5.18}$$

$$\text{Let}\quad g(k) = -\frac{Kp(1-k)}{k+Kp(1-k)}\frac{a}{4} \tag{3.5.19}$$

We obtain:

1. $[CO_2] = \dfrac{1-k}{\Sigma}$
2. $[CO] = \dfrac{k}{\Sigma}$
3. $[N_2] = \dfrac{3.76\lambda\left(1+\frac{a}{4}\right)}{\Sigma}$
4. $[O_2] = \dfrac{\lambda\left(1+\frac{a}{4}\right)+\frac{k}{2}-1+g(k)}{\Sigma}$
5. $[H_2] = \dfrac{ka}{2(k+Kp(1-k))\Sigma}$
6. $\Sigma = 4.76\lambda\left(1+\frac{a}{4}\right)+\frac{a}{2}+\frac{k}{2}+3g(k)$

These volume fractions depend only on four parameters: **λ, k, and Kp**, characterizing combustion, and **a** characterizing fuel.

Suppose that we know [CO] and $[CO_2]$ (their determination is relatively easy).

(1) and (2) can be combined to give k, by:

7. $k = \dfrac{[CO]}{[CO]+[CO_2]}$

and recombining with (2):

8. $\Sigma = \dfrac{1}{[CO]+[CO_2]}$

If a and K_p are known (which implies that the reaction freezing temperature is known), g(k) can be calculated, and λ is provided by (6). Then, (3), (4), and (5) allow to fully characterize reaction.

$$\lambda = \frac{[N_2]}{3.76\left(1+\frac{a}{4}\right)([CO]+[CO_2])} \qquad (3.5.20)$$

If Kp is not known, the $[O_2]$ value allows us to eliminate g(k) between (3) and (6), and thus determine λ.

In the case where, in addition, the value of a is uncertain, knowledge of $[H_2]$ allows us, as we can ascertain, to solve the problem.

3.5.3.2.1 Establishment of the Biard Chart

Now let us suppose fuel is known, and therefore the value of a. Determination of λ and k can be done by choosing two independent relations among five previous equations, since we set the K_p value by studying reaction at given temperature T.

Note that [CO] and $[H_2]$ are proportional, so we can often only consider their sum.

Some devices directly determine the amount of unburned hydrocarbons by measuring heat from catalytic oxidation of $CO + H_2$, whose heating values are close enough to be assimilated.

Detection of $[CO_2]$ and $[O_2]$ being relatively easy, it is usually from their equations that λ and k are determined.

System:

$$[CO_2] = \frac{1-k}{\Sigma}$$

$$[O_2] = \frac{\lambda\left(1+\frac{a}{4}\right)+\frac{k}{2}-1+g(k)}{\Sigma}$$

can be rewritten as:

$$\Sigma = 4.76\lambda\left(1+\frac{a}{4}\right)+\frac{a}{2}+\frac{k}{2}+3g(k)$$

$$\Sigma[CO_2] = 1-k$$

$$\Sigma[O_2] = \lambda\left(1+\frac{a}{4}\right)+\frac{k}{2}-1+g(k)$$

By expanding, we get: $\left[4.76\lambda\left(1+\frac{a}{4}\right)+\frac{a}{2}+\frac{k}{2}+3g(k)\right][CO_2] = 1-k$

$$\lambda = \frac{1-k-\left(\frac{a}{2}+\frac{k}{2}+3g(k)\right)[CO_2]}{4.76\left(1+\frac{a}{4}\right)[CO_2]} \qquad (3.5.21)$$

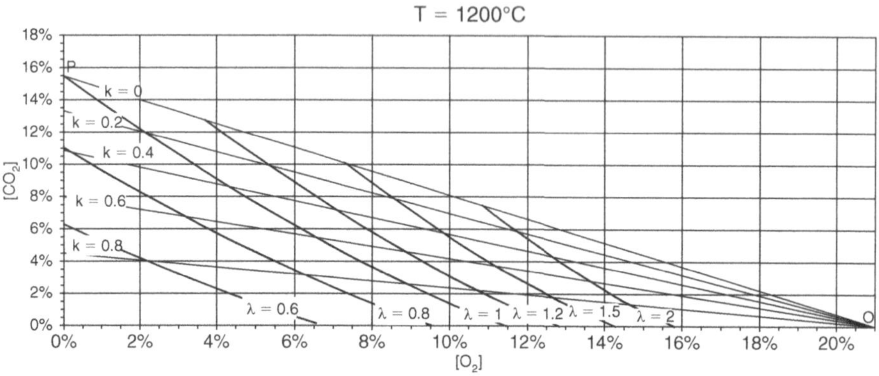

FIGURE 3.5.1 Biard chart for light fuel oil (T freezing = 1,200°C).

Replacing λ in the $[O_2]$ equation, we get:

$$(1-k)\left(4.76[O_2]-1\right)+\left(\frac{a}{2}+4.76-3.76\frac{k}{2}+1.76g(k)\right)[CO_2]=0 \quad (3.5.22)$$

For k = Const, this relationship defines in the plane ($[O_2]$,$[CO_2]$) a bundle of straight lines which all pass through point O of coordinates (0.21, 0) (Figure 3.5.1).

They cut the y-axis at coordinates:

$$\left[0;\frac{1-k}{\frac{a}{2}+4.76-3.76\frac{k}{2}+1.76g(k)}\right]$$

For k = 0, maximum point P has coordinates: $\left[0;\frac{1}{4.76+3.76\frac{a}{4}}\right]$.

Note that line k = 0 corresponds to perfect combustion, without unburned hydrocarbons. Point P represents then the neutral or stoichiometric combustion.

Elsewhere $[CO]=\frac{k}{\Sigma}$ and $[H_2]=\frac{ka}{2(k+Kp(1-k))\Sigma}$

We have:

$$\frac{[CO]}{[CO_2]}=\frac{k}{1-k}$$

$$\frac{[H_2]}{[CO_2]}=\frac{ka}{2(k+K_p(1-k))(1-k)}$$

Both expressions depend only on k, not λ, $[O_2]$ or $[CO_2]$.

We can therefore directly graduate lines k = const in $\frac{[CO]}{[CO_2]}$ or $\frac{[H_2]}{[CO_2]}$.

For λ = Const, the curve linking $[CO_2]$ and $[O_2]$ is given by parametric equations based on k:

$$[CO_2] = \frac{1-k}{\Sigma}$$

$$[O_2] = \frac{\lambda\left(1+\frac{a}{4}\right)+\frac{k}{2}-1+g(k)}{\Sigma} \qquad (3.5.23)$$

In the general case, they are decreasing curves concave upwards.

All these curves can be represented in the Biard chart, established for the given freezing temperature, example of which is given Figure 3.5.1.

3.5.3.2.2 Establishment of the Ostwald Chart

In some cases, it is possible to simplify the above expressions by a number of additional approximations.

In the temperature range of normal combustion in the presence of excess air (1,000–1,500 K), we have seen that constant K_p of mass action law can be approximated by the following equation:

$$K_p = \exp\left(3.387 - \frac{3,753}{T}\right), \text{ which varies here between 0.7 and 2.6}$$

Assuming that in K_p expression, k_1 and k_2 are very small compared to 1, we obtain: $K_p = \frac{k_1}{k_2}$.

Experience shows that we can consider that equilibrium is frozen when the flue gas temperature drops below a certain limit. We assume in what follows that it can be characterized by a K_p value close to 2.

Combustion can then be described by two parameters only: λ, or e, characterizing excess air, and $k = k_1 = 2k_2$.

Above equations are simplified, since function g(k) becomes linear and we can eliminate k in relationship between $[CO_2]$ and $[O_2]$. All curves λ = Const define in this case a second straight line bundle.

This double bundle has the following interesting features.

Line k = 0 passes through point O, of coordinates: $[CO_2]_O = 0$

$$[O_2]_O = \frac{1}{4.76} = 0.21 \quad \text{which corresponds to } \lambda = \infty$$

and through point P of coordinates

$$[CO_2]_P = \frac{1}{4.76 + 3.76\frac{a}{4}}$$

$$\text{Line slope equals to } \frac{-4.76}{4.76+3.76\frac{a}{4}}$$

- O is the point common to all the straight lines k = const, since it is determined by the air composition;
- Line k = 1 is the x-axis of equation $[CO_2]=0$.

Line $\lambda=1$ passes through point P above and meets the x-axis (k = 1) in Q, with coordinates: $[CO_2]_Q=0\,[O_2]_Q=\dfrac{1}{2\left(4.76+\dfrac{1}{4}\right)}=0.095$

and whose position is independent of a. Slope line is: $\dfrac{-8.52}{4.76+3.76\frac{a}{4}}$.

Any normal combustion must have composition ($[CO_2]$, $[O_2]$) located in triangle OPQ, closest to P.

This is called the Ostwald chart. It has appearance of Figure 3.5.2.

Assumption of linearity of iso-air ratio curves is better verified in the case of heavy fuel oil than for natural gas.

3.5.3.3 Composition of Flue Gases with Soot and without CO or HC

If one accepts that in the case of low reaction activation, there is essentially dry soot composed of carbon, the combustion equation becomes:

$$CH_a+\lambda\left(1+\frac{a}{4}\right)(O_2+3.76N_2)$$
$$\rightarrow(1-k)CO_2+kC+\frac{a}{2}H_2O+\left((\lambda-1)\left(1+\frac{a}{4}\right)+k\right)O_2+3.76\lambda\left(1+\frac{a}{4}\right)N_2 \tag{3.5.24}$$

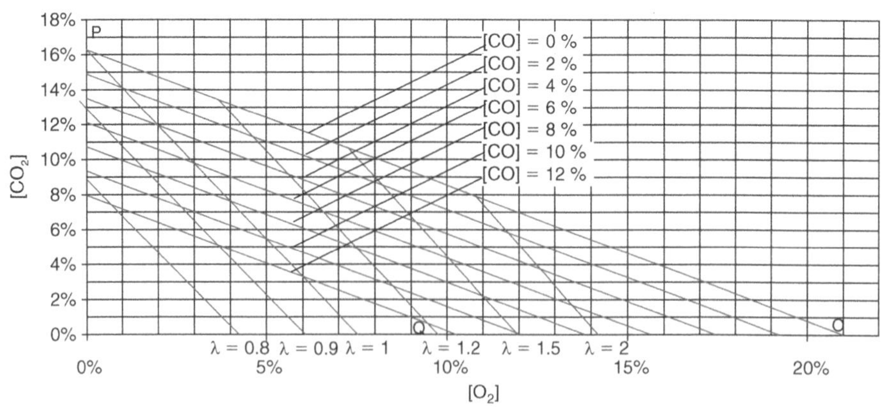

FIGURE 3.5.2 Ostwald chart for heavy fuel.

Analysis similar to the previous would allow one to determine different species concentrations.

3.5.3.4 Composition of Flue Gases without Soot and with CO and HC

In the third case, when unburned hydrocarbons are of type CO, HC (CH_4), and H_2, we can write the combustion equation as follows:

$$CH_a + \lambda\left(1+\frac{a}{4}\right)(O_2 + 3.76N_2) \rightarrow (1-x)CO_2 + yCO + (x-y)CH_4 + (1-t)\frac{a}{2}H_2O$$

$$+\frac{at-4(x-y)}{2}H_2 + \left((\lambda-1)\left(1+\frac{a}{4}\right)+x+\frac{y}{2}+\frac{at}{4}\right)O_2 + 3.76\lambda\left(1+\frac{a}{4}\right)N_2 \tag{3.5.25}$$

Mass action law gives here:

$$Kp = \frac{[CO][H_2O]}{[CO_2][H_2]} = \frac{y(1-t)\frac{a}{2}}{(1-x)\frac{at-4(x-y)}{2}} \tag{3.5.26}$$

Here also calculations could be carried out as in Section 3.5.3.2.

3.5.4 Energy Properties of Combustion Reactions

3.5.4.1 Thermodynamic Properties of Species in Reaction

One of the main objectives of combustion studies is determination of energy exchanges that occur when chemical reactions take place.

For this, it suffices to know internal energies or enthalpies of reactants and products, because energy exchange at stake does not depend on intermediate reactions, but only on initial and final states of chemical species in reaction.

We define combustion **heat of reaction** as change in the internal energy or enthalpy of the mixture, depending on whether reaction takes place at a constant volume or a constant pressure.

For reaction at constant volume, we have: $Q_v = (\Delta U_r)v = (U_p - U_r)v$

At constant pressure: $Q_p = (\Delta H_r)p = (H_p - H_r)p$

(3.5.27)

Evidently, heat of reaction calculation implies knowledge of internal energies U_r and U_p or enthalpies H_r and H_p of reactants and products. As mentioned previously, the principle of their calculation is based on assumption that different constituents react in the gaseous state, and each of them is comparable to ideal gas. Their mixture under these conditions is itself ideal gas.

If we note x_i molar concentrations and y_i mass concentrations of various constituents, we have, let us recall:

$$P_i = x_i P$$

$$M = \sum x_i M_i$$

3.5.4.1.1 In Mass Units

$$h = \sum y_i h_i$$

$$s = \sum y_i s_i (T, P_i) = \sum \left(y_i \int^{T} \frac{c_{pi}(t)}{t} dt - r y_i \ln P_i \right)$$

3.5.4.1.2 In Molar Units

$$H = \sum x_i H_i$$

$$s = \sum x_i s_i (T, P_i) = \sum \left(x_i \int^{T} \frac{c_{pi}(t)}{t} dt - R x_i \ln P_i \right)$$

Note that in equilibrium reaction, x_i and y_i depend on temperature, because of mass action law. Evaluation of h_i and s_i is based on reference common to all substances, at a temperature of 25°C and a pressure of 1 bar. The internal energy and enthalpy of different substances, known as heats of formation u_e or h_e, are determined in this reference state by methods that are outside this book scope; they are available in tables (e.g., JANAF Tables, Chase et al., 1985).

3.5.4.2 Enthalpies of Formation

For all elements, or simple substances, the enthalpy of formation (h_f) is chosen equal to 0 in the reference state. For compound substances, it is not zero, but its value is enthalpy consumed or released during chemical reaction that leads to compound development.

$$\text{If we have: } \nu_1 A_1 + \nu_2 A_2 \leftrightarrow \nu_3 A_3$$

with A_1 and A_2 simple substances of enthalpies of formation equal to 0,

$$h_{f^3} = \left(H_p - H_r\right)_p = H_p$$

Determination of the thermodynamic functions, temperature, and pressure of the reaction is then done using the following formula:

$$h(T) = h_f + \int_{298.15}^{T} c_p(t) dt \tag{3.5.28}$$

3.5.4.3 Entropies of Reference

Similarly, we can tabulate entropy values of different substances in this reference state. For this purpose, reasoning is based on the principle, derived from experimental observations, sometimes called Third Law of thermodynamics, which states that when the temperature approaches absolute zero, the entropy of pure substances approaches zero.

Reference entropies can thus be determined by integration between 0 and 298.15 K of equation:

$$ds = \frac{dh}{T} - v\frac{dP}{T}$$

For an ideal gas, for $P_0 = 1$ bar, $s_0 = \int_0^{298.15} \frac{c_p(t)}{t} dt.$

s_0 is usually read directly from tables, and calculating the entropy of an ideal gas at temperature T and pressure P is done by applying the following equation:

$$s = s_0 + \int_{298.15}^{T} \frac{c_p(t)}{t} dt - r\ln(P) \tag{3.5.29}$$

Enthalpies and entropies of formation can easily be found in the literature.

3.5.4.4 Heats of Reaction

Once enthalpies of formation of various chemical species are known, heat of reaction calculation is simple.

Let us use general equation (3.5.1) considered at the beginning of this chapter:

$$v_1A_1 + v_2A_2 \leftrightarrow v_3A_3 + v_4A_4$$

Let us note h_{fi} enthalpies of formation of A_i.

We obtain heat of reaction by writing:

$$-\Delta H_r = v_1h_{f1} + v_2h_{f2} - v_3h_{f3} - v_4h_{f4} \tag{3.5.30}$$

In equilibrium reaction, heat of reaction ΔH_r is the function of ν_i and thus x_i. Therefore, when the reaction is not complete or there is dissociation, the heat of reaction is different from that obtained under stoichiometric conditions, as shown in the following example.

Let us calculate, for example, the heat of reaction of octane for stoichiometric reaction:

$$C_8H_{18} + 12.5O_2 + 3.76 \times 12.5N_2 \rightarrow 8CO_2 + 9H_2O + 3.76 \times 12.5N_2$$

$$-\Delta H_r = h_{fC_8H_{18}} + 12.5h_{fO_2} - 8h_{fCO_2} - 9h_{fH_2O}$$

$$-\Delta H_r = -208{,}450 + 0 - 8(-393{,}520) - 9(-285{,}830) = 5{,}512{,}180 \text{ kJ / kmol}$$

Octane molar mass being 114.22 kg/kmol, its mass heat of reaction equals:

$$-Q_p = 48{,}259 \text{ kJ / kg}$$

If the reaction is not stoichiometric, but occurs in the absence of air, carbon monoxide formation decreases heat of reaction.

For richness equal to 1.25, for example, the reaction becomes:

$$C_8H_{18} + 10O_2 + 37.6N_2 \rightarrow 3CO_2 + 5CO + 9H_2O + 37.6N_2$$

$$-\Delta H_r = \Delta h_{fC_8H_{18}} + 10\Delta h_{fO_2} - 3\Delta h_{fCO_2} - 5\Delta h_{fCO} - 9\Delta h_{fH_2O}$$

$$-\Delta H_r = -208{,}450 + 0 - 3(-393{,}520) - 5(-110{,}530) - 9(-285{,}830)$$

$$= 4{,}097{,}230 \text{ kJ / kmol}$$

i.e., decrease of 25.7% over stoichiometric reaction.

3.5.4.5 Higher and Lower Heating Value

In most cases, reaction products are found in the gaseous state at the combustion end, but it is possible at low temperature, that some of them are liquid or even solidify, releasing heat of condensation or solidification.

Problem arises especially during hydrocarbon combustion, water appearing among products. Maximum energy release is obtained when water contained in flue gas is liquefied. The complete value of heat of reaction is named the Higher Heating Value or HHV. Where all water produced remains in the vapor state, it is given the name of Lower Heating Value or LHV. If only water fraction condenses, heat of reaction is an intermediate value between HHV and LHV.

Based on the above definitions, we identify four heating values which are particularly interesting: at constant pressure or volume, and lower or higher.

In fact, heating values at constant pressure and volume are substantially equal. However, vaporization enthalpies are far from negligible (for water, it is approximately 45 MJ/kmol at 0°C). So there are significant differences between fuel HHV and LHV values, and it is important to specify which one is used.

In conclusion, therefore, to determine energies involved in complete combustion reaction, you should use fuel heating value, specifying whether it is HHV or LHV.

Heating values of specific fuels commonly used in internal combustion engines are fairly close to each other. One can indeed show that the hydrocarbon LHV range is from 44 to 40 MJ/kg when we consider products of distillation of increasing molar weight.

For gaseous fuels, heating values are generally expressed in kJ/m^{3}N. They are obtained by dividing heat of reaction in kJ/kmol by volume occupied by normal kilomol (0°C, 1 bar), i.e., 22.414 m^3.

In Thermoptim, LHV values of gases are displayed in the compound gas editor screen.

3.5.4.6 Adiabatic Flame Temperature

If we assume that reaction takes place without heat exchange with surroundings, that is, adiabatically, it is possible to calculate the temperature reached by the mixture.

To do this, simply write that product enthalpy equals that of reactants in the initial state.

Temperature T_ad is thus given by:

$$H_p = H_{p0} + \int_{T_0}^{T_{ad}} Cp_p(t)\,dt = H_r \tag{3.5.31}$$

with

$$H_r = H_{r0} + \int_{T_0}^{T_r} C_{p_r}(t)\,dt$$

Values with index 0 are relative to the reference state (298.15 K, 1 bar), and T_r is the reactant temperature when combustion has been initiated.

We therefore should resolve:

$$\int_{T_0}^{T_{ad}} Cp_p(t)\,dt = H_{r0} - H_{p0} + \int_{T_0}^{T_r} Cp_r(t)\,dt$$

Since by definition $(-\Delta H_r) = H_{r0} - H_{p0}$

$$\int_{T_0}^{T_{ad}} Cp_p(t)\,dt = (-\Delta H_r) + \int_{T_0}^{T_r} Cp_r(t)\,dt \tag{3.5.32}$$

In the general case of equilibrium reaction in the presence of dissociation, this equation can be very complex, since ΔH_r depends on T_{ad} by mass action law.

In the case where reaction is complete, (3.5.32) corresponds to solving a polynomial with real coefficients in T_{ad}, which can be done iteratively, for example.

In practice, adiabatic temperature T_{ad} calculation has a number of pitfalls. It is particularly important to take into account variation in the mole number during reaction. For this, the best way is probably to think in molar units and to use chemical reaction coefficients.

With the general equation considered earlier in this chapter:

$$v_1 A_1 + v_2 A_2 \leftrightarrow v_3 A_3 + v_4 A_4$$

We have seen that heat of reaction calculation is very simple by writing:

$$-\Delta H_r = v_1 \Delta h_{01} + v_2 \Delta h_{02} - v_3 \Delta h_{03} - v_4 \Delta h_{04}$$

For this reaction, general equation (3.5.32) becomes:

$$(v_3 + v_4)\int_{T_0}^{T_{ad}} Cp_p(t)\,dt = (-\Delta H_r) + (v_1 + v_2)\int_{T_0}^{T_r} Cp_r(t)\,dt$$

where Cp represents the molar heat capacity.

It is also possible to take into account incomplete reactions (non-stoichiometric conditions, dissociation), provided equilibrium constant K_p values are set, iterating once temperature T_{ad} is known.

3.5.4.7 Combustion Efficiency

In a boiler or heat engine, total energy released by combustion is not always used: part is carried away by flue gases, and another corresponds to losses through walls, etc.

To characterize these losses, we introduce the combustion efficiency concept, one difficulty being that definitions vary according to authors and use types.

When it is possible to isolate combustion from other phenomena, it is quite natural to call the combustion efficiency ratio of useful energy at the combustion chamber outlet to energy theoretically available to the burner. Losses in the combustion chamber in this case essentially correspond to losses through walls and loss by incomplete combustion, which can be characterized by the concept of LHV (or possibly HHV) efficiency, the ratio of energy available in the burner to the fuel heating value.

$$\eta_c = \frac{\text{useful energy}}{\text{LHV}}$$

$$\eta_c = \frac{\text{LHV} - \text{losses by walls} - \text{losses by incomplete combustion}}{\text{LHV}}$$

$$\eta_{LHV} = \frac{\text{LHV} - \text{losses by incomplete combustion}}{\text{LHV}}$$

η_{LHV} is the combustion efficiency as calculated in Thermoptim.

When combustion cannot be isolated from other phenomena, things become more difficult. For example, in the automobile engine cylinder, combustion starts at the compression phase end and continues during early expansion phase. Precise definition of combustion efficiency becomes very complex.

In boilers, combustion takes place simultaneously with heat exchange by convection and radiation between combustion products and heated fluid. Losses in this case include three terms: incomplete combustion losses, losses through walls, and losses by flue gases, usually predominant, corresponding to the enthalpy of still hot gases released into air.

We then often define combustion efficiency as the ratio of useful energy actually transferred to the fluid to energy theoretically available to the burner. As first two loss types are generally lower than loss by flue gases, they are frequently neglected. Combustion efficiency is then:

$$\eta_c = \frac{\text{useful energy}}{\text{LHV}} = 1 - \frac{\text{loss by the fumes}}{\text{LHV}}$$

Combustion efficiency in this case depends crucially on flue gas temperature, their smoke losses being directly proportional. We understand why it is desirable to cool smoke as much as possible when it is possible to do so.

3.5.5 Emissions of Gaseous Pollutants

Major gaseous pollutants are sulfur, nitrogen, and carbon oxides. Additionally, CO_2 emissions, fatal combustion product, are now increasingly regarded as pollutants because of their impact on greenhouse effect.

SO_2 (sulfur dioxide) and SO_3 (sulfur trioxide) emissions depend only on fuel sulfur content, SO_3 formation resulting from complete sulfur oxidation. At low temperature, SO_3 combines with flue gas condensation to form highly corrosive sulfuric acid H_2SO_4. Reaction beginning at temperatures below 180°C, it is imperative that flue gases are discharged above this limit in cases where fuel is heavy fuel oil (3.5% S). Finally, stack height is determined by minimum levels of ground-level SO_2 concentration.

NO_x nitrogen oxides come in forms of NO (nitric oxide) and NO_2 (nitrogen dioxide). They come either from nitrogen presence in fuel or from unwanted oxidation of air nitrogen. It was shown that the nitrogen oxide formation rate from fuel impurities is much higher than that from air nitrogen, since nitrogen molecule binding energy is much greater than that which connects nitrogen atoms to carbon atoms or hydrogen.

To limit nitrogen oxide emission from combustion air, several possibilities exist: to lower excess air (though this increases temperature), making staged combustion, or flue gas recirculation, thereby lowering flame temperatures without increasing excess oxygen. One can indeed show that excess oxygen (and air) promotes nitrogen oxide onset, by studying reaction:

$$\frac{1}{2}O_2 + \frac{1}{2}N_2 \leftrightarrow NO$$

If we complete equation (3.5.14) of unburned hydrocarbon formation to reflect nitrogen dissociation, we can write:

$$CH_a + \lambda\left(1+\frac{a}{4}\right)(O_2 + 3.76N_2) \rightarrow (1-k_1)CO_2 + k_1CO + (1-k_2)\frac{a}{2}H_2O$$

$$+k_2\frac{a}{2}H_2 + \left((\lambda-1)\left(1+\frac{a}{4}\right) + \frac{k_1}{2} + k_2\frac{a}{4} - \frac{k_3}{2}\right)O_2 + 3.76\lambda\left(1+\frac{a}{4} - \frac{k_3}{2}\right)N_2 + k_3NO \qquad (3.5.37)$$

Constant K_p of mass action law for nitrogen dissociation reaction is very low (between 10^{-5} and 3×10^{-2} for temperatures between 1,000 and 2,200 K).

Under these conditions, k_3 is very small and can be neglected before oxygen, nitrogen, and total mole numbers.

$$\text{Therefore, at equilibrium: } [NO] = K_p [O_2]^{0.5} [N_2]^{0.5} \tag{3.5.38}$$

Since $[N_2]$ is almost constant and close to 78%, we obtain:

$$[NO] = 0.88K_p [O_2]^{0.5}$$

$[O_2]$ being increasing a function of excess air, [NO] is too.

This equilibrium value, however, is rarely reached, because NO formation kinetics is much slower than that of combustion. The concentration limit is thus lower than that which would suggest maximum temperatures were reached, but instead reaction freezes very early, so it is difficult to limit nitrogen oxide emissions.

Carbon monoxide (CO) is an incomplete combustion product, which indicates boiler malfunction.

3.5.6 Calculation of Combustion in Thermoptim

The combustion screen (Figure 3.5.3) is the most complex of those of processes, given the number of existing options. You should refer to beginning of Section 3.1.5 for the overview of process screens.

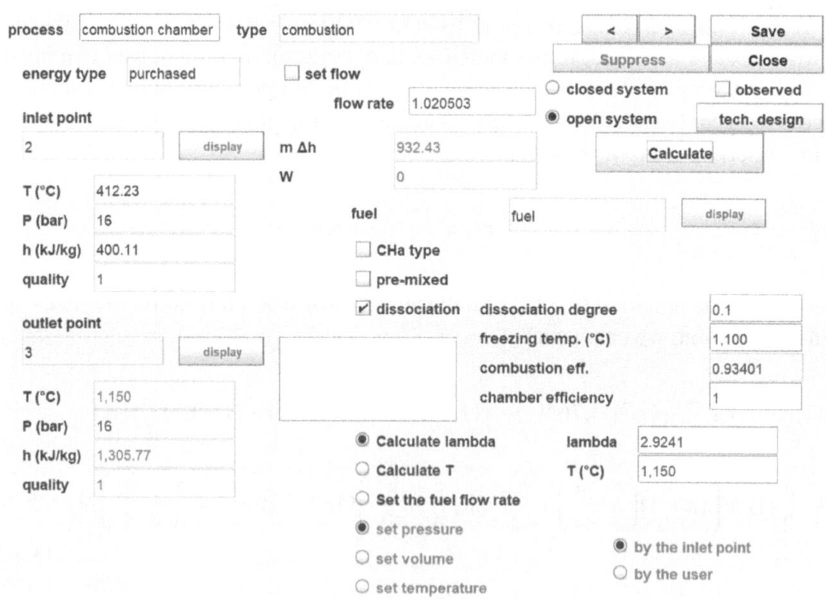

FIGURE 3.5.3 Combustion screen in Thermoptim.

3.5.6.1 Declaration of Fuel

In the case of excess air, non-stoichiometric complete combustion of fuel CH_a with atmospheric air is given by equation (3.5.11). Thermoptim uses a generalized equation of this type, where the oxidizer can be any compound gas including oxygen, and fuel is either given as CH_a or declared as pure or compound gas. Fuel definition is in the upper right of the screen (Figure 3.5.4).

If fuel is pure or compound gas, it may contain one or more of the following reactants: CO, H_2S, $C_nH_mS_pOq$, n, m, p, and q being decimal numbers less than 100. Inert gases taken into account are: Ar, CO_2, H_2O, N_2, and SO_2. Note that fuel is always assumed to be gaseous, which may cause slight error if it is in the liquid state.

Fuel and oxidizer can both contain reactive oxygen and inert gases, even if it is normal that fuel contains no oxygen, nor oxidizer fuel. It is thus possible to calculate complex combustion, such as that of the fuel mixture made before introduction into the combustion chamber (fresh charge of gasoline engine).

Reaction products are CO_2, H_2O, and SO_2 as well as CO and H_2 if there is dissociation, and fuel if the reaction is not complete.

Software analyzes chemical formulas of fuel and oxidizer components, and deduces reaction that takes place. Calculations can then be executed. Chemical formulas are obtained by decoding substance names.

When fuel is gas, name that appears on screen (here "fuel") should be that of the process (e.g. process-point) to determine its flow rate. The process itself is connected to point to specify the substance name and state.

The example below is taken from Getting Started example no. 2 "gas turbine." It relates to natural gas available at the LNG terminal of Gaz de France at Montoir de Bretagne (France) and illustrates how to declare fuel whose composition is given in Figure 3.5.5.

Each constituent, except for nitrogen, is fuel, whose chemical formula appears in the first part of the name that sometimes is followed by comment, separated from the formula by character "`". By analyzing the chemical formula of each constituent and its molar fraction, Thermoptim completely characterizes fuel.

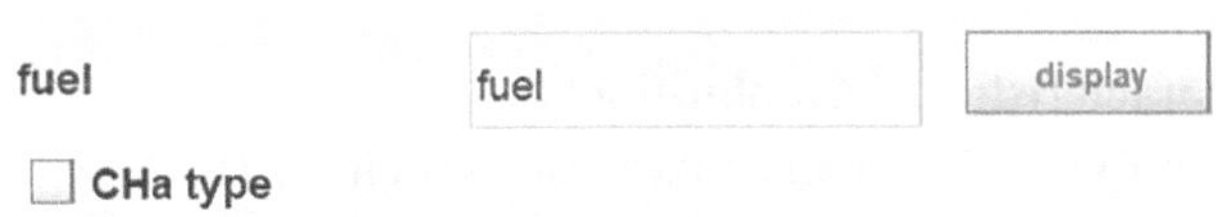

FIGURE 3.5.4 Fuel process link.

component name	molar fraction	mass fraction
CH4 ` methane	0.871	0.758966
C2H6 ` ethane	0.088	0.1437279
C3H8 ` propane	0.025	0.05987759
C4H10 ` n-butane	0.008	0.02525591
N2	0.008	0.01217253

FIGURE 3.5.5 Fuel composition.

This substance is associated to point, allowing specification of its temperature, which is taken into account for calculation of outlet gas temperature. The flow rate is associated with process-point, whose name serves as the parameter for combustion.

To find the oxidizer flow rate, software searches if there is a process whose outlet point is the combustion process inlet point, or if it is connected to node. Messages inform the user if a problem occurs, either because there are several processes that lead to the inlet point, or because no process or node is connected to it.

When fuel is given in form CH_a ("CH_a type" must be selected), it is necessary to enter values of a, and of its enthalpy of formation hf_0 (relative to formulation CH_a). It is then impossible to specify the flow rate.

3.5.6.2 Open Systems and Closed Systems

It is necessary to specify if combustion takes place in an open or closed system.

For open systems, pressure must be set to a certain value. It can be done either "by user," which means that this value will be that of the outlet, or "by inlet point," which corresponds to combustion at constant pressure equal to that of the upstream oxidizer.

For closed systems, one can choose between three possibilities: set volume, set pressure, or constant temperature combustion. For two first cases, two different modes of setting volume or pressure exist: "by user" or "by inlet point." In the third case, combustion temperature is constant and equal to that of the upstream oxidizer.

In the last two cases (set pressure and constant temperature), volume varies with combustion, so that part of energy liberated by combustion transforms directly into mechanical energy due to volume expansion.

The First Law of thermodynamics indicates indeed that internal energy variation (ΔU) is the algebraic sum of heat ($(-\Delta H_r \eta_{LHV} \eta_{th})$) and work ($W = -\int P dv$) received by the system.

Thermoptim calculates value W and displays it just below the value of energy liberated in combustion. This value is subsequently taken into account as useful energy when cycle balance is calculated. One can refer, for more understanding of this subject, to examples on diesel and spark ignition engines presented in portal.

3.5.6.3 Characteristics of Combustion

Dissociation of CO_2 to CO can be taken into account by checking the appropriate option. In the case of atmospheric air, combustion reaction is given by equations (3.5.14) and (3.5.15).

This equilibrium is independent of pressure and is a function of temperature. With assumption that combustion kinetics is sufficiently fast for equilibrium to be reached, mass action law can be written:

$$Kp = \frac{[CO][H_2O]}{[CO_2][H_2]} = \frac{k_1(1-k_2)}{(1-k_1)k_2} = f(T_f)$$

Thermoptim uses an approach of this type, but generalized.

If one chooses to take dissociation into account, a frame is displayed in which must be entered k_1, dissociation rate of CO_2, and T_f freezing temperature which is used for calculating constant K_p.

To account for any heat loss from the combustion chamber, not necessarily adiabatic, we introduce thermal efficiency η_{th}, initialized to 1 by default (right at the center of the screen). This efficiency differs from combustion efficiency η_{LHV}, calculated by software based on the dissociation rate and freezing temperature (see Section 3.5.4.7).

In the bottom left of the screen are located two fields, one corresponding to air factor λ, and other to temperature T_c at the end of combustion. It is possible to set either of these values, and calculate other, or to calculate both from fuel and oxidizer flows. The air ratio may be higher or lower than 1. If it is below 1, software considers this lack of combustion air leading to CO formation. If the air ratio is too low for all available carbon to be oxidized to CO, a message warns the user.

3.5.6.4 Calculation Options

The option "Calculate T" determines the outlet gas temperature T_c, from the value set for λ. If fuel is gas, the mass flow rate of the "fuel" process is adjusted in order that the relationship between volume flow rates of oxidizer and fuel equals the air factor.

The mass flow rate of the combustion process is set equal to the sum of fuel and oxidizer flow rates, which means that the combustion process behaves, at the hydraulic level, as a flow rate mixer.

The option "Calculate lambda" determines λ, from the value of T_c set. Flow rate calculation rules are analogous to those of the preceding option. If enthalpy released by stoichiometric combustion does not allow it to reach the desired temperature, a message warns the user.

The option "Set fuel flow rate" determines λ and T_c from oxidizer and fuel characteristics, which must be gas. The mass flow of the combustion process becomes equal to the sum of fuel and oxidizer flows. If fuel is type CH_a, nothing is done.

When combustion is calculated, the value of combustion efficiency η_{LHV} is determined.

3.5.6.5 Special Case of Premixed Reactants

As indicated above, it is possible to burn the reactant mix that is prepared before it is introduced to the combustion chamber (for example a fuel–air mixture prepared in a carburetor, which can be easily modeled in Thermoptim by a mixer node). In this case, there is no longer need to specify fuel that is already present in the oxidizer. In order that software knows that this is the case, it is necessary to choose option "CH_a type," and set values of a and h_{f0} to zero, or select option "premixed."

As it is clear in this case that notion of air factor λ loses its meaning, this parameter is used to represent the burned fraction of reactants ξ. If $\xi < 1$, we assume that only fraction ξ of mix has reacted, and that $(1 - \xi)$ has not reacted. Combustion gases are then considered as a dual gas mixture: first reaction products, including inert gases, and second fraction of initial mix that has not reacted. This allows one to start from the given mix, and to separate its combustion in several phases, for example constant volume, then constant pressure, and then

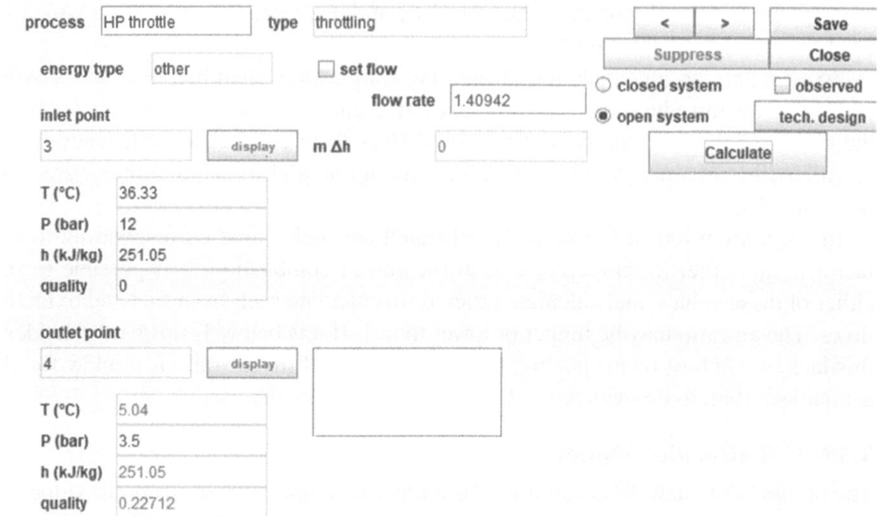

FIGURE 3.6.1 Throttling process screen in Thermoptim.

constant temperature. One can refer, for more understanding of this subject, to the example on spark ignition engines cited above.

3.6 THROTTLING OR FLASH

We call throttling the adiabatic expansion of fluid in a fixed wall device such as filter, valve, or refrigerator expansion valve. The process being adiabatic, $Q=0$. Since walls are fixed, $\tau=0$.

If kinetic energies are negligible, which is usually the case, the First Law is then written $\Delta h=\tau+Q=0$. Throttling is an isenthalpic process.

In Thermoptim, the throttling process screen (Figure 3.6.1) is one of the simplest, as only one action can be performed. You should refer to beginning of Section 3.1.5 for the overview of process screens. Process calculation is to determine the temperature and entropy of the outlet point knowing its pressure and enthalpy.

3.7 WATER VAPOR/GAS MIXTURE PROCESSES

In Section 2.6.2, we presented the definitions and conventions for moist mixtures—in practice, mixtures of a compound gas and water whose composition may vary due to condensation or vaporization of this water. In that same section, we outlined the principles governing the calculation of these mixtures. The main practical application of these studies is air conditioning for controlling temperature and humidity in indoor environments, but the problem is more general.

This section presents the basic treatments that moist mixtures undergo, illustrating each with examples pertaining to air. Air processing cycles involve different processes discussed below (Wang et al., 2000).

To represent these processes, professional practice has established the use of Carrier or psychrometric diagrams, or Mollier diagrams for moist mixtures of the kind that are built into Thermoptim (see Figures 2.6.21 and 2.6.22).

The main processes that moist mixtures can undergo are: mixing, heating, cooling (with or without water condensation), humidification by water or steam, dehumidification, and finally, for air conditioning applications, it is often useful to determine the supply conditions for obtaining desired comfort.

In what follows, we present these basic operations and associated technologies, recall the equations that come into play, and explain how they can be determined with Thermoptim and represented in moist mixture diagrams.

3.7.1 Moist Process Screens

To study changes that a moist mixture may undergo, a "moist" process has been introduced. In fact, it corresponds to six different processes, which are distinguished by their category. The screen appears as shown in Figure 3.7.1, but it varies slightly depending on the category. You should refer to the beginning of Section 3.1.5 for an overview of process screens.

This is a "supply" moist process, used to calculate the supply conditions for maintaining desired comfort, given water and thermal loads.

Before detailing how to use this screen, which will be done in Section 3.7.7, let us highlight some points valid for all moist processes.

First, Thermoptim calculates moist mixture processes by representing them through their dry gas and specific or relative humidity. The reason is simple: this

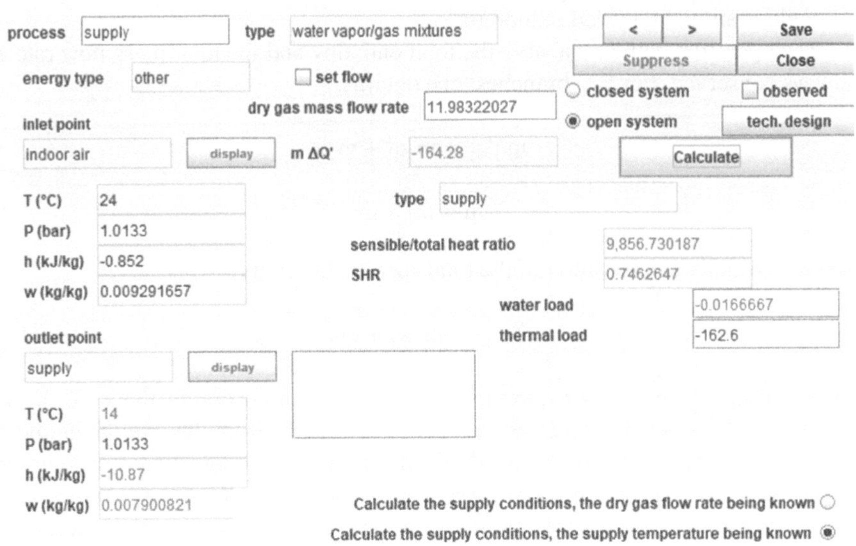

FIGURE 3.7.1 Supply process in Thermoptim.

mode of representation eliminates the need to introduce a new substance for each value of relative humidity.

It should be noted that this way of working is quite exceptional in Thermoptim: for all other processes, calculations are made from the exact composition of the substance considered. If coupling must be made between moist processes and other processes, care should be taken to link them with moist mixtures of appropriate composition.

Additionally, since quantities are preferentially expressed in specific units, flows that appear on the moist processes are dry gas flows.

The concept of effectiveness is often used to describe actual processes versus theoretical or ideal processes. In practice, these generally correspond to changes whose final state is saturated (full humidification, cooling to saturation).

The upper right corner of the screen recalls the general characteristics of the process, while the left shows the inlet and outlet points, indicating their temperature, pressure, enthalpy, and specific humidity.

The settings of the various categories of moist processes and calculation methods are located in the central and lower right parts.

It should also be noted that the moist processes are only accessible from the simulator screen: they have no equivalent in the diagram editor.

3.7.2 Moist Mixers

3.7.2.1 Principle and Equations

A moist mixer is used to determine the particular properties of a mixture of several moist gases. This operation takes place, for example, when two pipes containing mixtures of different humidities come together to form a single stream. In practice, it most often involves air mixtures in air conditioning systems, for example, a mixture of outdoor air and recycled indoor air.

The operation being adiabatic, the total enthalpy and the total mass flow rate are always conserved. For two branches, one obtains:

$$\dot{m}q'_{mix} = \dot{m}_1 q'_1 + \dot{m}_2 q'_2$$

$$\dot{m} = \dot{m}_1 + \dot{m}_2$$

There is usually conservation of the total specific humidity:

$$\dot{m}w_{tot} = \dot{m}_1 w_1 + \dot{m}_2 w_2 \quad (3.7.1)$$

except where there is supersaturation, in which case a portion of the water is condensed. In this case, the liquid water must be subtracted from the specific humidity.

The three previous equations show that in a (w, q') coordinate system, and when there is no supersaturation, the point representative of the mixture is the barycenter of the two points representing the moist mixtures, the coefficients being equal to their mass flow rates. It follows that the three points are aligned, and that the determination of the mixture can be done simply graphically in a

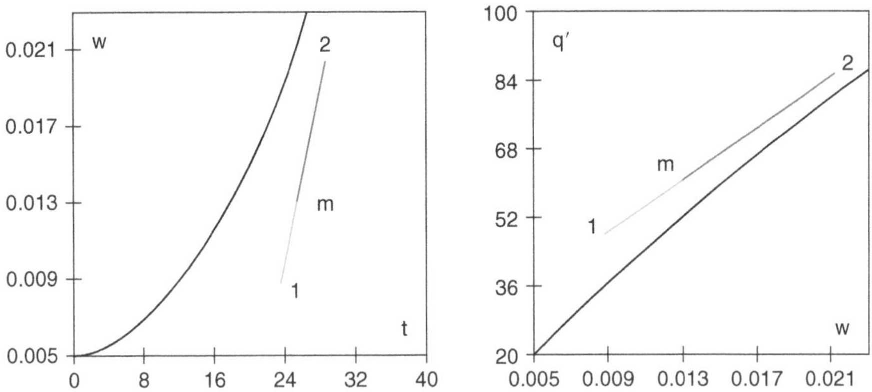

FIGURE 3.7.2 Moist mix in Carrier and Mollier diagrams.

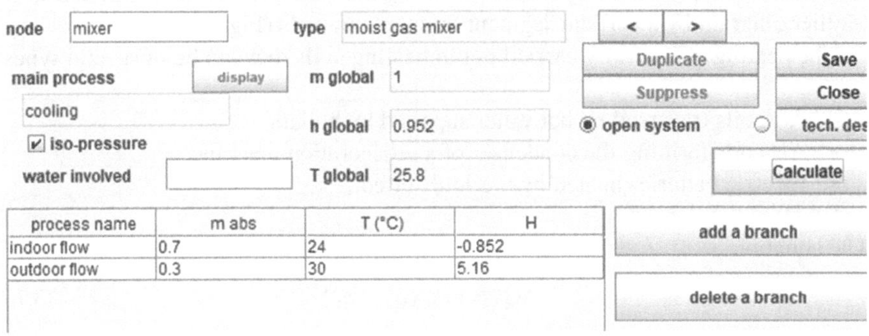

FIGURE 3.7.3 Moist gas mixer.

moist mixture diagram using this system of axes. In the Carrier (w, t) diagram, this is only true as a first approximation.

The representation of the mixture of two moist air flows 1 and 2 in the diagram is given in Figure 3.7.2. The mixture m is located at the junction of the two light and dark aligned segments.

The enthalpy being conserved, we begin by examining the saturation conditions q'mix. If the saturated specific humidity wsat exceeds wtot, then $w_{mix} = w_{tot}$. Otherwise, $w_{mix} = w_{sat}$.

The dry temperature is, in the first case, that corresponding to q'_{mix} and w_{mix}, and in the second case, that of saturation.

3.7.2.2 Calculation of a Mixture in Thermoptim

The moist mixer node can determine the properties of a moist mixture of several dry or moist gases.

A moist mixer (Figure 3.7.3) differs from a simple mixer as follows:

- First, it accepts as branches both moist and non-moist processes. Since the flow of moist processes is referred to the dry gas, the moist mixer makes the necessary corrections.
- Second, its main stream must be a moist process. If this is not the case, a message warns the user, and the calculations are performed as if it were a simple mixer.

In general, the outlet point is located on the mixing line. In the case of supersaturation, a message informs the user and Thermoptim searches for the mixing point on the saturation curve, ensuring conservation of enthalpy. Excess water is then displayed in the upper right.

The values that are displayed are not expressed relative to dry gas.

3.7.3 Heating a Moist Mixture

3.7.3.1 Principle and Equations

The heating of a moist mixture is a constant moisture operation. In the Carrier diagram, it is represented by a horizontal segment oriented to the right, and in the Mollier diagram by a vertical segment oriented upwards (Figure 3.7.4).

The heating is generally carried out in heating coils that can be of several types:

- Air coils traversed by hot water supplied by boilers.
- Air coils forming the condenser of a refrigeration machine.
- Electric batteries heated by the Joule effect.

The equation is:

$$\Delta Q' = m_{air}\left(q_2' - q_1'\right) \tag{3.7.2}$$

After processing, it reduces to:

$$\Delta Q' = m_{air}\,(1 + w)\left(h_{mm2} - h_{mm1}\right) \tag{3.7.3}$$

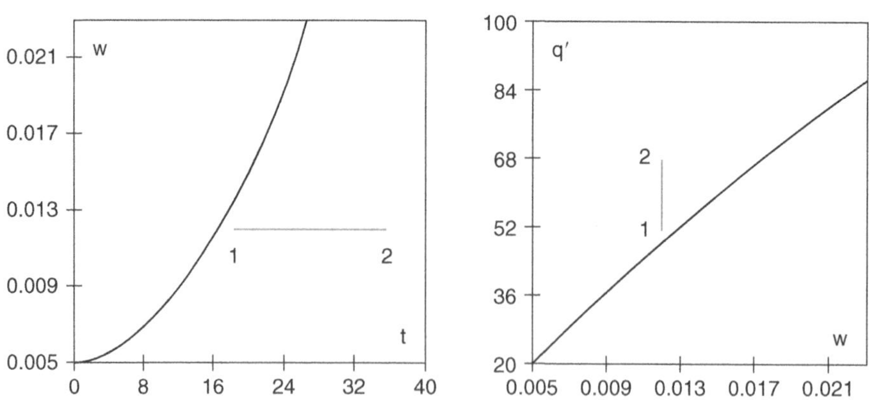

FIGURE 3.7.4 Moist heating in Carrier and Mollier diagrams.

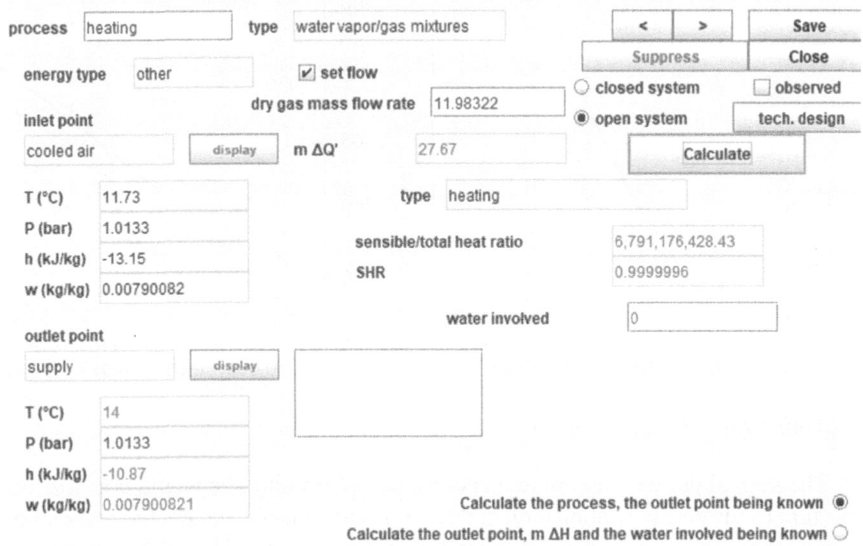

FIGURE 3.7.5 Moist heating process.

where h_{mm} is the enthalpy of the moist mixture.

If $\Delta Q'$ is known, we reverse the equation, which gives t_2.

3.7.3.2 Calculation of Heating in Thermoptim

This process (Figure 3.7.5) can perform various calculations of evolution of a moist mixture. Depending on circumstances, the water content of the inlet and outlet points can be the same or different.

Two actions are possible here:

- Calculate the process, assuming the inlet and outlet points are known, in which case $\Delta Q'$ and the water involved are determined.
- Determine the state of the outlet point, with the enthalpy change and the water involved being known.

3.7.4 Cooling of Moist Mixture

3.7.4.1 Principle and Equations

To cool a moist mixture, it is passed through a special heat exchanger called a cooling coil, which can be cooled by chilled water or by direct evaporation of a refrigerant. The mixture, being in contact with the cold surfaces, sees its temperature decrease. Depending on circumstances, there may or may not be condensation. If there is no condensation, specific humidity remains constant and cooling can be represented by a horizontal segment oriented to the left in the Carrier diagram, and vertically oriented downwards in the Mollier diagram. If condensation occurs, which is very often the case, the segment is oriented to the bottom left in both diagrams (Figure 3.7.6).

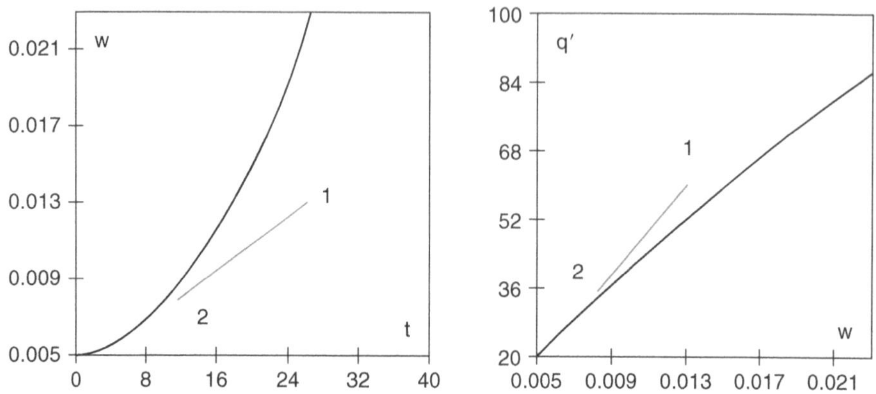

FIGURE 3.7.6 Moist cooling in Carrier and Mollier diagrams.

Theoretical perfect cooling in a cooling coil of infinite size would cool the moist mixture to the coil saturation temperature. It is customary to characterize a real process by taking this cooling as the theoretical reference, introducing effectiveness ε of the cooling coil and its average surface temperature t_s.

In the U.S., another approach is generally retained: the bypass factor method is based on the assumption that part of the total flow that goes through the coil is never in contact with the cold surface—it bypasses it. Thus, the final state of the air exiting the cooling coil corresponds to the mixture between both streams. Calling b the bypass factor, we could easily show that $\varepsilon = 1 - b$.

The advantage of both methods is that values of ε and b of a given coil are coil characteristics largely independent of conditions of use.

To calculate point 2, we first search for the saturation conditions $w_{sat}(t_s)$:

$$w_{sat}(t_s) = \frac{M_{H_2O}P_{sv}}{M_{dg}P - P_{sv}} \quad \text{and} \quad q'(t_s, w_{sat})$$

We calculate the final moist properties in view of effectiveness:

$$\varepsilon = \frac{w_2 - w_1}{w_{sat} - w_1}, \text{ which gives:}$$

$$w_2 = \varepsilon w_{sat} + (1 - \varepsilon) w_1 \tag{3.7.4}$$

$$q_2' = \varepsilon q'(t_s, w_{sat}) + (1 - \varepsilon) q_1'$$

We then search t_2 such that $q_2' = q'(t_2, w_2)$.

A problem may arise when, in the psychrometric diagram, the line from point 1 intersects the saturation curve at two points. Indeed, in this case, the point calculated from effectiveness may be in the saturated zone. This is called early condensation. Two possibilities exist: if one sets the outlet point specific humidity w_2, the end point lies on the saturation curve for $w_{sat} = w_2$; if we set the effectiveness, it is on the saturation curve for $q_{sat}' = q_2'$. We must therefore ensure that the point is not found in the saturated zone.

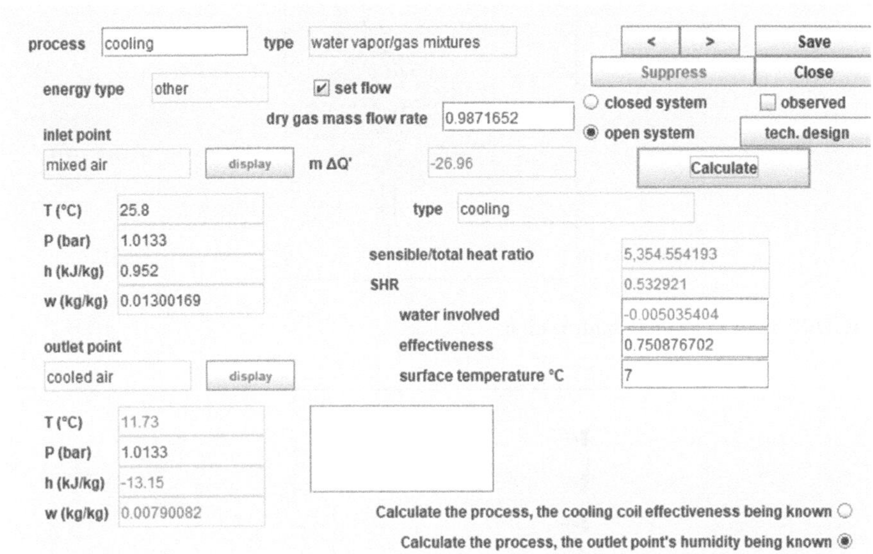

FIGURE 3.7.7 Moist cooling process.

3.7.4.2 Calculation of Cooling in Thermoptim

This process (Figure 3.7.7) allows one to study the cooling of a moist mixture on a cooling coil, with or without condensation.

The cooling conditions are specified on the right of the screen: surface temperature and effectiveness of the cooling coil, if both are known. If the effectiveness is not known, the outlet point humidity must be given to calculate the process.

Two actions are possible here:

- With the effectiveness being set, calculate the outlet point moisture. If early condensation occurs (when the line connecting the inlet point to the point where the saturation curve at the surface temperature intersects the saturation curve at two points), the outlet point is sought on the saturation curve, with the enthalpy being calculated from the effectiveness of the battery.
- With the humidity of the outlet point being set, calculate its temperature and the coil effectiveness. In the case of early condensation (see above), the outlet point is searched on the saturation curve for the specific humidity desired.

3.7.5 Humidification of a Gas

3.7.5.1 Principle and Equations

To humidify a gas, we can proceed in two ways:

- Injecting steam into the gas, usually slightly superheated, or spraying water in very fine droplets with a diameter of 5–20 μm by ultrasound or compressed air, making sure to avoid condensation on the walls (Figure 3.7.8).

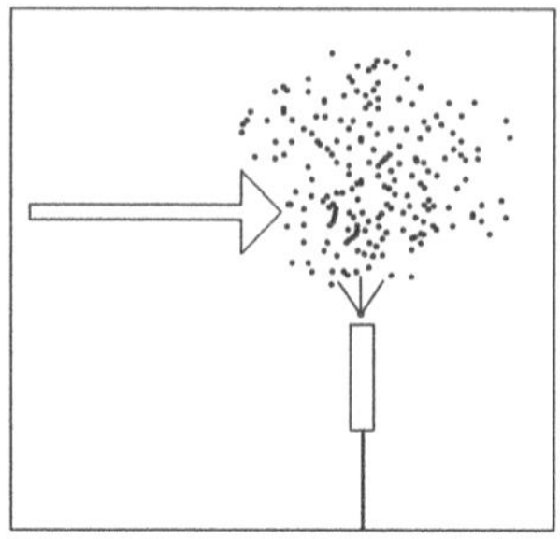

FIGURE 3.7.8 Steam humidification.

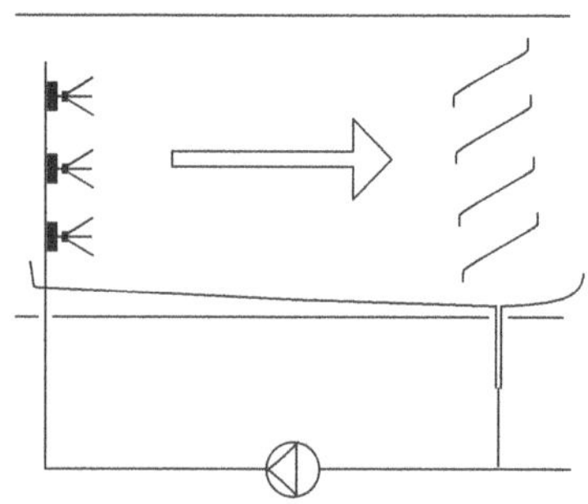

FIGURE 3.7.9 Humidification by water spray.

- Spraying water to form rain that wets the gas (Figure 3.7.9), where the contact surface between gas and water is very important and the gas is far from the saturated state, the water vaporizes and the gas humidity increases. In this case, the vaporization of water requires heat input, which can be supplied by both water and gas, or gas alone, in which case it is called adiabatic humidification. With this type of device there is, however, a risk of bacterial growth and it is therefore less used.

Humidification is mainly used in air conditioning, for example, to humidify very dry outside air in winter before blowing it into a warm interior, or in summer to provide passive cooling.

As has been presented for the cooling of a moist mixture, with the reference being the theoretical wetting leading the gas to saturation, the real humidification is characterized by the previously defined effectiveness ε.

3.7.5.1.1 Cases of Non-Adiabatic Humidification (Vapor or Liquid Water)

If we know the humidification effectiveness ε, we first find the boundary saturation conditions w_{sat}:

$$\text{such as } w_{sat} = w_1 + \frac{m_{water}}{m_{air}} \quad \text{and} \quad q'(t, w_{sat}) = q_1' + (w_{sat} - w_1) h_{H_2O} \qquad (3.7.5)$$

$$\text{with } w_{sat} = \frac{M_{H_2O}}{M_{dg}} \frac{P_{sv}(t)}{P - P_{sv}(t)}$$

We then calculate the final moist conditions in view of effectiveness:

$$w_2 = \varepsilon w_{sat} + (1-\varepsilon) w_1$$
$$q_2' = \varepsilon q'(t, w_{sat}) + (1-\varepsilon) q_1' \quad (3.7.6)$$

We then seek t_2 such that $q_2' = q'(t_2, w_2)$.

If we know the water absorbed by the fluid m_{water}, w_2 is directly calculated:

$$w_2 = w_1 + \frac{m_{water}}{m_{air}}$$

We then seek t_2 such as $q'(t_2, w_2) = q_1' + (w_2 - w_1) h_{water}$

If we seek the boundary saturation conditions, we deduce the effectiveness ε.

Humidification by water or steam spraying is reflected in the Carrier diagram by a line segment tilted to the right or left according to the enthalpy of the fluid injected, and in the Mollier diagram by a segment inclined toward the right. Figure 3.7.10 shows a humidification with steam superheated at 115°C.

3.7.5.1.2 Case of Adiabatic Humidification

We begin by examining the saturation conditions at the wet-bulb temperature $w_{sat}(t')$:

$$w_{sat}(t') = \frac{M_{H_2O}}{M_{dg}} \frac{P_{sv}(t')}{P - P_{sv}(t')}$$

$$\text{and } q'(t', w_{sat})$$

We calculate the final moist conditions in view of effectiveness:

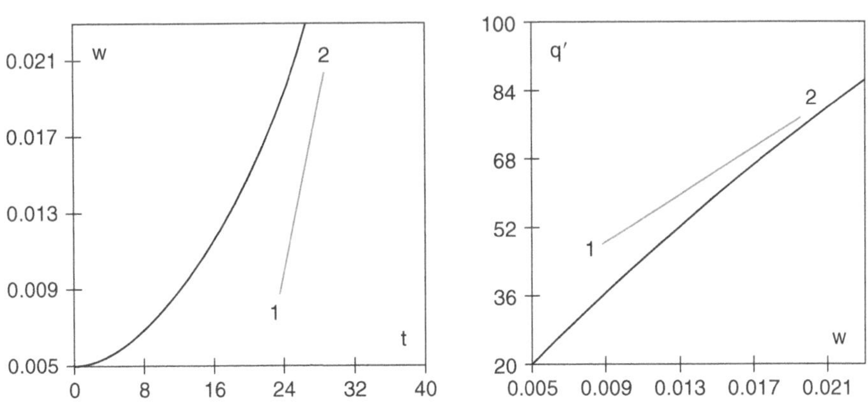

FIGURE 3.7.10 Spray humidification in Carrier and Mollier diagrams.

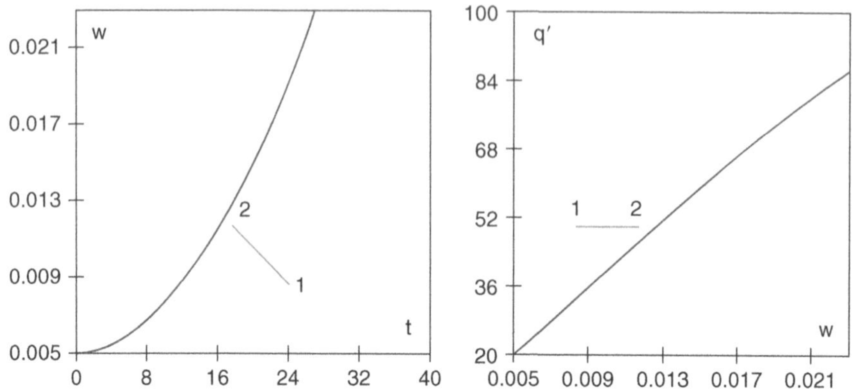

FIGURE 3.7.11 Adiabatic humidification in Carrier and Mollier diagrams.

$$w_2 = \varepsilon w_{sat} + (1-\varepsilon)w_1 \tag{3.7.7}$$

$$q'_2 = \varepsilon q'(t_s, w_{sat}) + (1-\varepsilon)q'_1$$

We then seek t_2 such that $q'_2 = q'(t_2, w_2)$

Adiabatic humidification is reflected in the Carrier diagram by a line segment tilted to the left corresponding to $t' =$ constant, and in the Mollier diagram by a nearly horizontal segment, with q' being approximately constant (Figure 3.7.11).

3.7.5.2 Calculation of Humidification in Thermoptim

Two distinct but similar types of processes are used to study either water or steam humidification, or adiabatic humidification (Figure 3.7.12).

The humidification conditions are specified on the right of the screen: water or steam temperature and pressure, and humidifier effectiveness, if known. If effectiveness is not known, the outlet point humidity must be given to calculate the process.

Two actions are possible here:

- With the effectiveness being set, calculate the outlet point moisture. If early condensation occurs (when the line connecting the inlet point to the point of the saturation curve at the surface temperature intersects the saturation curve at two points), the outlet point is sought on the saturation curve, with the enthalpy being calculated from the humidifier effectiveness.
- Calculate the outlet point to achieve the desired humidity. When there is early condensation (see above), the outlet point is searched on the saturation curve for the specific humidity desired. The effectiveness of the humidifier is then calculated.

3.7.6 Dehumidification of a Mixture by Desiccation

3.7.6.1 Principle and Equations

To dehumidify a moist mixture by desiccation, it passes over a bed of solid hygrophilic adsorbent, which extracts moisture from the gas by exothermic physical

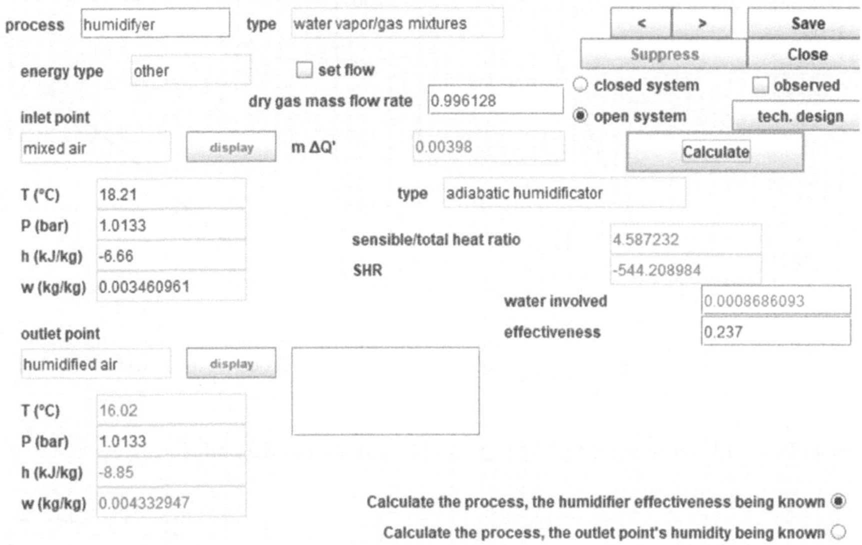

FIGURE 3.7.12 Adiabatic humidificator screen.

(adsorption) or physico-chemical (chemisorption) effects. The regeneration of the desiccant is achieved by the opposite effect through warming. Among the systems that accomplish desiccation, one of the most used takes the form of a rotary regenerative heat exchanger which simultaneously dehumidifies an air stream and regenerates part of the desiccant.

The same equations govern both effects.

The equation of adsorptive dehumidification or desiccant regeneration is provided by the enthalpy balance, which is written here, with L_s being the heat of sorption:

$$(1+w_2)\left(h_{mm}(t_2)-h_{mm}(0°C)\right)-(1+w_1)\left(h_{mm}(t_1)-h_{mm}(0°C)\right)+L_s(w_2-w_1)=0 \tag{3.7.8}$$

Knowing w_2, we simply invert this expression for t_2, taking care that the composition of the moist mixture varies between points 1 and 2.

Knowing t_2, w_2 is obtained by expressing the above equation as a function of dry gas, which leads to a simple first-order polynomial equation in w_2.

Dehumidification by adsorption results in the Carrier diagram as a line segment tilted to the right, and in the Mollier diagram as a segment close to horizontal, with q' increasing slightly (Figure 3.7.13). That the operation is carried out at substantially constant enthalpy does not mean it is free in terms of energy, since the desiccant must be regenerated.

3.7.6.2 Calculation of Desiccation in Thermoptim

This process (Figure 3.7.14) allows the study of dehumidification by desiccation or regeneration of the desiccant. Different types of desiccant materials are proposed,

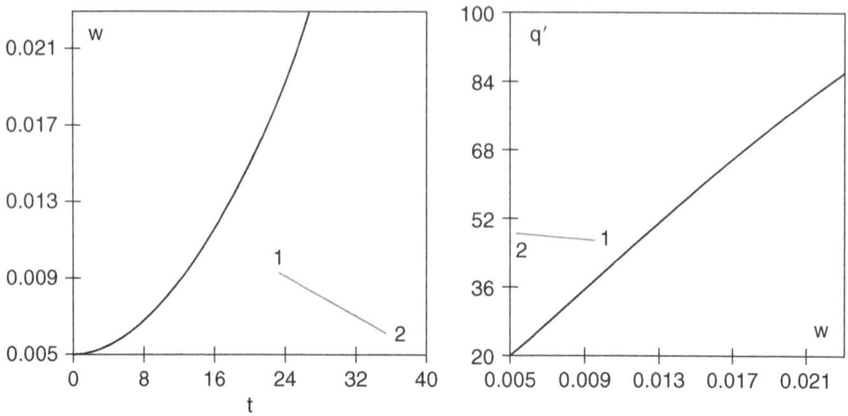

FIGURE 3.7.13 Desiccation process in Carrier and Mollier diagrams.

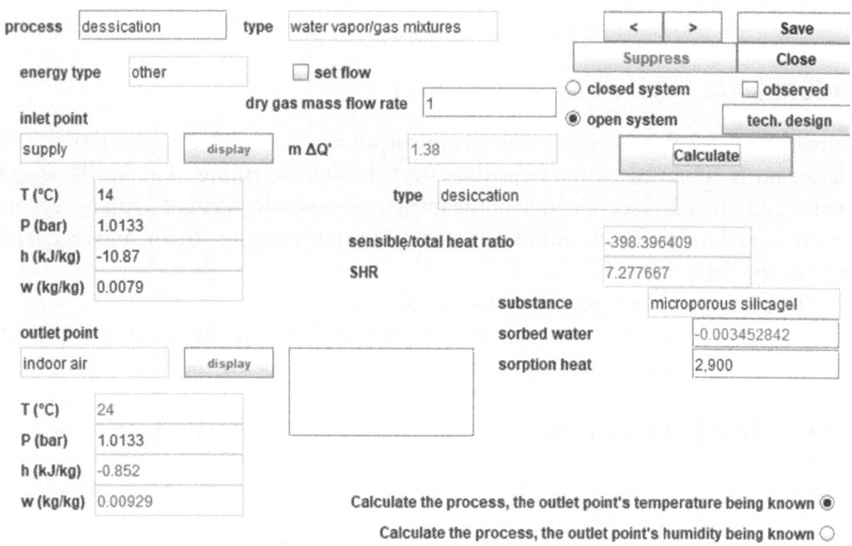

FIGURE 3.7.14 Desiccation process.

with their name and their heat of sorption being displayed on the right of the screen in kJ/kg. If the user wishes, they can change the latter.

Two actions are possible here:

- With the outlet point temperature being known, calculate its moisture.
- Calculate the outlet point temperature to obtain the desired humidity.

Note that the same process calculates both desiccation and desiccant regeneration, according to the relative values of the inlet point and outlet point. For regeneration, the software does not check, however, whether the temperature is sufficient.

3.7.7 Determination of Supply Conditions

3.7.7.1 Principle and Equations

In air conditioning a space, we often experience the following problem: we want to maintain the atmosphere inside a building at a given dry bulb temperature and relative humidity. External climatic conditions are known: dry bulb temperature and relative humidity. We must evacuate the internal and external thermal loads of known magnitude, and a quantity of water corresponding to the internal gains. Depending on countries, calculations are expressed slightly differently: in the U.S., for example, sensible $\dot{Q}_s$ and latent $\dot{Q}_l$ loads are considered (Agami Reddy, 2001), while in France we talk of enthalpy $\dot{Q}_s$ and water $\dot{m}_{eau}$ loads (AICVF, 1999). This simply means that in the first case, water to be extracted is directly converted in energy terms (BTU/hr or kW), while in the second it is expressed in kg/s.

L_{water} being the latent heat of evaporation at indoor air temperature, we have:

$$\dot{Q}_l = \dot{m}_{eau} L_{water}$$

In addition, for reasons of hygiene and comfort, the supply temperature should generally not be less than a given value, and the proportion of recirculated air should not exceed a limit. You then have to determine the supply conditions, that is to say the flow rate $\dot{m}_{air}$ and state of the air blown, so that the atmosphere obtained is the desired one.

1. If the air flow is known, equations are:

$$\text{Water balance: } w_{su} = w_l + \frac{\dot{m}_{water}}{\dot{m}_{air}} \tag{3.7.9}$$

$$\text{Enthalpy balance: } q'(t_{su}, w_{su}) = q'_l + \frac{\dot{Q}_s + \dot{m}_{water} h_{water}(t_l)}{\dot{m}_{air}} \tag{3.7.10}$$

$h_{water}(t_l)$ being the enthalpy of water at temperature t_l.

$$\text{or (U.S. notations) } q'(t_{su}, w_{su}) = q'_l + \frac{\dot{Q}_s + \dot{Q}_l}{\dot{m}_{air}}$$

We therefore determine w_{su}, and then seek t_{su} such that $q' = q'(t_{su}, w_{su})$.

2. If the supply temperature is known, the method is as follows:

$$\text{Water balance: } \dot{m}_{air} = \frac{w_{su} - w_l}{\dot{m}_{water}} \tag{3.7.11}$$

$$\text{Enthalpy balance: } q'(t_{su}, w_{su}) = q'_l + \frac{\dot{Q}_s + \dot{m}_{water} h_{water}(t_l)}{\dot{m}_{air}}(w_{su} - w_l) \tag{3.7.12}$$

With t_{su} being known, w_{su} can be calculated by:

$$q'(t_{su}, w_{su}) = h_{dg}(t_{su}) - h_{dg}(0°C) + w_{su}\left(h_{H_2O}(t_{su}) - h_{H_2O}(0°C)\right) + w_{su} L_{0water}$$

$$w_{su} = \frac{q_1' - w_1\left(\dfrac{Q_s}{m_{water}} + h_{lwater}\right) - h_{dg}(t) + h_{dg}(0°C)}{h_{H_2O}(t_{su}) - h_{H_2O}(0°C) + L_{0water} - \dfrac{Q_s}{m_{water}} - h_{lwater}(t_1)} \tag{3.7.13}$$

Note that eliminating $\dot{m}_{air}$ from equations (3.7.9) and (3.7.10) yields:

$$\theta = \frac{q_{su}' - q_1'}{w_{su} - w_1} = \frac{\dot{Q}_s + \dot{m}_{water} h_{water}(t_1)}{\dot{m}_{water}}$$

θ is called the enthalpy–moisture ratio. It expresses that variations of specific enthalpy and humidity are proportional whatever the air flow rate value $\dot{m}_{air}$: all points lying on a straight line called the condition line or load line satisfy the sensible and latent loads, i.e., are solutions of equations (3.7.9) and (3.7.10).

The position of point "su" on the line determines the value of $\dot{m}_{air}$. It depends on various factors, such as the maximum allowable temperature difference to avoid any inconvenience (usually 6°C–12°C according to the technique used).

3.7.7.2 Calculation of Supply Conditions in Thermoptim

This process connects the point describing the known indoor conditions (inlet point) to the point corresponding to the supply conditions sought (outlet point). Its screen is given at the beginning of this section (Figure 3.7.1).

Let us point out that the determination of supply conditions does not really correspond to what we called a Thermoptim process until now: this is strictly speaking a misnomer. The problem is finding the thermodynamic state of air to be blown into the space to be conditioned in order to compensate for thermal and water loads.

There are two possible options in this case:

- Determine the temperature and the specific humidity of the supply point when the dry gas flow rate is known.
- Set the supply temperature, so that Thermoptim determines the specific humidity of the supply point, as well as the dry gas flow rate required.

Fields on the right of the screen allow you to enter the water (rate of moisture gain) and thermal (sensible rate of heat gain) loads to be removed, respectively, in kg/s and kW, which defines the supply line. It is assumed that the water load must be dissipated at the inside point temperature. Note that units here are not exactly those generally used in the U.S. where the water load is usually directly expressed in kW.

Thermoptim then calculates the value of the sensible/total heat ratio (also known as sensible heat factor SHF or sensible heat ratio SHR) which is the ratio of the rate of sensible heat gain $\dot{Q}_s$ (in kW) for the space to the rate of total energy gain for the space $\dot{Q}_s + \dot{Q}_l$ (in kW). As this ratio is proportional to the slope of the supply or condition line, the ASHRAE psychrometric diagram includes a protractor which facilitates drawing the condition line.

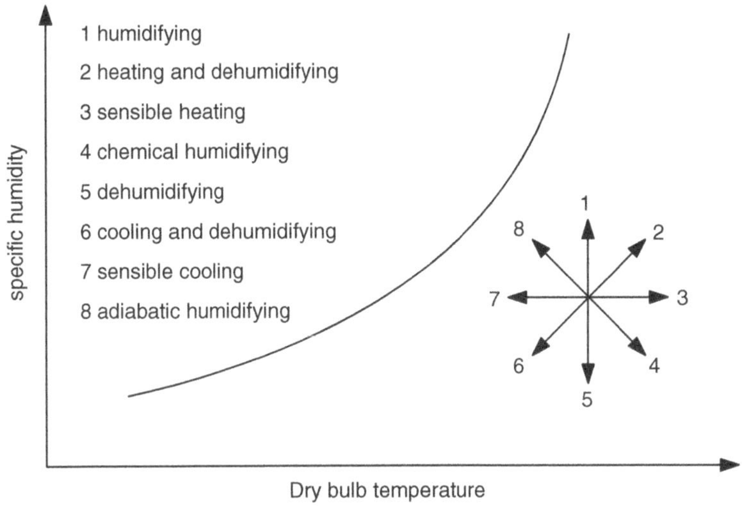

FIGURE 3.7.15 Air conditioning processes in a psychrometric diagram.

In France, the SHR is replaced by the ratio, directly expressed in the psychrometric diagram units, between changes in enthalpy and specific humidity. Known as the "slope ratio γ," it is equal to the enthalpy–moisture ratio θ.

$$SHR = \frac{Q_s}{Q_s + Q_l} = \frac{\Delta h_s}{\Delta h_{tot}}$$

3.7.8 Air Conditioning Processes in a Psychrometric Diagram

Figure 3.7.15 summarizes how the different air conditioning processes can be represented in a psychrometric diagram.

3.8 CONCLUSION

This chapter presented the physical phenomena governing the operation of key components in energy conversion technologies and their calculation methodology in particular within the Thermoptim modeling environment.

The components examined include compression (volumetric and dynamic), expansion (turbines), combustion, throttling, and the processing of wet mixtures. For each type, the theoretical foundations were established, followed by practical calculation methods and performance characterization. Special emphasis was placed on irreversibilities, definitions of efficiency (isentropic and polytropic), and off-design behavior.

The analysis of combustion phenomena covers complete and incomplete reactions, pollutant formation mechanisms, and energy balances. The processing of moist mixtures includes air conditioning processes.

Performance is characterized using dimensionless parameters and performance maps essential for design and optimization, providing engineers with the tools for accurate modeling of energy technologies.

Examples illustrating how the concepts presented in this chapter can be applied using Thermoptim are available on this book's website.[1]

NOTE

1 Available at: https://server.s4e2.com/crc/amtc/chapters/chapter3/_resources/index.html.

BIBLIOGRAPHY

Agami Reddy, T. (2001). "Psychrometrics and comfort." In Kreider, J.F. (ed.), *Handbook of heating, ventilation, and air conditioning*. CRC Press, London, pp. 10–50. ISBN 0-8493-9584-4.

Boehm, R.F. (2000). "Pumps and fans." In Kreith, F. (ed.), *The CRC handbook of thermal engineering*. CRC Press, Boca Raton, FL, pp. 4-170–4-188. ISBN 0-8493-9581-X.

Chase, M.W. Jr. et al. (1985). "JANAF thermochemical tables." *Journal of Physical and Chemical Reference Data,* 14(1).

Cohen, R., Groll, E., Harden, W.H., Hickman, K.E., Mistry, D.K. & Muir, E. (2000). "Compressors." In Kreith, J.F. (ed.), *The CRC handbook of thermal engineering*. CRC Press, Boca Raton, FL, pp. 4-37–4-58. ISBN 0-8493-9581-X.

Conan, J.G. (1988). *Réfrigération industrielle.* Eyrolles, Paris.

de Vlaminck, M. & Wauters, P. (1988). *Thermodynamique et turbines.* Ciaco, Paris.

Giampaolo, T. (2010). *Compressor handbook: Principles and practice.* The Fairmont Press, Lilburn/CRC Press, Boca Raton, FL. ISBN 0-88173-615-5.

Glassman, I. (1996). *Combustion,* 3rd ed. Academic Press, San Diego, CA.

Goodger, E.M. (1977). *Combustion calculations: Theory, worked examples and problems.* The MacMillan Press, London.

Kuo, K.K. (2000). *Principles of combustion.* John Wiley & Sons, Hoboken, NJ.

Lebrun, J. (1985). *Thermodynamique appliquée: Introduction aux machines thermiques.* Polycopiés de l'Université de Liège.

Li, D. & Groll, A. (2005). "Transcritical CO_2 refrigeration cycle with ejector-expansion device." *International Journal of Refrigeration,* 28, 766–773.

Spalding, D. (1979). *Combustion and mass transfer.* Pergamon Press, Oxford.

Vivier, L. (1965). *Turbines à vapeur et à gaz.* Albin Michel, Paris.

Wang, S.K., Lavan, Z. & Norton, P. (2000). *Air conditioning and refrigeration engineering.* CRC Press, Boca Raton, FL. ISBN 0-8493-0057-6.

Williams, F.A. (1985). *Combustion theory: The fundamental theory of chemically reacting flow system*, 2nd ed. Addison-Wesley Publishing Company, Redwood City, CA.

4 Heat Exchangers

INTRODUCTION

Heat exchangers are devices for transferring heat between two fluids at different temperatures. In most cases, the two fluids are not in direct contact, and the transfer occurs through a separating wall. Within the wall, the heat transfer mechanism is conduction, while on each surface in contact with the fluids, convection almost always predominates. In many cases, fluids remain in a single phase, whether gas or liquid. However, there are three main types of heat exchangers involving phase change: evaporators (liquid-to-vapor transition), condensers (vapor-to-liquid transition), and evaporative condensers (simultaneous phase change in both fluid streams).

In practice, heat exchangers play a critical role in many energy systems where heat is transferred, particularly when thermal energy recovery is desired. They can be essential to the performance of energy systems involving internal heat exchange, such as regeneration or combined cycles, whose optimization requires minimizing the internal irreversibilities related to internal temperature differences.

4.1 PRINCIPLES OF OPERATION OF A HEAT EXCHANGER

The study of heat exchangers in an energy system can be conducted at very different levels of complexity depending on the objectives pursued:

- The most basic approach, very simple to implement, is to perform an enthalpy balance of the heat exchanger by determining, for example, the outlet temperature of a fluid when the inlet temperatures and flow rates of both fluids are known, as well as one outlet temperature.
- The problem can be posed in reverse if the two outlet temperatures are known and only one inlet temperature. The approach remains simple. Indeed, it can be shown that a heat exchanger has 5 degrees of freedom, and once five quantities are set among the inlet and outlet temperatures and flow rates, the sixth is directly deduced.
- It is also common to characterize a heat exchanger by either an effectiveness or a minimum temperature difference between the two fluids (called pinch). Solving this problem is also fairly straightforward if four quantities are known. Note that very often in the study of energy systems, heat exchanger calculations are not explored further. They indeed already provide sufficient information to conduct preliminary analysis of many cycles, especially at the functional level, apart from specific technological considerations. The theory shows, as we see in this chapter, that this approach allows us to characterize an exchanger by what we call the UA, the product of the overall heat transfer coefficient U between the two fluids and its exchange area A.

DOI: 10.1201/9781003606017-5

- To proceed further and determine the required exchange surface, calculations become much more complex, requiring thorough knowledge of the internal operation of heat exchangers in order to estimate as precisely as possible the value of U. There is actually a very important qualitative leap between the basic arithmetic of balancing the heat exchanger and what we call its component sizing, which will be discussed in Chapter 7.

In a heat exchanger, the fluid flows can be arranged in multiple configurations.

It can easily be shown that thermodynamically, the most efficient heat exchanger is the counter-flow heat exchanger (Figure 4.1.1), but considerations other than thermodynamic effectiveness are taken into account when designing a heat exchanger: the maximum permissible temperatures for one fluid, or more often considerations of size, weight, or cost.

Consequently, the configurations of exchangers encountered in practice are relatively numerous.

However, these configurations can be grouped into three main geometries:

- Counter-flow, in which fluids flow parallel and in opposite directions.
- Parallel-flow, in which fluids flow parallel and in the same direction.
- Cross-flow, in which fluids flow in perpendicular directions.

We will denote the hot fluid by index h, and the cold fluid by index c. Besides the geometric configuration, exchanger sizing or performance depends on many parameters and variables:

- Mass flow rates $\dot{m}_h$ and $\dot{m}_c$ passing through them.
- Inlet temperatures T_{hi} and T_{ci} and outlet temperatures T_{ho} and T_{co} of both fluids.
- Heat transfer coefficients U_h and U_c of each fluid.
- Thermal resistance of the wall $\frac{e}{\lambda}$.
- The surface area A of the exchanger.
- The pressures of both fluids, assumed nearly constant.
- The thermophysical properties of fluids, which are used to determine coefficients U_h and U_c. These are essentially the specific heat capacity c, density ρ, thermal conductivity λ, and viscosity μ.

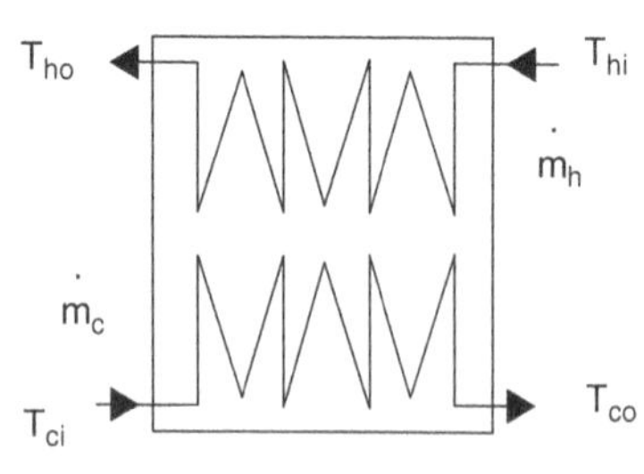

FIGURE 4.1.1 Heat exchanger sketch.

In what follows, we will assume that the heat transfer coefficients U_h and U_c and thermophysical properties of fluids maintain constant values throughout the entire heat exchanger.

If this assumption is not verified, then in order to study the performance of the exchanger, it is necessary to divide it into small volume elements in which these properties can be considered constant. The calculations then become much more cumbersome.

The NTU method that we will use to calculate heat exchangers only provides access to the product UA of U and the exchange surface A, which already allows us to solve many thermal coupling problems in energy systems. In this chapter, we will not examine in detail the calculation of the exchange coefficients U. However, this point will be addressed in Chapter 7 during the study of the behavior of systems under off-design conditions.

Finally, we always assume that the heat exchanger is globally adiabatic, meaning there is no heat exchange with the surroundings.

4.1.1 Heat Flux Exchanged

It can be shown, where fluid flow is pure (parallel or counter-flow), that the heat flux exchanged between the two fluids is given by the formula:

$$\phi = UA\Delta T_{ml} \tag{4.1.1}$$

$$\text{with} \quad \Delta T_{ml} = \frac{\Delta T_0 - \Delta T_L}{\ln \frac{\Delta T_0}{\Delta T_L}} \tag{4.1.2}$$

where A is the exchange surface, U is the overall heat transfer coefficient (see the next section), ΔT_{ml} the Log Mean Temperature Difference (LMTD), and ΔT_0 and ΔT_L are the temperature differences between fluids at one end and the other end of the exchanger, respectively, with the convention: $\Delta T_0 > \Delta T_L$.

Care must be taken that we refer to temperature differences between the two fluids at the entrance and exit of the exchanger (positions 0 and L), not the differences between inlet temperatures ($T_{hi} - T_{ci}$) and outlet temperatures ($T_{ho} - T_{co}$) of the fluids.

In the example in Figure 4.1.2, we have:

$$\Delta T_0 = T_{hi} - T_{co}$$

$$\Delta T_L = T_{ci} - T_{ho}$$

When the flow configuration is more complex, we introduce a correction factor F (less than 1), which is given by charts that are calculated or determined experimentally, most often provided by manufacturers.

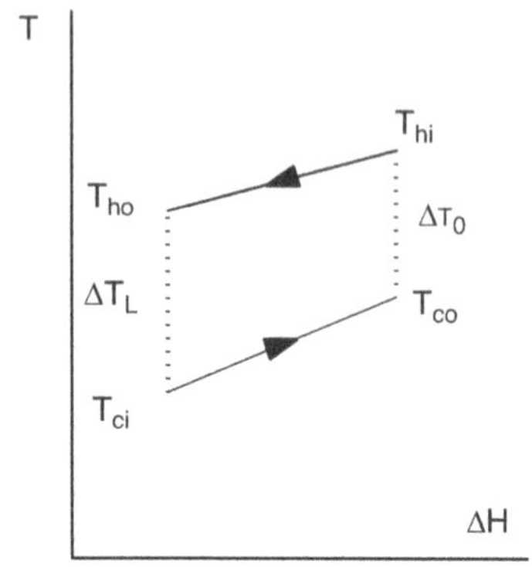

FIGURE 4.1.2 LMTD notations.

Equation (4.1.1) becomes:

$$\phi = UAF\Delta T_{ml} \tag{4.1.3}$$

This formula shows that to transfer a given flux ϕ, if we wish to reduce the irreversibilities and therefore ΔT, the product UA must be as large as possible, which can be achieved either by increasing the surface area, though this directly affects cost, or by increasing the U value, which is an objective sought by all heat exchanger designers.

In this chapter, the design of the heat exchanger will be limited to determining the product UA of the exchanger surface A and the overall heat transfer coefficient U, as the accurate estimation of the latter depends on the internal constructive details of the exchanger.

In general, the design of heat exchangers is a compromise between conflicting objectives, the two main ones being:

- A large exchange area is desirable to increase the effectiveness of heat exchangers, but it results in high costs.
- Small fluid flow sections increase the values of the heat transfer coefficients U_h and U_c defined in the next section, and thus reduce the required area, but they also increase pressure drops.

4.1.2 Heat Exchange Coefficient U

We only address the precise determination of U and the study of pressure drops in heat exchangers in Chapter 7, not because these difficult issues lack interest, quite the contrary, but because, as for other components, we are not currently seeking to complete their detailed internal design: we limit ourselves to dealing with their integration into energy technology systems. We will only present here the principles governing the calculation of U.

In most heat exchangers, the rate of heat transfer on each side of the wall is governed by convection. The convective heat transfer coefficient, denoted h, relates the heat flux to the temperature difference between the wall and the bulk fluid:

$$Q = hA(T_b - T_w)$$

where:

A is the heat transfer surface area [m^2],

T_b is the bulk fluid temperature [K], and

T_w is the wall temperature [K].

The coefficient h depends on fluid properties, flow velocity, flow regime (laminar or turbulent), and channel geometry. It is therefore not a property of the fluid alone, but of the combined system of fluid and geometry.

When heat is exchanged between two fluids separated by a wall, the total resistance results from convection on the hot side, conduction through the wall, and convection on the cold side. The overall heat transfer coefficient U relates the heat flux to the temperature difference between the two fluids:

$$Q = UA(T_h - T_c)$$

where: U is the overall heat transfer coefficient $\left[W/m^2/K\right]$, and T_h and T_c are the bulk temperatures of the hot and cold fluids [K].

Thermal science teaches us that for a flat plate heat exchanger, the overall heat transfer coefficient U is such that its inverse, called thermal resistance, is the sum of thermal resistances between the two fluids:

$$\frac{1}{U} = \frac{1}{h_h} + \frac{e}{\lambda} + \frac{1}{h_c} \tag{4.1.5}$$

where h_h is the heat transfer coefficient between the hot fluid and the wall, h_c is the heat transfer coefficient between the cold fluid and the wall, e is the wall thickness, and λ is the thermal conductivity of the wall.

If the wall is composed of several layers of different materials, or if deposits have covered the wall, this formula becomes:

$$\frac{1}{U} = \frac{1}{h_h} + \sum_{k=1}^{n} \frac{e_k}{\lambda_k} + \frac{1}{h_c} \tag{4.1.6}$$

The overall thermal resistance is the sum of (n + 2) thermal resistances, the largest of which most significantly limits the heat transfer.

This form highlights the weakest link: the smallest convection coefficient, or a poorly conducting wall, will dominate the resistance and limit U. Improving U requires either increasing the limiting convection coefficient (e.g., by using fins or turbulence promoters) or reducing the wall resistance.

For example, in the case of a gas/liquid heat exchanger, the convection coefficient can be 10–100 times lower on the gas side than on the liquid side. The thermal resistances of the wall and the liquid side are generally negligible, and U is approximately equal to h.

In this case, we seek to increase the gas-side exchange surface by using fins. The previous formulas become more complicated, and since the cold- and hot-side surfaces are not the same, we introduce a hot-side overall exchange coefficient U_h, such that its inverse is:

$$\frac{1}{U_h} = \frac{1}{\eta_{0,h} h_h} + \frac{e}{\frac{A_w}{A_h}\lambda} + \frac{1}{\frac{A_c}{A_h}\eta_{0,c} h_c} \tag{4.1.7}$$

with A_h and A_c total exchange areas on hot and cold sides, respectively, A_w exchanger wall surface, and $\eta_{0,h}$ and $\eta_{0,c}$ overall fin effectiveness on warm and cold sides, respectively (see Section 4.1.3).

There is also a cold side overall exchange coefficient U_c, with of course:

$$U_c A_c = U_h A_h$$

4.1.3 Fin Effectiveness

The efficiency of a fin depends on its geometry and the balance between conduction along the fin and convection from its surface. For a thin rectangular fin of thickness e (e << l, Figure 4.1.3), height h, thermal conductivity λ, and subjected to a convection coefficient h, the effectiveness can be evaluated by introducing the fin parameter:

$$\eta_a = \frac{\tanh(ml)}{ml}$$

$$\text{with } m = \sqrt{\frac{2h}{\lambda e}} \tag{4.1.8}$$

This factor expresses the ratio between the actual heat flux and the maximum flux that would be obtained if the entire fin were at the base temperature.

In practice, high-conductivity materials (e.g., copper, aluminum) lead to high efficiency, while thick fins with low conductivity are less effective. Fin design must therefore balance additional area against material and pressure drop penalties.

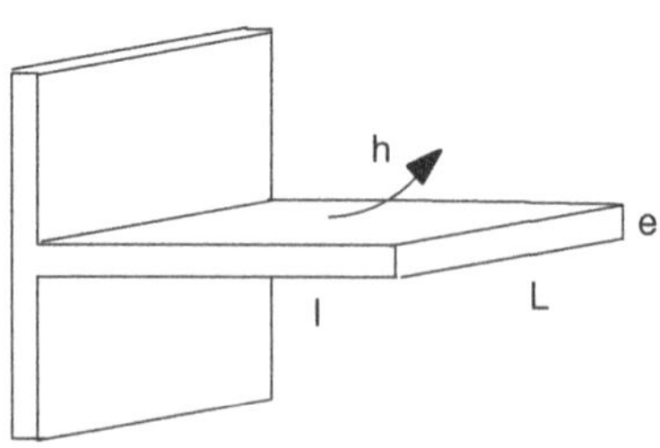

FIGURE 4.1.3 Rectangular fin.

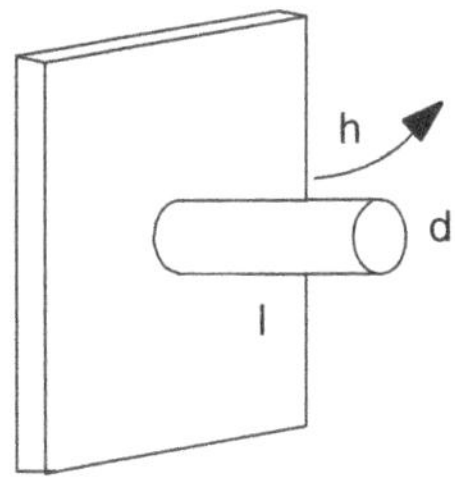

FIGURE 4.1.4 Needle fin.

For fins of different shapes, the value of m changes. For example, for circular needle fins (Figure 4.1.4):

$$m = \sqrt{\frac{4h}{\lambda d}} \tag{4.1.9}$$

For other geometries, we refer to the literature (Incropera & Dewitt, 1996).

The total flux exchanged equals the sum of the fluxes exchanged on the one hand by the fins (area A_a) and on the other hand by the wall between the fins (area $A - A_a$). This defines an overall effectiveness η_0 equal to the ratio of the flux actually exchanged to the flux that would be exchanged if the entire surface were at the base temperature:

$$\eta_0 = \frac{hA_a\eta_a + h(A - A_a)}{hA} = 1 - \frac{A_a}{A}(1 - \eta_a) \tag{4.1.10}$$

4.1.4 Heat Transfer Correlations

The values of convection coefficients h_h and h_c depend on fluid thermophysical properties and exchange configurations. Further explanations will be given in Chapter 7 on heat exchanger sizing.

To determine the convective coefficient h, correlations are used to relate the Nusselt number Nu to the Reynolds number Re and the Prandtl number Pr:

$$Nu = \frac{hd_h}{\lambda}, \quad Re = \frac{\rho V d_h}{\mu} \quad \text{and} \quad Pr = \frac{\mu c_p}{\lambda}$$

The hydraulic diameter d_h equals the ratio of four times the flow area S to the wetted perimeter p: $d_h = 4S/p$ (if there are isolated walls, the perimeter of heat exchange must be considered).

4.1.4.1 Forced Convection in Tubes

For laminar flow in a circular tube with constant wall temperature:

$$Nu = 3.66$$

For fully developed turbulent flow, one widely used correlation is the Dittus–Boelter equation:

$$Nu = 0.023Re^{0.8}Pr^{n}$$

with n = 0.3 for heating and n = 0.4 for cooling.

4.1.4.2 External Flow

For cross-flow over tubes, empirical charts and correlations are available, depending on the Reynolds number, arrangement (in-line or staggered), and the number of rows.

For flow perpendicular to tubes, several options exist, such as the Colburn correlation:

$$Nu = 0.33Re^{0.6}Pr^{0.33} \tag{4.1.12}$$

In general, the exponent of the Reynolds number is between 0.5 and 0.8, and that of the Prandtl number between 0.33 and 0.4.

4.1.4.3 Natural Convection

When flow is induced by buoyancy, correlations are based on the Grashof number Gr and the Prandtl number, combined into the Rayleigh number $Ra = Gr \cdot Pr$. Typical forms are:

$$Nu = CRa^{n}$$

with constants depending on geometry (vertical plates, horizontal cylinders, etc.).

4.1.4.4 Two-Phase Flow

The exchange coefficients in two-phase flow are always much higher than for single-phase flows. Their estimation is much more complicated. Examples of correlations will be provided in Chapter 7.

4.1.4.5 Typical Values of Heat Transfer Coefficients

For order-of-magnitude estimates, the following ranges are useful:

- Natural convection in air: $h \approx 2-10\ W/m^2/K$
- Forced convection in air: $h \approx 10-100\ W/m^2/K$
- Water (forced convection in tubes): $h \approx 500-10{,}000\ W/m^2/K$
- Condensation of steam on vertical surfaces: $h \approx 5{,}000-100{,}000\ W/m^2/K$
- Boiling of water: $h \approx 5{,}000-100{,}000\ W/m^2/K$

These values illustrate the wide variation of heat transfer intensity depending on the fluid and regime. In general, gas-side coefficients are much lower than those for liquids or phase-change processes.

4.2 PHENOMENOLOGICAL MODELS FOR THE CALCULATION OF HEAT EXCHANGERS

To size a heat exchanger, it is possible to use relations (4.1.1) or (4.1.3), but they assume the LMTD is known, which is rarely the case.

4.2.1 Number of Transfer Units Method

A simpler and more general method is the NTU method developed by Kays and London (1984). This method has the advantage of requiring knowledge of fluid inlet temperatures, not those at the outlet.

By definition, NTU is defined as the ratio of the product UA of the heat exchanger to the minimum heat capacity rate.

$$NTU = \frac{UA}{\left(\dot{m} \cdot c_p\right)_{min}} \tag{4.2.1}$$

We call R the ratio (less than or equal to 1) of the heat capacity rates:

$$R = \frac{(\dot{m}c_p)_{min}}{(\dot{m}c_p)_{max}} \leq 1 \tag{4.2.2}$$

and **ε** the **effectiveness** of the exchanger, defined as the ratio between the heat flux actually transferred and the maximum possible flux:

$$\varepsilon = \frac{\phi}{\phi_{max}} \tag{4.2.3}$$

We can show that ϕ_{max}, which would be obtained for a counter-flow heat exchanger of infinite length, is:

$$\phi_{max} = (\dot{m}c_p)_{min} \Delta T_i$$

where ΔT_i is the difference between inlet temperatures of both fluids. Figure 4.2.1 shows the temperature profile in a heat exchanger. Note that the abscissa used here is the enthalpy exchanged, not the length of the exchanger, which explains, with the assumption that the heat capacities are constant, why this profile is linear rather than exponential.

Note also that ΔT_{max} and ΔT_{min} are the temperature differences within each of the two fluids, not between them as in Figure 4.1.2.

$$\text{As } \phi = (\dot{m}c_p)_{min} \Delta T_{max} = (\dot{m}c_p)_{max} \Delta T_{min} \tag{4.2.4}$$

$$\text{We have: } \varepsilon = \frac{\Delta T_{max}}{\Delta T_i} \tag{4.2.5}$$

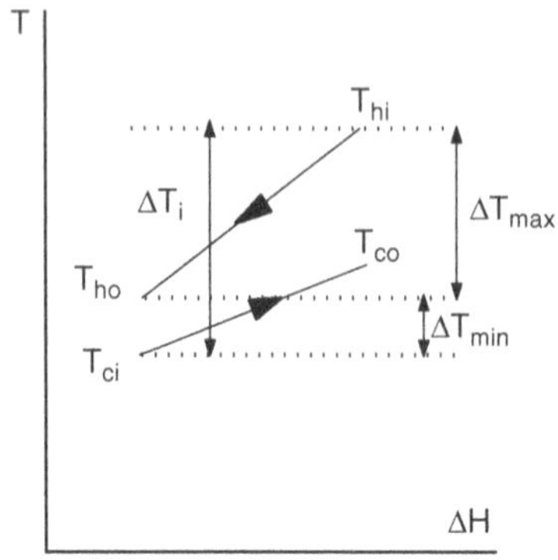

FIGURE 4.2.1 Effectiveness notations.

$$\text{Since } \phi = UA\Delta T_{ml} = (\dot{m}c_p)_{min}\,\Delta T_{max} = (\dot{m}c_p)_{max}\,\Delta T_{min}$$

$$NTU = \frac{UA}{\dot{m}c_{pmin}} = \frac{\Delta T_{max}}{\Delta T_{ml}}$$

NTU measures the ability of the exchanger to change the temperature of the fluid whose heat capacity rate is lower.

This is indeed an indicator of the equipment quality: the more efficient the heat exchanger, the more it allows heating or cooling a fluid with a small temperature difference relative to the other fluid.

When U varies along the surface of the exchanger,

$$NTU = \frac{1}{\dot{m}c_{pmin}} \int_0^A U\,dA$$

For instance, in the example of Figure 4.2.1, where $(\dot{m}c_p)_h < (\dot{m}c_p)_c$, we have:

$$(\dot{m}c_p)_{min} = (\dot{m}c_p)_h$$

$$\Delta T_i = T_{hi} - T_{ci}$$

$$\Delta T_{max} = T_{hi} - T_{ho} \quad \text{and} \tag{4.2.6}$$

$$\varepsilon = \frac{T_{hi} - T_{ho}}{T_{hi} - T_{ci}}$$

With these definitions, it is possible to show that there is a general relation of type:

$$\varepsilon = \mathbf{f}(\mathbf{NTU, R, flow\ pattern}) \tag{4.2.7}$$

In practice, it suffices to have a set of relationships corresponding to flow patterns representative of the exchangers studied, and the design of a heat exchanger is based on the one hand on balance equations (4.2.4) and on the other hand on the internal equation (4.2.7).

If we know the flow rates of both fluids, their inlet temperatures, and the heat flux transferred, the procedure is as follows:

- We start by determining the outlet temperatures of fluids from equation (4.2.4).
- We deduce the fluid heat capacity rates $\dot{m}c_p$ and their ratio R.
- The effectiveness ε is calculated from equation (4.2.5).
- The value of NTU is determined from the appropriate (NTU, ε) relationship.
- UA is calculated from equation (4.2.1).

If UA is known, equation (4.2.1) gives the value of NTU, and ε can be determined from the appropriate (NTU, ε) relationship. Equation (4.2.5) allows calculation of the heat transferred.

In both cases, the enthalpy balance provides the outlet temperatures. This method therefore connects the UA product to the exchanger performance, and the surface area is deduced from an estimate of the value of the exchange coefficient U.

As shown in equation (4.2.6), ε measures the "temperature" effectiveness of the exchanger, i.e., the fraction of the maximum temperature difference between the fluids that is achieved by the fluid with the lower heat capacity rate.

The value of the method proposed by Kays and London lies in its generality due to the formal appearance of $(\dot{m}c_p)_{min}$ and $(\dot{m}c_p)_{max}$.

Phenomenological (ε, NTU) models provide only the exchanger UA product. As we have seen in Section 4.1.2, the calculation of U depends on many factors not taken into account by these models. We will show in Chapter 7 the types of sizing models that can be built to assess U.

Given its advantages, this method has been very successful and is now widely used. As we shall see, it is often even generalized to size heat exchangers other than those we have considered here, where we assumed that heat transfer coefficients U_h and U_c and fluid thermophysical properties keep constant values.

Experience shows that, at a thermodynamic level, (ε, NTU) models generally represent well the behavior of many types of exchangers, even more complex than those we have studied so far. This is especially what we have done in Thermoptim for thermocouplers, whose mechanism complements heat exchangers by allowing components other than "exchange" processes to connect to one or more "exchange" processes to represent thermal couplings (see Section 4.3.5).

4.2.2 Relationship between NTU and ε

In the pages that follow, we give relations between ε and NTU as functions of R, and their inverses which can be expressed simply, as well as the corresponding graphs

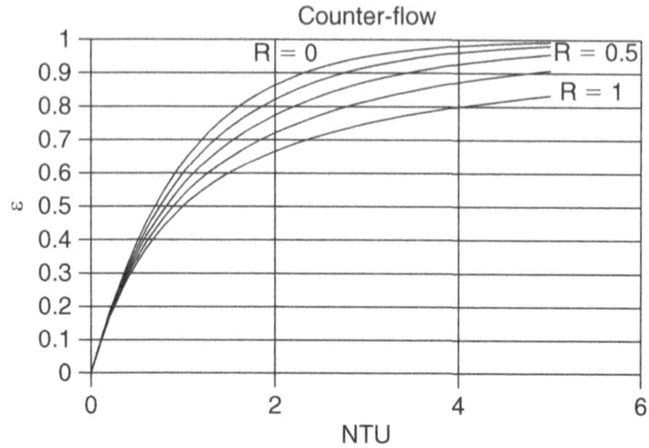

FIGURE 4.2.2 Counter-flow heat exchanger chart.

in which curves represent ε as a function of NTU, for R varying from 0 to 1, top to bottom, in steps of 0.25.

4.2.2.1 Counter-Flow Heat Exchangers

For a counter-flow heat exchanger, analytical expressions are, for $R \neq 1$:

$$NTU = \frac{1}{1-R}\ln\frac{1-\varepsilon R}{1-\varepsilon} \tag{4.2.8}$$

$$\varepsilon = \frac{1-\exp(-NTU(1-R))}{1-R\exp(-NTU(1-R))}$$

$$\text{For } R = 1 \quad \varepsilon = \frac{NTU}{1+NTU}$$

The corresponding chart is given in Figure 4.2.2.

4.2.2.2 Parallel-Flow Heat Exchangers

For a parallel-flow heat exchanger:

$$NTU = -\frac{1}{1+R}\ln(1-\varepsilon(1+R)) \tag{4.2.9}$$

$$\varepsilon = \frac{1-\exp(-NTU(1+R))}{1+R}$$

The corresponding chart is given in Figure 4.2.3.

4.2.2.3 Cross-Flow Heat Exchangers

For a $((\dot{m}c_p)_{min})$ cross-flow heat exchanger with $(\dot{m}c)$ mixed fluid:

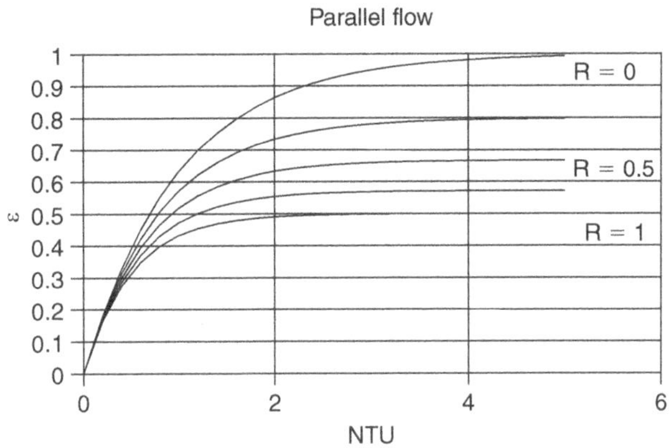

FIGURE 4.2.3 Parallel-flow heat exchanger chart.

$$NTU = -\frac{1}{R}\ln[1 + R\ln(1-\varepsilon)] \quad (4.2.10)$$

$$\varepsilon = 1 - \exp(-\Gamma/R) \quad \text{with} \quad \Gamma = 1 - \exp(-RNTU)$$

The corresponding chart is given in Figure 4.2.4.

For a $((\dot{m}c_p)_{max})$ cross-flow heat exchanger with $(\dot{m}c)$ mixed fluid:

$$NTU = -\ln\left[1 + \frac{1}{R}\ln(1-\varepsilon R)\right] \quad (4.2.11)$$

$$\varepsilon = \frac{1}{R}[1 - \exp(-\Gamma R)] \quad \text{and} \quad \Gamma = 1 - \exp(-NTU)$$

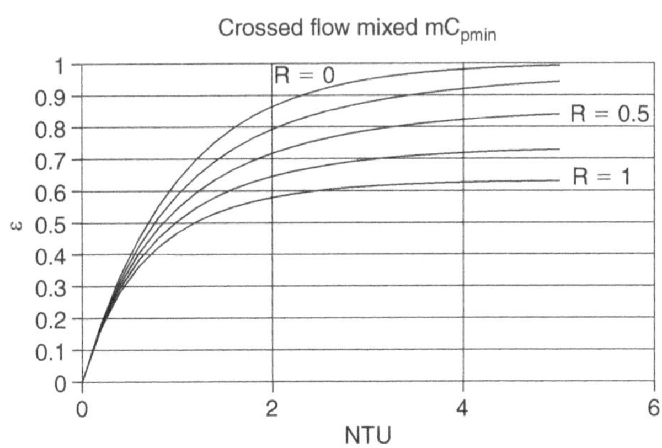

FIGURE 4.2.4 Cross-flow heat exchanger chart (mCp_{min}).

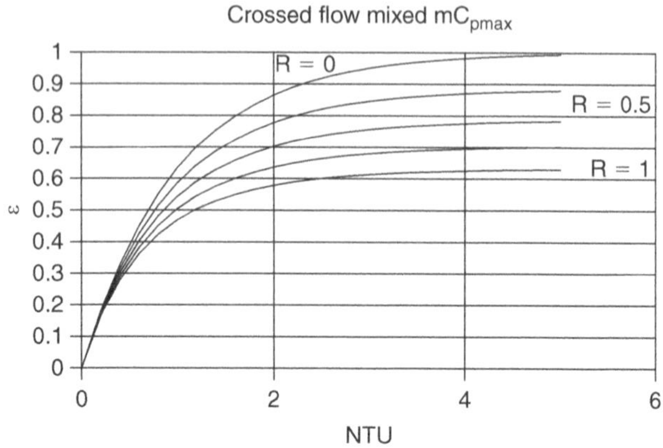

FIGURE 4.2.5 Cross-flow heat exchanger chart (mCp_{max}).

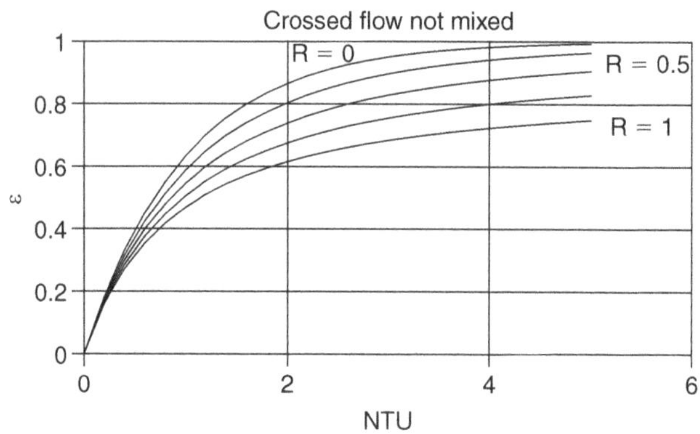

FIGURE 4.2.6 Cross-flow heat exchanger chart (not mixed).

The corresponding chart is given in Figure 4.2.5.

For a non-mixed fluid cross-flow heat exchanger (Holman, 1989), the corresponding chart is given in Figure 4.2.6.

$$\varepsilon = \left[1 - \exp\left(\frac{-\Gamma NTU^{0.22}}{R}\right)\right] \quad \text{with} \quad \Gamma = 1 - \exp(-RNTU^{0.78}) \qquad (4.2.12)$$

4.2.2.4 Shell and Tube Heat Exchangers

For a shell and tube heat exchanger, the following relationships are valid for one shell pass, regardless of the number of tube passes:

$$NTU = \frac{1}{\sqrt{1+R^2}} \ln\left[\frac{2 - \varepsilon(1 + R + \sqrt{1+R^2})}{2 - \varepsilon(1 + R - \sqrt{1+R^2})}\right] \qquad (4.2.13)$$

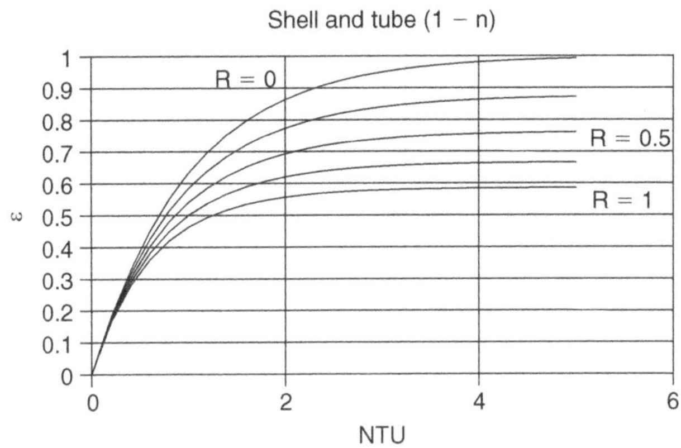

FIGURE 4.2.7 Shell and tube heat exchanger chart (1 – n).

$$\text{Setting} \quad \Gamma = \sqrt{1+R^2}$$

$$\varepsilon = \frac{2}{1+R+\Gamma\dfrac{1+\exp(-NTU\Gamma)}{1-\exp(-NTU\,\Gamma)}}$$

The corresponding chart is given in Figure 4.2.7.

For a multiple-pass exchanger, with the assumption that the fluid is mixed between passes and the NTU is distributed evenly between identical channels placed in an overall counter-flow arrangement, we will establish relationship (4.2.26) (Section 4.2.4.1), which simplifies here to give, where p is the number of passes and ε_p is the effectiveness of each pass, depending on NTU_p:

$$NTU = pNTU_p \qquad \varepsilon = \frac{1-\left[\dfrac{1-R\varepsilon_p}{1-\varepsilon_p}\right]^p}{R-\left[\dfrac{1-R\varepsilon_p}{1-\varepsilon_p}\right]^p} \tag{4.2.14}$$

For example, for a shell and tube heat exchanger with three shell passes and two tube passes, we get the chart given in Figure 4.2.8.

If we seek an overall effectiveness ε, setting $A^p = \dfrac{1-R\varepsilon_p}{1-\varepsilon_p}$, then:

$$A = \left[\frac{1-R\varepsilon}{1-\varepsilon}\right]^{1/p} \qquad \varepsilon_p = \frac{\left[\dfrac{1-R\varepsilon}{1-\varepsilon}\right]^{1/p}-1}{\left[\dfrac{1-R\varepsilon}{1-\varepsilon}\right]^{1/p}-R} \tag{4.2.15}$$

NTU_p is then deduced, followed by NTU.

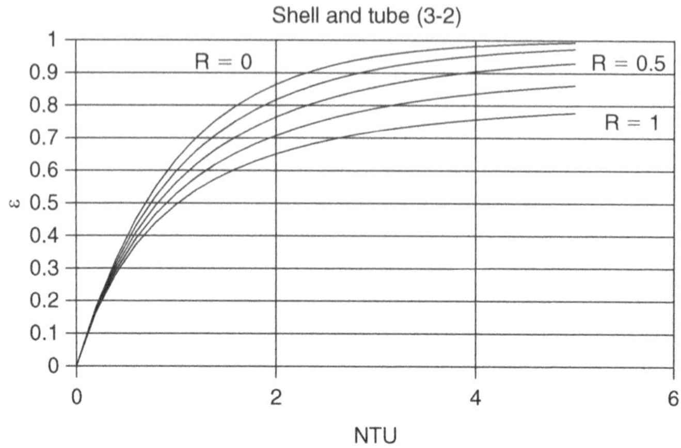

FIGURE 4.2.8 Shell and tube heat exchanger chart (3–2).

If we seek an NTU value, we deduce NTU_P, then ε_P, and finally ε.

4.2.3 Matrix Formulation

In the case where $(\dot{m}c_p)_h < (\dot{m}c_p)_c$, (4.2.6) can be rewritten as:

$$T_{ho} = (1-\varepsilon)T_{hi} + \varepsilon T_{ci} \tag{4.2.16}$$

Moreover, equation (4.2.4) give $\phi = (\dot{m}c_p)_{min}(T_{hi} - T_{ho}) = (\dot{m}c_p)_{max}(T_{co} - T_{ci})$

Given equation (4.2.2):

$$\begin{aligned} T_{co} - T_{ci} &= R(T_{hi} - T_{ho}) = R(T_{hi} - (1-\varepsilon)T_{hi} - \varepsilon T_{ci}) \\ T_{co} &= R\varepsilon T_{hi} + (1 - R\varepsilon)T_{ci} \end{aligned} \tag{4.2.17}$$

(4.2.16) and (4.2.17) can also be expressed in matrix form

$$\begin{pmatrix} T_{ho} \\ T_{co} \end{pmatrix} = \begin{pmatrix} 1-\varepsilon & \varepsilon \\ R\varepsilon & 1-R\varepsilon \end{pmatrix} \begin{pmatrix} T_{hi} \\ T_{ci} \end{pmatrix} \tag{4.2.18}$$

If we had considered a system where $(\dot{m}c_p)_h > (\dot{m}c_p)_c$, the matrix system would have been:

$$\begin{pmatrix} T_{ho} \\ T_{co} \end{pmatrix} = \begin{pmatrix} 1-R\varepsilon & R\varepsilon \\ \varepsilon & 1-\varepsilon \end{pmatrix} \begin{pmatrix} T_{hi} \\ T_{ci} \end{pmatrix} \tag{4.2.19}$$

With $R = \dfrac{(\dot{m}c_p)_c}{(\dot{m}c_p)_h}$

If we define:

$$\varepsilon_h = \inf\left(\varepsilon, \frac{(\dot{m}c_p)_c}{(\dot{m}c_p)_h}\varepsilon\right) \quad \text{and} \quad \varepsilon_c = \inf\left(\varepsilon, \frac{(\dot{m}c_p)_h}{(\dot{m}c_p)_c}\varepsilon\right)$$

we obtain a generalized matrix formulation that remains always valid.

$$\begin{pmatrix} T_{ho} \\ T_{co} \end{pmatrix} = \begin{pmatrix} 1-\varepsilon_h & \varepsilon_h \\ \varepsilon_c & 1-\varepsilon_c \end{pmatrix} \begin{pmatrix} T_{hi} \\ T_{ci} \end{pmatrix}$$

$$\begin{pmatrix} T_{ho} \\ T_{co} \end{pmatrix} = A \begin{pmatrix} T_{hi} \\ T_{ci} \end{pmatrix}$$

This relationship can be modified to express not the outlet temperatures as functions of inlet temperatures, but temperatures from one end of the exchanger in relation to those of the other. We can readily obtain:

$$\begin{pmatrix} T_{hi} \\ T_{co} \end{pmatrix} = \begin{pmatrix} \dfrac{1}{1-\varepsilon_h} & -\dfrac{\varepsilon_h}{1-\varepsilon_h} \\ \dfrac{\varepsilon_c}{1-\varepsilon_h} & 1-\dfrac{\varepsilon_c}{1-\varepsilon_h} \end{pmatrix} \begin{pmatrix} T_{ho} \\ T_c \end{pmatrix} = B \begin{pmatrix} T_{ho} \\ T_c \end{pmatrix} \tag{4.2.21}$$

A third formulation is used to express the temperatures of the hot fluid in terms of those of the cold fluid:

$$\begin{pmatrix} T_{ho} \\ T_{hi} \end{pmatrix} = \begin{pmatrix} \dfrac{1-\varepsilon_h}{\varepsilon_c} & 1-\dfrac{1-\varepsilon_h}{\varepsilon_c} \\ \dfrac{1}{\varepsilon_c} & 1-\dfrac{1}{\varepsilon_c} \end{pmatrix} \begin{pmatrix} T_{co} \\ T_{ci} \end{pmatrix} = C \begin{pmatrix} T_{co} \\ T_{ci} \end{pmatrix} \tag{4.2.22}$$

4.2.4 Heat Exchanger Assemblies

When we cannot achieve the desired heat transfer in a single exchanger, it is possible to combine various heat exchangers, either in series or in series–parallel.

4.2.4.1 Assembly in Series

Given their poor performance, parallel-flow heat exchangers are not put in series. Consider an assembly of heat exchangers in series.

Let us call T_{hi}^j and T_{ho}^j the inlet and outlet temperatures of the hot fluid in exchanger j, T_{ci}^j and T_{co}^j those of the cold fluid, respectively. Given the serial arrangement, we have:

$$T_{co}^j = T_{ci}^{j+1} \quad \text{and} \quad T_{hi}^j = T_{ho}^{j+1}$$

Let as before

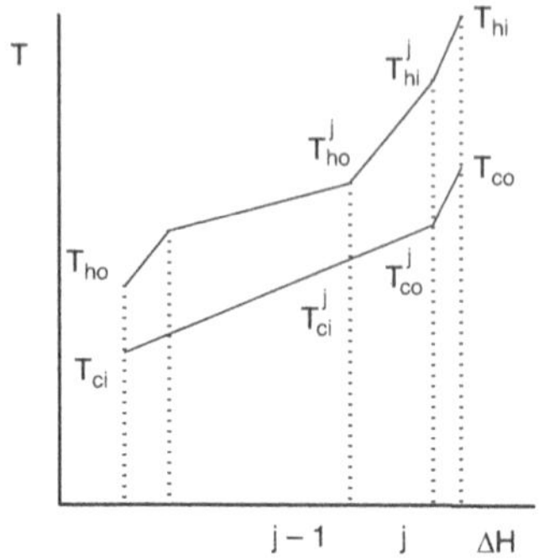

FIGURE 4.2.9 Assembly in series.

$$\varepsilon_h^j = \inf\left(\varepsilon^j; \frac{(\dot{m}c_p)_{ci}}{(\dot{m}c_p)_{hi}}\varepsilon^j\right)$$

$$\varepsilon_c^j = \inf\left(\varepsilon^j; \frac{(\dot{m}c_p)_{hi}}{(\dot{m}c_p)_{ci}}\varepsilon^j\right)$$

We have, using formulation (4.2.21):

$$\begin{pmatrix} T_{hi}^j \\ T_{co}^j \end{pmatrix} = \begin{pmatrix} \dfrac{1}{1-\varepsilon_h^j} & -\dfrac{\varepsilon_h^j}{1-\varepsilon_h^j} \\ \dfrac{\varepsilon_c^j}{1-\varepsilon_h^j} & 1-\dfrac{\varepsilon_c^j}{1-\varepsilon_h^j} \end{pmatrix} \begin{pmatrix} T_{ho}^j \\ T_{ci}^j \end{pmatrix} = B_j \begin{pmatrix} T_{ho}^j \\ T_{ci}^j \end{pmatrix} \quad (4.2.23)$$

$$\begin{pmatrix} T_{ho}^{j+1} \\ T_{ci}^{j+1} \end{pmatrix} = B_j \begin{pmatrix} T_{ho}^j \\ T_{ci}^j \end{pmatrix} \quad \text{if} \quad i < n$$

$$\begin{pmatrix} T_{hi} \\ T_{co} \end{pmatrix} = B_n \begin{pmatrix} T_{ho}^n \\ T_{ci}^n \end{pmatrix} \quad \text{if} \quad j = n$$

$$(4.2.24)$$

$$\begin{pmatrix} T_{hi} \\ T_{co} \end{pmatrix} = B_n \begin{pmatrix} T_{ho}^n \\ T_{ci}^n \end{pmatrix} = B_n B_{n-1} \begin{pmatrix} T_{ho}^{n-1} \\ T_{ci}^{n-1} \end{pmatrix}$$

$$\begin{pmatrix} T_{hi} \\ T_{co} \end{pmatrix} = \prod_{i=n}^{i=1} B_j \begin{pmatrix} T_{ho} \\ T_{ci} \end{pmatrix}$$

*Knowing the hot and cold inlet temperatures in the system, this formula directly gives the corresponding outlet temperatures.

The inverse of the matrix B_j equals:

$$B_j^{-1} = \begin{pmatrix} 1-\dfrac{\varepsilon_h^i}{1-\varepsilon_c^i} & \dfrac{\varepsilon_h^i}{1-\varepsilon_c^i} \\ -\dfrac{\varepsilon_c^i}{1-\varepsilon_c^i} & \dfrac{1}{1-\varepsilon_c^i} \end{pmatrix}$$

$$\det\left(B_j\right) = \frac{1-\varepsilon_c^j}{1-\varepsilon_h^i}$$

The product $B_j B_k = B_{jk}$, which represents the heat exchanger obtained by connecting exchangers j and k in series, in that order, is such that:

$$\varepsilon_c^{jk} = 1 - \frac{\left(1-\varepsilon_c^j\right)\left(1-\varepsilon_c^k\right)}{1-\varepsilon_h^j \varepsilon_c^k}$$

$$\varepsilon_h^{jk} = 1 - \frac{\left(1-\varepsilon_h^j\right)\left(1-\varepsilon_h^k\right)}{1-\varepsilon_h^j \varepsilon_c^k}$$

It is clear that $B_j B_k \neq B_k B_j$

Moreover, if we subtract the second line of (4.2.23) from the first, we get

$$T_{ho}^{j+1} - T_{ci}^{j+1} = \frac{1-\varepsilon_c^j}{1-\varepsilon_h^j}\left(T_{ho}^j - T_{ci}^j\right) \tag{4.2.25}$$

For n exchangers:

$$T_{ho}^{n+1} - T_{ci}^{n+1} = \prod \frac{1-\varepsilon_c^j}{1-\varepsilon_h^j}\left(T_{ho}^1 - T_c^0\right)$$

For a counter-flow arrangement of n exchangers containing fluids with identical and constant thermophysical properties, whose individual effectiveness is ε_i, we deduce that the overall effectiveness ε can be expressed as

$$\varepsilon = \frac{1 - \prod_{i=1}^{i=n} \dfrac{1-R\varepsilon_i}{1-\varepsilon_i}}{R - \prod_{i=1}^{i=n} \dfrac{1-R\varepsilon_i}{1-\varepsilon_i}} \tag{4.2.26}$$

If $R = 1$, this expression becomes

$$\varepsilon = \frac{\sum_{i=1}^{i=n} \frac{\varepsilon_i}{1-\varepsilon_i}}{1-\sum_{i=1}^{i=n} \frac{\varepsilon_i}{1-\varepsilon_i}} \tag{4.2.27}$$

These relationships lead to the following addition rule: the overall NTU corresponding to overall effectiveness ε equals the sum of NTU_j of each module corresponding to its partial effectiveness εj.

4.2.4.2 Assembly in Series–Parallel

With the same assumptions as above regarding the constancy of fluid thermophysical properties, two cases may arise, depending on whether the fluid that passes through heat exchangers in series is the hot or cold fluid.

Let us suppose here that the cold fluid passes horizontally in series through all exchangers while the hot fluid is divided into n parallel branches.

Relation (4.2.20) gives here

$$\begin{pmatrix} T_{ho}^j \\ T_{co}^j \end{pmatrix} = A_j \begin{pmatrix} T_{hi}^j \\ T_{ci}^j \end{pmatrix}$$

$$\text{with} \quad T_{ci}^j = T_{co}^{j-1}$$

We can obtain a recurrence matrix relation as follows:

$$\begin{pmatrix} T_{ho}^j \\ T_{hi} \\ T_{co}^j \end{pmatrix} = \begin{pmatrix} 0 & 1-\varepsilon_h^j & \varepsilon_h^j \\ 0 & 1 & 0 \\ 0 & \varepsilon_c^j & 1-\varepsilon_c^j \end{pmatrix} \begin{pmatrix} T_{ho}^{j-1} \\ T_{hi} \\ T_{co}^{j-1} \end{pmatrix}$$

$$\begin{pmatrix} T_{ho}^j \\ T_{hi} \\ T_{co}^j \end{pmatrix} = D_j \begin{pmatrix} T_{ho}^{j-1} \\ T_{hi} \\ T_{co}^{j-1} \end{pmatrix}$$

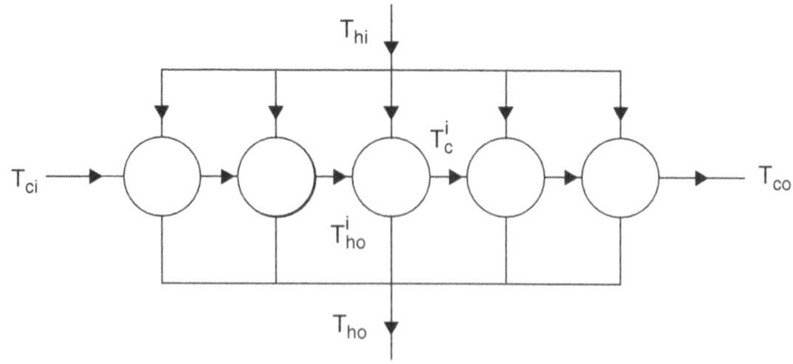

FIGURE 4.2.10 Assembly in series–parallel.

$$D_jD_k = D_{jk} = \begin{pmatrix} 0 & 1-\varepsilon_h^j+\varepsilon_c^k\varepsilon_h^j & \varepsilon_h^j\left(1-\varepsilon_c^k\right) \\ 0 & 1 & 0 \\ 0 & \varepsilon_c^j+\varepsilon_c^k\left(1-\varepsilon_c^j\right) & \left(1-\varepsilon_c^j\right)\left(1-\varepsilon_c^k\right) \end{pmatrix}$$

$$1-\varepsilon_c^{jk} = \left(1-\varepsilon_c^j\right)\left(1-\varepsilon_c^k\right)$$

$$\varepsilon_h^{jk} = \varepsilon_h^j\left(1-\varepsilon_c^k\right) = \frac{\varepsilon_h^j}{\left(1-\varepsilon_c^j\right)}\left(1-\varepsilon_c^j\right)\left(1-\varepsilon_c^k\right) = \frac{\varepsilon_h^j}{\left(1-\varepsilon_c^j\right)}\left(1-\varepsilon_c^{jk}\right)$$

The hot fluid outlet temperature T_{ho} is obviously the barycentric average of the T_{ho}^j, weighted by the heat capacity rates in each branch.

The relationship providing the overall effectiveness is here:

$$1-\varepsilon = \prod_{i=1}^{i=n}(1-\varepsilon_i) \tag{4.2.28}$$

knowing that ε should always be calculated for the fluid passing through heat exchangers in series. This relationship can be transformed by noting that $NTU = -\ln(1-\varepsilon)$ for $R = 0$, and leads to the following addition rule: the overall NTU of a set of exchangers in series–parallel is obtained by summing the NTU_i related to the fluid flowing through them in series, calculated for a heat capacity rate ratio $R = 0$.

$$NTU(\varepsilon,R=0) = \sum_{i=1}^{i=n} NTU_i(\varepsilon,R=0) \tag{4.2.29}$$

4.2.5 Relationship with the LMTD Method

The log-mean temperature difference method (LMTD) starts from equation (4.1.3):

$$\phi = UAF\Delta T_{ml}$$

It is possible to switch from one calculation method to another using the relationship:

$$\Delta T_{ml} = \frac{\varepsilon \Delta T_i}{NTU}$$

4.2.6 Heat Exchanger Pinch

A heat exchanger pinch is defined as the minimum temperature difference between both fluids. For a simple exchanger without phase change, it is:

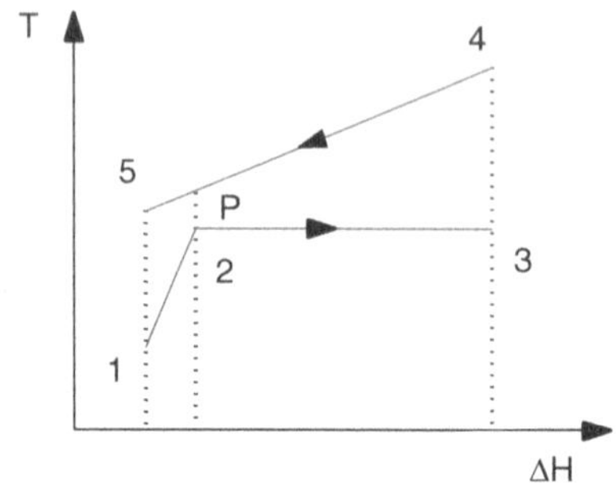

FIGURE 4.2.11 Pinch point.

$$\left(T_{ho} - T_{ci}\right) \quad \text{if} \quad (\dot{m}c_p)_h \leq (\dot{m}c_p)_c$$

$$\left(T_{hi} - T_{co}\right) \quad \text{if} \quad \left(\dot{m}c_p\right)_h \geq \left(\dot{m}c_p\right)_c$$

Consider the graphical representation of a double exchanger consisting of an economizer followed by an evaporator, in the (T, ΔH) diagram where temperature is the ordinate and enthalpies involved are the abscissa (Figure 4.2.11).

The fluid that evaporates is represented by the lower curve (1-2-3), which presents an inflection point 2 corresponding to the onset of boiling. The hot fluid cools from 4 to 5, decreasing its sensible heat.

On segments $(1-2)$ and $(4-5)$ we have $\Delta H = \dot{m}c_p\Delta T$, and thus $\Delta T = \frac{1}{\dot{m}c_p}\Delta H$

The slopes of segments (1–2) and (4–5) equal the inverse of the heat capacity rates of the corresponding fluids. For segment (2–3), we obviously have $\Delta T = 0$, with vaporization taking place at constant temperature for a pure component or azeotrope.

At point 2 on the diagram, P represents the minimum temperature difference between the two fluids, called the pinch. This point plays a fundamental role in the design of heat exchangers, as it represents the smallest temperature difference in the device and thus corresponds to the most constrained area of the system.

The importance of the pinch is most evident in problems of complex heat exchanger network design, to the point that powerful optimization methods based on this concept have been developed. We present in Section 3.4.5 of this set's volume 2 the Thermoptim method derived from the pinch method, which aims to find the best integration of the entire network.

The exchanger, sized according to the pinch, operates outside this point with temperature differences greater than the minimum pinch between the two fluids. The irreversibilities depend directly on the temperature difference ΔT between the hot and cold fluids. They are even lower when the slopes of hot and cold fluids are close to one another, meaning that the R ratio of their heat capacity rates is close to unity. The heat exchanger that has the least irreversibility is one whose heat capacity rates are equal and whose temperature difference equals the minimum allowable pinch.

4.3 CALCULATION OF HEAT EXCHANGERS IN THERMOPTIM

In Thermoptim, a heat exchanger is not represented by a particular component, but by a connection established between two "exchange" processes representing the hot and cold fluids. This way of working has many advantages. In particular, it is possible to design fluid networks at the hydraulic level, calculate their thermodynamic state, and thermally connect them only at a later time. For example, when attempting to optimize the design of a network of heat exchangers, it is not necessary to define a priori fluid pairings. In addition, it is clear that a heat exchanger aims precisely at linking changes in two fluids, which is consistent with the mode of representation adopted in the software.

We begin by presenting the "exchange" processes, and then we show how they can be connected to form a heat exchanger.

4.3.1 "Exchange" Processes

An "exchange" process is used to calculate the heating or cooling of a fluid between two states represented by upstream and downstream points.

Its screen (Figure 4.3.1) is similar to that of the compression process, with fewer options. You should refer to the beginning of Section 3.1.5 for an overview of the process screens.

The middle-right part displays the following:

- A field that sets the minimum accepted pinch for this fluid. This parameter can be used in sizing heat exchangers.

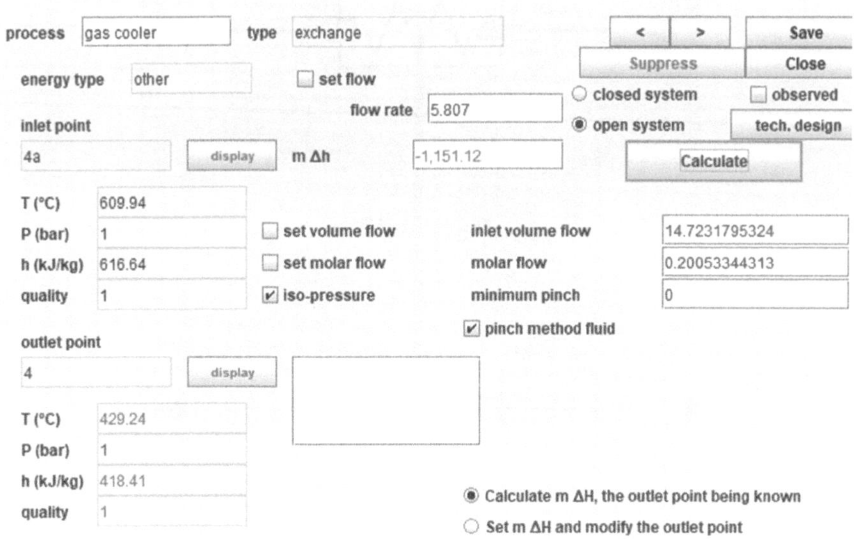

FIGURE 4.3.1 Exchange process.

- An option "pinch fluid method" used to indicate that the process must be taken into account in optimization calculations by thermal integration (see Volume 2 of the reference manual of the software).

Two complementary calculation methods can be chosen according to the option selected in the bottom right of the screen:

- "Calculate $\dot{m}\Delta H$, the outlet point being known" assesses the enthalpy change Δh involved in the process.
- "Set $\dot{m}\Delta H$ and modify the outlet point" recalculates downstream point temperature to ensure that the enthalpy change Δh involved in the process equals the value entered in the $\dot{m}\Delta H$ field.

4.3.2 Creation of a Heat Exchanger in the Diagram Editor

To represent the heat exchanger in the diagram editor, we use non-oriented links connecting two components of type "exchange."

An "exchange" component has at its center a small maskable exchanger connection port (Figure 4.3.2), which appears (as a small blue or red circle) only when the mouse is located above it, or when it is connected with another such component to represent a heat exchanger.

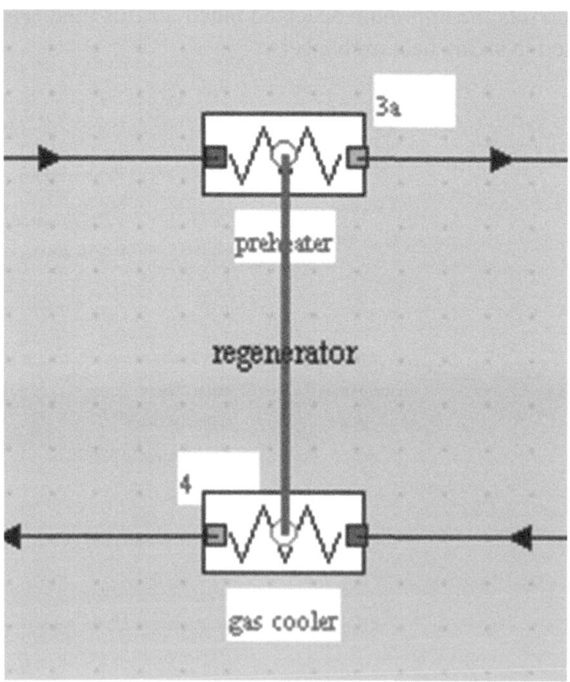

FIGURE 4.3.2 Exchanger link.

To make the connection, click on this port of one component, drag the mouse to the same port of another component, and then release the button. During the connection, the name of the exchanger is requested. Once the heat exchanger settings are entered, the port is blue for the cold fluid and red for the hot fluid.

A heat exchanger is created in the simulator when an exchanger link exists in the diagram and the two "exchange" components that it connects correspond to two "exchange" processes in the simulator that are sufficiently well defined (it is particularly necessary that their inlet and outlet temperatures are different, that they have been calculated so that the value of ΔH is not zero, and that one is warming while the other cools). By default, the exchanger is initialized with a "counter-flow" type.

4.3.3 Heat Exchanger Screen

The connection is done in two stages, one for the hot part, and one for the cold part.

The screen (Figure 4.3.3) contains information about the hot fluid in the middle left part, while that for the cold fluid is in the right part. To connect a fluid, if not done in the diagram editor, you must double-click the name field, and then choose the process from the list. Besides the values of temperatures, flow rates, heat capacities, and enthalpies involved, constraints exist on the temperatures and flow rates used to manage the heat exchanger calculations, allowing distinction between the variables of the problem, those set and those to be calculated.

Possible types of exchangers are: counter-flow, parallel, cross-flow (mixed or not mixed), and multi-pass shell and tube (p-n).

In the bottom left, there are three buttons to optionally specify the absence or presence of implicit constraints on temperatures (see below). In the lower right are options for defining the calculation method ("design" or "off-design").

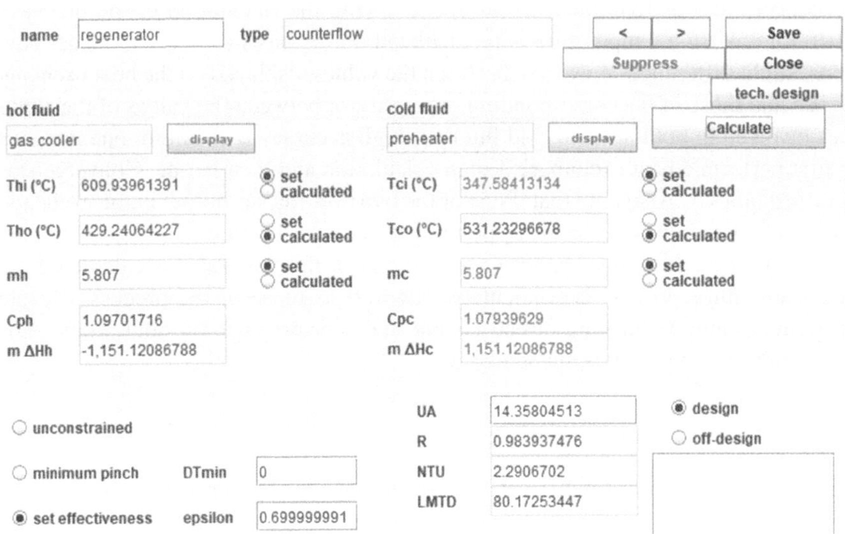

FIGURE 4.3.3 Heat exchanger screen.

4.3.4 Simple Heat Exchanger Design

A heat exchanger connects two "exchange" processes. One of them is a hot fluid which corresponds to a substance that cools, while the other is a cold fluid that heats up. Once the coupling of the processes is made, the design problem is as follows: it is necessary on the one hand to conserve enthalpy in the heat exchanger, and on the other hand to respect certain temperature constraints.

Given that there are four temperatures (two for each fluid) and two flow rates, the problem has five degrees of freedom once enthalpy is conserved. It can furthermore be shown that at least one of the two flow rates must be set, otherwise the problem is indeterminate.

For temperatures, one can either set explicit constraints (for example the fluid inlet temperatures), or implicit constraints (the value of the heat exchanger effectiveness, or the minimum pinch value). To set an effectiveness value, enter it in the ε field and click on button "set effectiveness." To set a "minimal pinch," click on the corresponding button. The minimal pinch value is initialized, when the heat exchanger is created, or when one double-clicks in this field, with a value equal to the half sum of the minimum pinches defined in both hot and cold processes (see Section 4.3.1).

For the problem to have a solution, it is necessary to set a total of five constraints, among which there is at least a flow rate. If one of them is implicit (effectiveness or minimal pinch), four must be explicit (three temperatures and one flow rate, or two temperatures and two flow rates), or otherwise all five must be explicit (a single flow rate or a single temperature is free). These conditions are necessary but not sufficient. This is why the software analyzes all the constraints proposed. If there is a solution, it is found. Otherwise, a message tells the user that the calculation is impossible.

Note that the design of heat exchangers is always made with the implicit hypothesis that the thermophysical properties of the fluids remain constant in the heat exchanger, but this hypothesis is not made during the calculation of the processes. Therefore, when a temperature is recalculated on the basis of heat exchanger equations, slight differences may exist between the values displayed on the heat exchanger screen and those of the corresponding processes, or between the values of the enthalpies involved in both hot and cold fluids. If high accuracy is required, one may need to first perform a heat exchanger design calculation and then iterate. Convergence is obtained quickly. Also note that if one of the two flow rates is not set in the exchanger, it is recalculated, even if the corresponding process is "set flow."

When a heat exchanger is designed and one of the flow rates is calculated, that of the exchange process is recalculated, as well as those of the processes located upstream as long as they are directly connected. The propagation stops when a node or a combustion process is encountered.

4.3.5 Thermocouplers

Thermocouplers complement heat exchangers by allowing components other than exchange processes to be connected to one or more exchange processes in order to represent a thermal coupling. This mechanism does not replace the exchanger mechanism: two exchange processes cannot be directly connected by a thermocoupler.

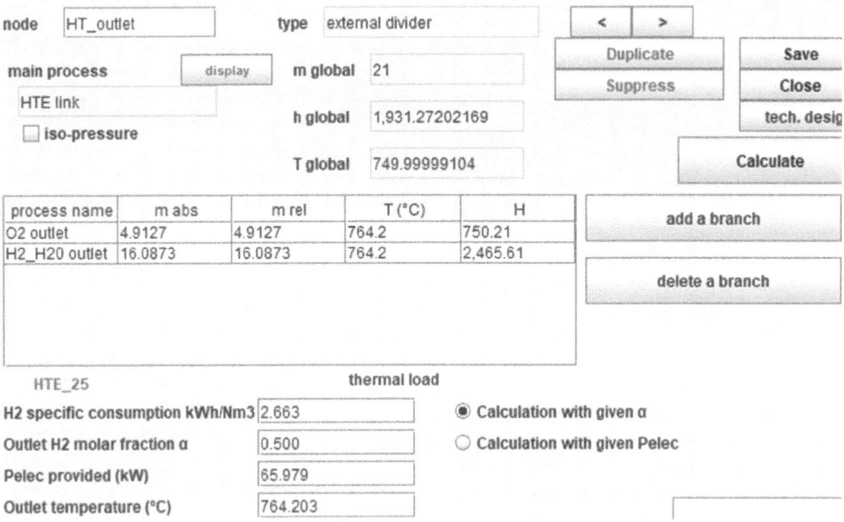

FIGURE 4.3.4 High-temperature electrolyzer screen.

This mechanism has a number of benefits, because it can be used to represent many heat transfers that do not constitute a heat exchanger in the traditional sense, such as, for example, cooling the walls of the combustion chamber of a reciprocating engine, cooled compression, and above all supply or removal of heat from multi-functional external components.

Figure 4.3.4 illustrates this: a high-temperature electrolyzer involves an endothermic reaction that requires a heat input, which is provided by a high-temperature heat source. The representation of this heat exchange is possible thanks to the thermocoupler mechanism: the external node calculates the thermal energy that must be exchanged, and the thermocoupler recalculates the corresponding "exchange" process, which updates its downstream point.

The types of thermocouplers used by an external component appear in the lower right corner of the screen. Double-click on one of the types (here a single one: "thermal load") to open the screen of the corresponding thermocoupler.

Given that thermocouplers are a type of heat exchanger, it is valuable, as stated in Section 4.3.1, to define them by values such as effectiveness ε, UA, NTU, or LMTD, that can be calculated using similar equations. The component sends to each of its thermocouplers the equivalent values for flow rates, inlet and outlet temperatures, and thermal energy transferred, which they must take into account in their calculations. Specific methods are provided in the external class code and are not user-modifiable.

However, there are limits to the similarities with exchangers: for example, temperature crossovers unacceptable in an exchanger may occur in a thermocoupler, leading to absurd values.

It is therefore best to transmit values that are unlikely to lead to this type of situation. One possible solution is to assume that the thermocoupler is isothermal for calculations of characteristics similar to exchanger characteristics. For example,

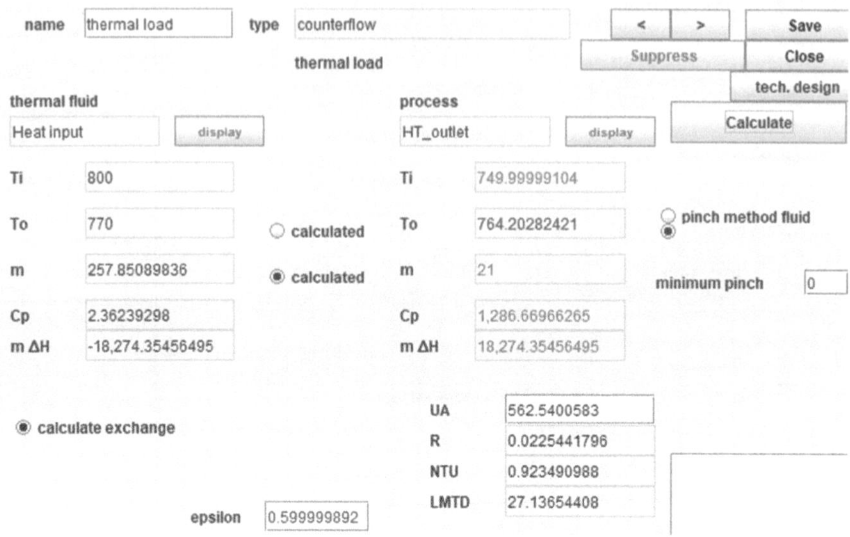

FIGURE 4.3.5 High-temperature electrolyzer thermocoupler screen.

a combustion chamber may be assumed to be at mean temperature between its upstream and downstream points when calculating the cooling. This assumption is not absurd and may prevent a temperature crossover between the cooling fluid and the gases that pass through the component.

The electrolyzer thermocoupler screen is given in Figure 4.3.5.

For external processes and for external nodes that accept several thermocouplers, the potential complexity of the calculations prevents the connected exchange process from controlling the thermocoupler calculations. Its thermal load is always set by the external component. This is why there are fewer options for calculating a thermocoupler than for a heat exchanger: the user can only choose between calculating the outlet temperature of the exchange process (at a given flow rate) or calculating the flow rate when the temperature is known.

Note that on the thermocoupler screen, the external component fluid can be selected as a pinch fluid and a minimum pinch value can be entered.

Finally, note that the off-design calculation method is not available for thermocouplers, as their behavior can deviate significantly from that of a conventional heat exchanger.

4.4 CONCLUSION

This chapter presented the principles and calculation methods for heat exchangers, devices that transfer heat between two fluids at different temperatures.

Following the introduction of basic principles, the fundamental equation for heat flux was established using the LMTD and the overall heat transfer coefficient (U), including the effects of fins and convection correlations.

The core of this chapter focused on the NTU method by Kays and London, which relates the heat exchanger effectiveness to design parameters. Detailed NTU–effectiveness relationships were provided for various configurations: counter-flow, parallel-flow (co-current), cross-flow, and shell and tube heat exchangers.

Matrix formulation techniques were introduced to analyze complex heat exchanger networks. The concept of pinch point was explained as a critical design constraint.

The practical implementation within Thermoptim concluded this chapter, establishing a bridge between theoretical foundations and engineering practice for system analysis and detailed thermal design.

Examples illustrating how the concepts presented in this chapter can be applied using Thermoptim are available on this book's website.[1]

NOTE

1 Available at: https://server.s4e2.com/crc/amtc/chapters/chapter4/_resources/index.html.

BIBLIOGRAPHY

Incropera, F.P. & DeWitt, D.P. (1996). *Fundamentals of heat and mass transfer*, 4th ed. John Wiley & Sons, Hoboken, NJ.

Kakac, S. & Liu, H. (2002). *Heat exchangers: Selection, rating, and thermal design*, 2nd ed. CRC Press, Boca Raton, FL. ISBN 0-8493-0902-6.

Kays, W.M. & London, A.L. (1984). *Compact heat exchangers*. McGraw-Hill.

Shah, R.K. & Bell, K.J. (2000). "Heat exchangers." In Kreith, F. (ed.), *The CRC handbook of thermal engineering*. CRC Press, Boca Raton, FL, pp. 4-118–4-170. ISBN 0-8493-9581-X.

Smith, E.M. (1997). *Thermal design of heat exchangers: A numerical approach*. Wiley, Chichester.

5 External Class Development

INTRODUCTION

A few years after the publication of the first edition of our book on energy systems, the dual analytical/systems modeling approach that makes possible the new pedagogical paradigm initiated at École des Mines de Paris was already used by more than 100 higher education institutes. The reason it was adopted so widely is that it meets a need, as students become increasingly opposed to traditional thermodynamics teaching methods.

Based on the experience of its users and the expectations they raised, it became desirable to extend the capabilities of the tool so that it could solve problems more difficult than those for which it was originally designed: advanced modeling of certain energy systems was beyond the capabilities of Thermoptim versions released before late 2002, their limits being in two main dimensions:

- First, the area covered by the package: only predefined core components available on the diagram editor palette could be used in energy systems studied. Even though they allowed representation of a very large number of conventional energy technologies, some could not be modeled in Thermoptim, including new systems with low environmental impact. Similarly, some users had expressed the wish to modify models available in the tool by introducing their own equations, which was then impossible.
- Second, the modeling refinement: phenomenological models built certainly allowed one to study thermodynamic cycles of the technologies studied, but not to perform specific component sizing or to simulate off-design behavior, the latter two issues being much more complex than the first.

New features have been incorporated into the software to allow it to exceed these limits.

We limit ourselves here to a brief presentation of these extensions. More detailed explanations are available through links on this book's website.[1]

5.1 GENERAL, EXTERNAL SUBSTANCES

5.1.1 Introducing Custom Components

5.1.1.1 General Principle of Thermoptim External Classes

Thermoptim's graphical environment allows users to visually build models of energy systems, from the simplest (such as a refrigerator) to the most complex (such as power

DOI: 10.1201/9781003606017-6

plants). This approach simplifies modeling, makes it easier to use and maintain models, and automatically ensures consistent connections between components.

However, Thermoptim's core only includes a certain number of the most common standard components (compressors, pumps, turbines, heat exchangers, combustion chambers, fluids, etc.). Since its developers cannot anticipate every possible use case, advanced users can create their own elements using what are known as **external classes**.

In practice, an external class is a small piece of Java code added to the software. It defines a new element (such as a new fluid, a solar collector, or a custom component) that integrates seamlessly into Thermoptim's interface, as if it were part of the original software.

This mechanism must adhere to certain constraints:

1. Thermoptim's core is protected: its code is intentionally obfuscated (renamed and compressed) to prevent copying or modification, meaning internal classes cannot be directly accessed or extended.
2. To work around this, some parts of the software remain accessible via a bridge called the extThopt package.
3. External classes connect to these entry points and are automatically recognized by Thermoptim when the software is launched.

Thermoptim thus opens the door to additional modules, but in a controlled, safe, and limited manner.

5.1.1.2 External Classes Enable Users To

1. **Extend Thermoptim:** Create new fluids or original components (e.g., a solar collector, a chemical absorber) as transparently as a core component.
2. **Control Thermoptim Externally:** Access its thermodynamic libraries or coordinate calculation sequences to perform optimizations or solve coupled equation systems.

5.1.1.3 Implementation in Simple Terms

Technically, adding an external class is relatively straightforward for a programmer: you create a Java class that inherits from an abstract parent class provided by Thermoptim (such as ExtProcess for components or ExtSubstance for fluids). Interaction is achieved through:

1. Methods for calculations (equations of the new model).
2. A graphical panel (JPanel) integrated into Thermoptim's interface, where you define your custom interface (buttons, fields, etc.).

Thermoptim automatically loads these classes at startup by scanning two files where they are stored (extThopt2.zip and extUser2.zip). Once loaded, they are sorted and integrated into the software's internal lists.

External components are of four types, depending on the number of inlet and outlet streams:

- External processes have a single stream as the inlet and one as the outlet, whose substance can be either the same as the inlet or another substance.
- External mixers are connected to multiple inlet streams, usually associated with different substances, and only one at the outlet.
- External dividers are connected to a single inlet stream and several at the outlet, usually associated with different substances.
- Finally, components connected to multiple streams at the inlet as well as at the outlet, which form a multipole, are represented by the combination of an external inlet mixer and an external outlet divider, connected by a process-point, as direct connection between two nodes is not possible in Thermoptim for reasons of logical consistency.

5.1.1.4 Challenges and Solutions

Creating an external class may seem daunting because the communication rules between the protected core and extensions are strict:

1. Some methods are obfuscated (and thus inaccessible).
2. Only a few are left visible, and their syntax must be followed precisely. This ensures consistency and security but can be intimidating for beginners.

Fortunately, in practice, it is quite easy to start from an existing example close to what you want to achieve and modify it to meet your needs. You simply adapt the code defining the visual interface (the JPanel) and the calculations, which is usually not too difficult.

This is especially true since developing a new external class can be done in a user-friendly development environment like Eclipse, which greatly facilitates loading into Thermoptim's interface through an emulation mechanism that allows dynamic testing of the class as it is being developed.

When programming in Java, the most delicate part requiring some experience is defining the class architecture. Since this architecture is largely predefined by Thermoptim's extension mechanism, the adaptation work is generally quite limited. For example, for a component like SolarCollector, you override the methods for specific calculations (efficiency, temperature, etc.) and add your parameters to the panel.

External classes are distributed according to open-source principles to facilitate their implementation and help build libraries of components that are reusable and easily customizable if the need arises, by a community as large as possible.

The model library under development already contains a score of documented models (solar collector, diffuser, nozzle, absorber and desorber for the lithium bromide–water pair, shift CO reaction, fuel cells, reformer) and will be completed in the coming years.

We would like to point out that the external classes we provide have not been fully tested, and users should carefully verify that the implemented models meet their specific needs. They are distributed in the spirit of open source, *as is*, and without any warranty of proper functioning or suitability for particular purposes.

It is obvious that the design of such external classes requires more knowledge of thermodynamics and Thermoptim software environments than those necessary for simply assembling predefined components. Package documentation specifies rules to follow and provides a number of examples but cannot go into details of the underlying problem exposed in this book.

In this chapter, we restrict ourselves to some indications on the following classes:

- A simple substance (Dowtherm A) and an external mixture (pair LiBr-H_2O).
- A simple external process modeling a flat plate solar collector.
- One example of a class involving moist mixtures with external nodes: a direct contact cooling tower.

Examples of Thermoptim external controllers will be explained in the case studies.

For further details, refer to Volume 3 of the Thermoptim reference manual. Some code excerpts are given below, but it is recommended that you obtain the complete Java classes available in the Thermoptim-UNIT portal model library.

5.1.2 Simple Substance: Example of Dowtherm A

It is common that one wishes to use in Thermoptim a substance whose properties are very simple to model, such as a thermal fluid not available in the core. We illustrate how to proceed with class DowThermA.java, which represents an oil sold by Dow Chemical that may in particular serve as a thermal fluid for high-temperature solar collectors (100°C–400°C). The substance remains permanently in the liquid state, and calculations are simple: for example, that of the enthalpy as a sole function of temperature, and its inversion.

5.1.2.1 Creation of the Class

To create an external substance, simply subclass extThopt.ExtSubstance.

The very simple constructor initializes molar mass, critical pressure and temperature, and the range of validity of the model:

```
public DowThermA (){
    //data from FLUIDFILE, Dow Chemical calculation software
    //données extraites de FLUIDFILE, logiciel de calcul de Dow Chemical
    super();
    type=getType();
    M=166; PC=31.34; TC=770.15;
    Tmini=288.15;  Tmaxi=680;
    chemForm="C12H10O (73.5% vol), C12H10 (26.5% vol)";
    aCv=0.003037892;//valeurs du Cp [kJ/(kg K)], fonction affine Cp=a*t+b avec
    bCv=1.466457638;//Cp [kJ/(kg K)] values, Cp=a*t+b with t=T-273.15
    aV=-0.430370424;//valeurs du volume [m^3/kg], fonction parabolique inversé
    bV=1041.622415;//volume [m^3/kg] values, reverse parabolic function V=1/(c
    cV=-0.001179466;
    aVisc=1.080162;//valeurs de la viscosité [10^-5 Pa·s], fonction mu=c*t*t+a
    bVisc=-641.5157547;//viscosity [10^-5 Pa·s] values, function mu=c*t*t+a*t+
    cVisc=-7.95033E-04;
    dVisc=234997.5416;
    aCond=-0.00016;//valeurs de la conductivité [W/(m K)], fonction affine lam
    bCond=0.1419;//conductivity [W/(m K)] values, function lambda=a*t+b with t
}
```

Method getCompType() returns the type of substance as it will appear in Thermoptim screens:

```
public String getCompType(){
    return "Dowtherm A";
}
```

Method getClassDescription() allows entry of a brief description of the class, which will appear in the display screen of external classes. Provide some indications on the model chosen and, if possible, refer to more detailed documentation.

```
public String getClassDescription(){
    return "data from FLUIDFILE software by Dow Chemical\n\nauthor : R. Gicqu
}
```

5.1.2.2 Methods for Calculating Substance Properties

The calculation methods are inherited from the parent class, with the parameter values having been provided in the constructor. Calculations of substance properties can be done as indicated below for enthalpy.

5.1.2.2.1 Direct Calculation of Enthalpy

Calculations of specific heat capacity, enthalpy, and entropy as functions of temperature are performed using a linear correlation for specific heat, while the volume is determined from a quadratic function of the density.

```
public double calcH (double T, double P,double x) {
    double V=calcV(T, P,x);
    double TT=T-273.15;
    return bCv*TT+aCv/2.*TT*TT+V*(P-1)*100;
}
```

5.1.2.3 T-Inversion of Enthalpy

The T-inversion of enthalpy or entropy is obtained by a bisection search over the temperature interval defined in the constructor.

```
public double getT_from_hP(double $h,double $P){
    double T=Util.dicho_T (this,$h,$P, "calcH",Tmini, Tmaxi,0.0001);
    return T;
}
```

5.1.3 Coupling to a Thermodynamic Properties Server

When the substance that you want to use in Thermoptim is not as simple as the one we have been studying, writing all the methods for calculating the properties can become a very complex and laborious task. This is particularly true when dealing with mixtures of real fluids.

Considering both that the primary purpose of Thermoptim is to model energy systems, from simple to complex, rather than being software for calculating the thermodynamic properties of substances, and, second, that there are a number of software applications specializing in this type of calculation, we found that the best solution was to combine these two types of environments. Therefore, we have developed an external class sub-mechanism that allows Thermoptim to perform calculations using various thermodynamic property servers.

5.1.3.1 Characteristics of External Mixtures

The names of pure external substances appear directly in the substance selection list screen where they are used to instantiate classes.

For external mixtures, things are much more complex because the same external class can be used to define multiple systems, and each system generates as many compositions as the user wants, except of course if it is a pure substance, in which case both mole and mass fractions are equal to 1. In a file placed in the "mixtures" directory, each class defines all the external systems it offers, which can be modified from Thermoptim by the user.

Like compound gases, the list of external mixtures is stored in a file called "mel_ext.txt," which is updated whenever a new mixture is created and is used to initialize the substance selection screen.

To instantiate an external mixture, you need to:

- Ensure that the system is well defined in the correct file of folder "mixtures."
- Check that the external class is present in extThopt2.zip or extUser2.zip.
- Create the mixture into the editor from a point screen, and then save.

5.1.3.2 Example of an External Mixture: System LiBr-H_2O

The LiBr-H_2O system is a simple example of external mixture definition. The equations of this mixture are given in Section 2.6.3.3.

This mixture, which was modeled in class H_2OLiBrMixture.java, is presented in more detail in Volume 3 of the Thermoptim reference manual.

The constructor in this case performs the same initializations as before and additionally creates Vector vMixtures, which here defines a single system.

Methods allow one to define the class description, the software used, the unit used to define the composition, the proposed system, and the identifier of the class.

The updating of the composition is performed by the updateComp() method, and calculation of the substance state by the CalcPropCorps() method, calling the methods that ensure the resolution of equations (2.6.14) to (2.6.16).

5.1.3.3 Vapor Diagrams for External Mixtures

In order to allow plotting of cycles using external mixtures, vapor diagrams for external mixtures can also be partially plotted in (T,s) and (h,P) coordinates.

The only lines which are displayed are saturation curves as well as either isobars or isotherms depending on the diagram (the diagram in Figure 5.1.1 corresponds to a highly accurate equation of state for CO_2 which can be used to model supercritical CO_2 cycles).

5.2 FLAT PLATE SOLAR COLLECTORS

A flat plate solar collector receives a radiative flux of variable intensity (less than about 1,200 W/m^2), part of which is reflected and absorbed by the glazing. It is crossed by a fluid that heats up and recovers part of the available power, the rest being lost by heat exchange with the surroundings. The collector performance is characterized by two parameters: a radiative flux transmission coefficient representative of reflection losses and absorption, and a heat exchange coefficient with the outside air.

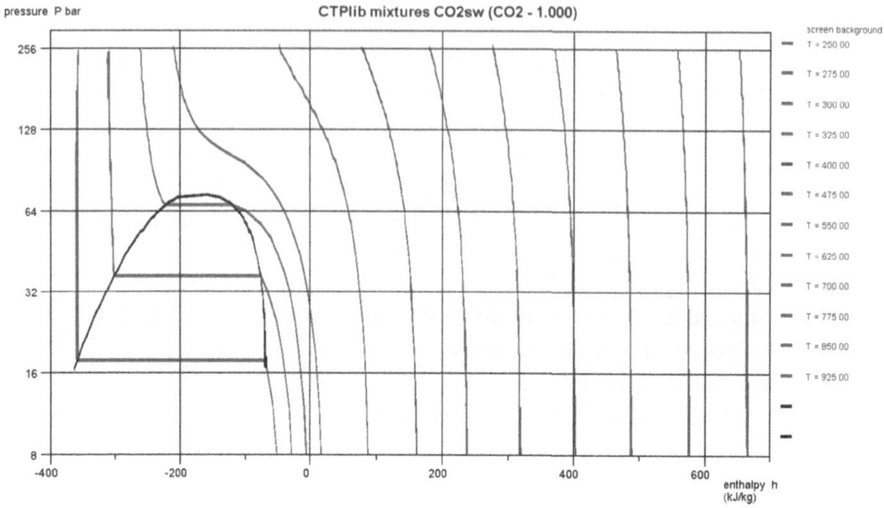

FIGURE 5.1.1 Highly accurate model for supercritical CO_2 cycle analysis.

To be easily comprehensible, the model presented here (SolarCollector.java) is relatively simple in thermodynamic terms and easily calculable with these assumptions. However, it is complicated enough to clearly show the real problems that arise when creating an external component and solutions to resolve them.

5.2.1 Design of the External Component

5.2.1.1 General

Model parameters are:

- the glazing transmittivity τ
- the thermal loss coefficient U (W/m^2/K)
- the incident solar flux G (W/m^2)
- the collector surface A (m^2)
- the outside temperature T_{out} (°C)

The model input data are as follows (provided by other system components):

- The thermal fluid temperature at the collector inlet T_i (°C)
- The flow of the thermal fluid (kg/s)

The outputs are:

- The thermal fluid temperature at the collector outlet T_o (°C)
- The thermal power received by the thermal fluid Q (W/m^2)
- The collector effectiveness

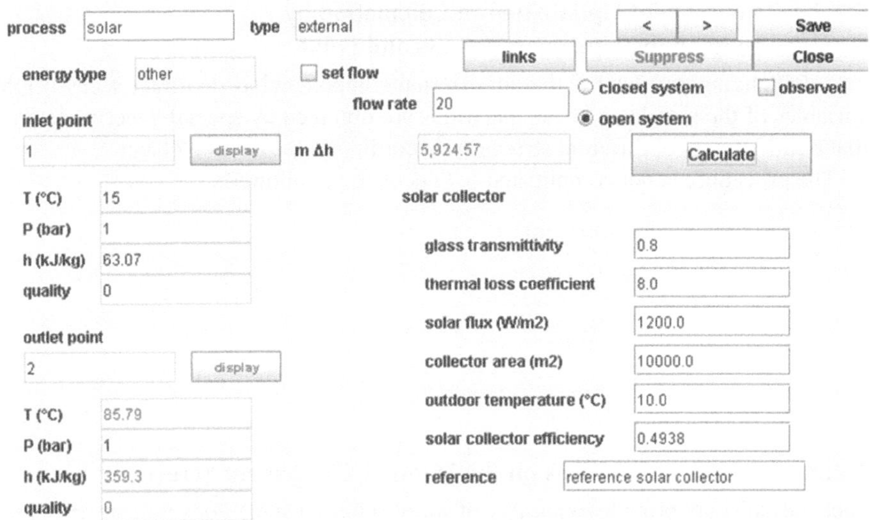

FIGURE 5.2.1 GUI of the external component.

5.2.1.2 Graphical User Interface (GUI)

A graphical interface for the component can be deduced (Figure 5.2.1). It allows building the bottom third of the screen, the rest being defined as a Thermoptim standard.

Parameters correspond to the first five rows added; the calculation results automatically appear in the sixth row, in the downstream point state and in the field ΔH. The input data are supplied by the upstream process of the system in which the component is inserted: thermal fluid flow rate and upstream point state.

5.2.2 Physical Model

With the previous notations, the model equation is:

$$Q = \dot{m}Cp(T_o - T_i) = \dot{m}Cp\left(\tau\frac{G}{U} - T_i + T_{out}\right)\left[1 - \exp\left(-\frac{UA}{\dot{m}Cp}\right)\right]$$

Specifically, the sequence of calculations is as follows:

1. Component update before calculation from the upstream process and point.
2. Reading of settings on the external component screen.
3. Calculation of the thermal power in play and the downstream point state.
4. Update of the external component screen.

Let us now consider the problems encountered in practice at each of these steps.

5.2.2.1 Component Update before Calculation by Loading the Upstream Process and Point

The challenging part here is that an external component has no direct access to the variables of the simulator: these quantities are obtained by special generic methods that build Vectors of different structures according to the desired object.

The procedure is not complicated but needs to be followed:

```
String[] args=new String[2];
args[0]="process";//type of the element (see method getProperties(Strir
args[1]=tfe.getCompName();//name of the process (see method getProperti
Vector vProp=proj.getProperties(args);
Double f=(Double)vProp.elementAt(3);
double flow=f.doubleValue();
String amont=(String)vProp.elementAt(1);//gets the upstream point name
getPointProperties(amont);//direct parsing of point property vector
Ti=Tpoint;
```

5.2.2.2 Reading of Settings on the External Component Screen

Package extThopt provides a number of simple but robust methods for converting the String displayed in the JTextField controls used in the GUI to double, and vice versa for displaying double in these fields. They are implemented as static methods of class extThopt.Util:

```
P=Util.lit_d(P_value.getText());
A=Util.lit_d(A_value.getText());
tau=Util.lit_d(tau_value.getText());
K=Util.lit_d(K_value.getText());
Tout=Util.lit_d(Tex_value.getText())+273.15;
```

5.2.2.3 Calculation of the Thermal Power in Play and the Downstream Point State

We begin by estimating the Cp of the thermal fluid by numerically deriving the enthalpy function:

```
double H=Hpoint;
lecorps.CalcPropCorps(Tpoint+1, Ppoint, Xpoint);//this method recalculates
getSubstProperties(nomCorps);
double Cp=(Hsubst-H);
```

We then estimate the value of the outlet temperature in order to determine the thermal power absorbed Q:

```
double DT0=tau*P/K-Ti+Tout;
double T=Ti+(DT0)*(1-Math.exp(-K/1000*A/flow/Cp));
double DT=T-Tpoint;
double Q=Cp*DT*flow;
```

We determine the value of the enthalpy of the downstream point, and then invert that function to determine the exact value of the outlet temperature (ExtSubstance method):

```
double hAval=Q/flow+Hpoint;
Tpoint=lecorps.getT_from_hP(hAval,Ppoint);
getSubstProperties(nomCorps);
Xpoint=Xsubst;
```

5.2.2.4 Update of the External Component Screen

Thermoptim method setupPointAval() allows updating the downstream point from the values loaded in a Vector built here by the method getProperties() of ExtProcess. Then the value of the collector efficiency is determined and displayed.

```
tfe.setupPointAval(getProperties());
eff_value.setText(Util.aff_d(Q/P*1000/A, 4));
```

5.2.3 Saving and Loading Model Parameters

It is possible to save in the normal Thermoptim project files (and then re-read) the settings of external components using two specific methods:

```
public String saveCompParameters()
public void readCompParameters(String ligne_data)
```

The only constraint is that the entire external component setting must fit into one line, with a format compatible with that used in the core of the package: the various backup fields are separated by tabs (Figure 5.2.2).

extThopt.Util provides a generic method to attribute to the value of a parameter a code allowing it to be identified:

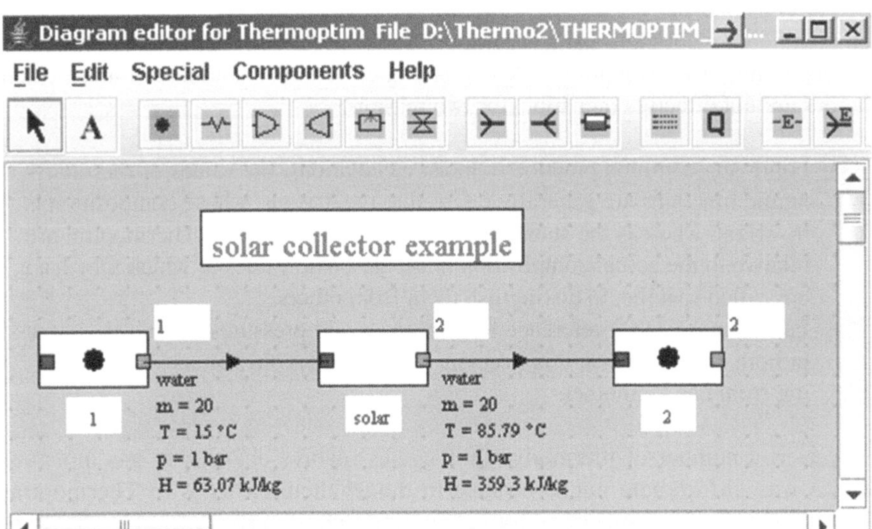

FIGURE 5.2.2 Simulation of a water solar collector.

```
public static String extr_value(String ligne_data, String
search)
```

The backup is done in the form "parameter = value" and the search is performed in the form:

```
value = Util.lit_d(Util.extr_value(ligne_data, "parameter"));
```

If the configuration of the component is too complex to be saved in this way, nothing prevents a user from using it to save the name of a specific setting file and then re-reading this file as desired.

Examples of use of this class are given in the Thermoptim-UNIT portal.

5.3 CALCULATION OF MOIST MIXTURES IN EXTERNAL CLASSES

5.3.1 Introduction

As explained in Section 2.6.2, we call a moist mixture, and by abuse of language **moist gas**, a mixture of a gas that does not condense, which we call the **dry gas**, and **water** that could condense.

Since its inception, Thermoptim has had functions for calculating properties of moist gases and points, but these are generally used for particular calculations decoupled from standard thermodynamic cycle calculations, such as for air conditioning treatment (see Section 3.7). It is for this reason that moist processes have no icon in the diagram editor.

We assume in what follows that the reader is sufficiently familiar with both external class use and moist gas thermodynamics. The definitions, notations, and equations are those in Section 2.6.2.

For convenience, we use the term **standard** when we refer to the Thermoptim environment except moist calculations.

The standard Thermoptim cycle computing environment and that of moist calculations are not directly compatible for two reasons:

- Following common practice in moist calculations, the values of thermodynamic functions are generally referred to the dry gas, whose composition is invariant, whereas the standard calculations performed in Thermoptim are relative to the actual composition of the gas. The values to which they lead are called specific, to distinguish them from others.
- Furthermore, since reference temperatures and pressures are not the same in both environments, it is necessary to perform conversions when switching from one to another.

However, a number of thermodynamic cycles involve changes in gas moisture, and it was unfortunate not to be able to model them easily with Thermoptim. That is why the core functions for calculating the properties of moist gases and points have been made available from external classes. This section explains how to use them.

A moist gas can be represented in Thermoptim in two equivalent ways:

- Either directly as a compound substance comprising at least two components: H_2O and another gas, pure or compound,
- Or as a dry gas of known specific humidity.

The first way has the advantage that the composition of moist gas is available at any time. However, it implies creating a new substance for each value of moisture for the same dry gas. The second presentation is itself much more concise, since it only uses the invariant gas and the humidity value. This is why it is used by moist gas calculation functions, while the first is the rule in the standard Thermoptim cycle computing environment.

Let us recall that we call **relative humidity** ε the ratio of the partial pressure of water vapor divided by its saturation vapor pressure at the temperature of the mixture, and that, by definition, with index dg corresponding to the dry gas, **absolute or specific humidity** w is equal to the ratio of the mass of water contained in a given volume of moist mixture to the mass of dry gas contained in this volume, namely:

$$w = \frac{y_{H_2O}}{y_{dg}} = \frac{M_{H_2O} x_{H_2O}}{M_{dg} x_{dg}}$$

This relationship allows the calculation of w when we know the moist gas composition.

5.3.2 Methods Available in the External Classes

Generally, moist gas calculations are performed (from external classes) at one point, but a method allows one to directly modify the humidity of a gas. The gas humidity is entered by giving either its absolute humidity w or its relative humidity ε. Both methods are defined in class ExtProcess, from which derive all external processes and nodes.

Method **updatepoint("pointName," ...)** can perform moist calculations while method **getPointProperties("pointName")** recovers in addition to the standard properties (P, T, h, s ...) the moist properties of a point with the following values: Wpoint for absolute humidity w, Epsipoint for relative humidity ε, Qprimepoint for specific enthalpy q′, Tprimepoint for adiabatic temperature t′ (°C), Trpoint for dew temperature tr (°C), VPrimepoint for specific volume vs, Condpoint for condensates, and M_secpoint for the molar mass of dry gas.

5.3.2.1 Search of a Point Humidity

When the point state has been calculated, w is known and method **getPointproperties("pointName")** allows one to access it directly.

When the gas composition is determined by programming, it may be necessary to recalculate w, which can be done by the following formula, with fractH2OFuel being x_{H_2O}:

$$// \, w = \frac{M_{H_2O} x_{H_2O}}{M_{dg} x_{dg}}$$

```
// Calculate the gas absolute humidity
   double inlet_w=18.01528*fractH2OFuel/fuelM/(1-fractH2OFuel);
```

5.3.2.2 Updating a Point Moist Properties

Method **updatepoint**(String name, boolean updateT, double T, boolean updateP, double P, boolean updateX, double x, boolean melHum, String task, double value) is a generic method for update and recalculation of point state variables, which has been generalized to allow for moist calculations.

If boolean melHum is false, the point update is for T, P, or x, depending on whether booleans updateT, updateP, and updateX are true or false: this is a standard property update without moist calculations.

If boolean melHum is "true," only moist calculations are performed, even if updateT, updateP, and updateX are "true."

These calculations are defined by two parameters **task** and **value**.

task is a String specifying the type of calculation to perform, and **value** is a double providing the value of the variable to modify.

The detail of the possible combinations for task and value is given in Volume 3 of the Thermoptim reference manual.

5.4 COOLING TOWERS

The performance of steam systems operating in a closed circuit is very sensitive to conditions under which steam may be condensed. Indeed, a condenser is a heat exchanger of a particular type, involving phase change, whose thermal equilibrium determines the condensation temperature of steam, and thus its condensation pressure.

The cycle low pressure, which directly defines the power that the turbine can produce, is thus set by the condenser.

Heat power transferred to the condenser is typically removed by cooling water flowing inside the tubes, which must be renewed (open-loop cooling) or cooled by contact with ambient air (air cooling).

For systems that operate in open loop, cooling water is drawn and released directly in the sea or a river after passing through the condenser, while in others, condenser cooling water circulates in a closed circuit cooled with air, usually in a cooling tower.

In both cases, the amount of water required is very important, which is a burden, especially in the context of global warming and depletion of water resources. Innovative binary cycles presented in Chapter 5 of this set's volume 2 are proposed today to reduce this constraint.

As heat exchangers have been studied in Chapter 4, we will not further develop water-cooled condensers. In this section, we present cooling towers, which have not been previously addressed.

5.4.1 Principle of Operation of Cooling Towers

A cooling tower is a heat exchanger of a particular type that discharges heat to the surrounding air in the form of both sensible heat and latent heat due to the increase of its moisture.

By working this way, it is possible to cool a fluid at a temperature a few degrees above the ambient air wet bulb temperature (and possibly below its dry bulb temperature), at the cost of a water consumption of about 5% of that which water cooling would require. Both economically and environmentally, cooling towers are very interesting systems, particularly in hot and dry climates.

Note that moist gas equations are presented in Section 2.6.2.

There are two main types of cooling towers, called direct or open-cycle, and indirect contact or closed-cycle.

In a direct contact tower, hot water is cooled in **contact** with ambient air either by spraying fine droplets or by runoff along flow surfaces. With both fluids in contact, heat is exchanged by convection, and part of the water vaporizes, thereby increasing the humidity. If it is not saturated, air starts to cool in an almost adiabatic process before warming up along the saturation curve. Therefore, water may exit at a temperature lower than ambient air.

An indirect contact tower comprises two circuits known as external and internal. In the latter, the cooling fluid, which can be arbitrary, remains confined in a tube bundle around which the external cooling circuit water flows. It warms on contact and then is cooled by exchange with ambient air by the same mechanism as in a direct contact tower.

5.4.2 Phenomenological Model

The theoretical modeling of cooling towers is quite complex given the multiplicity of transfers that take place.

The ASHRAE (2000) Handbook (Chapter 36) presents the theory proposed by Baker and Shryock (1961), which considers (Figure 5.4.1) three media exchanging mass and energy:

- The flow of water that cools.
- The air flow which is heated and whose moisture increases.
- An interstitial film.

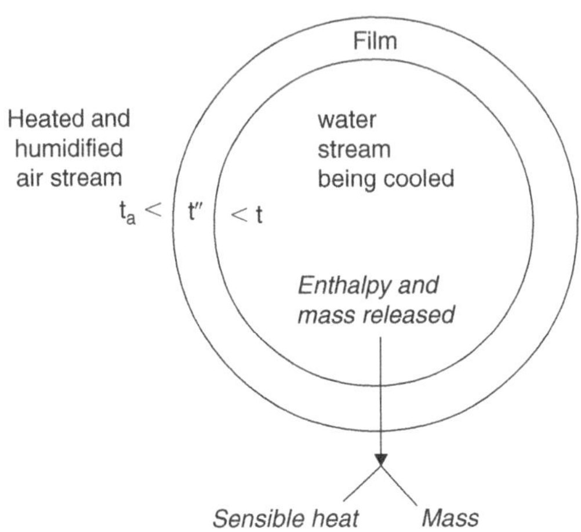

FIGURE 5.4.1 Sketch of transfers.

The assumptions are:

- Air leaving the tower is nearly saturated.
- The interstitial film is saturated moist air at the water temperature.
- The Lewis number is equal to 1.
- Thermal resistance on the liquid side is negligible compared to that on the air side (for indirect contact towers, it is assumed that the air-liquid film temperature gradient is equal to that of air-water cooling).

The water balance of a small tower element of surface ΔA is written with our usual notations:

$$\Delta Q_{water} = \dot{m} C_{p,water} \Delta T_{water}$$

This energy is used, first, to heat air (ΔQ_s), and, second, to evaporate water absorbed by air (ΔQ_L).

$$\Delta Q_s = k_s \Delta A (t'' - t_a)$$

where k is the heat exchange coefficient in the presence of mass transfer, t″ is the film temperature (assumed equal to that of water), and t_a that of air.

$$\Delta Q_L = k' \Delta A L_{water} (w''_{sat} - w_{asat})$$

k′ being the conductance referred to mass transfer, w''_{sat} the film saturation moisture, and w_{asat} that of air.

We have thus:

$$\Delta Q_{water} = k_s \Delta A (t'' - t_a) + k' \Delta A L_{water} (w''_{sat} - w_{asat})$$

In the turbulent regime, with Sc and Pr being the Schmidt and Prandtl numbers and Le the Lewis number, it can be shown that k_s and k′ are related by the following equation:

$$\frac{k_s}{k' Cp_{air}} = \left[\frac{Sc}{pr}\right]^{2/3} = Le^{2/3} \tag{5.4.1}$$

It turns out that for a water–air mixture, Le ≈ 1 as long as $t_{water} < 50°C$, as shown by Merkel in 1921 when he proposed to consider that for cooling towers Le = 1.

With this assumption, the above equation is simplified, showing the specific enthalpy q′:

$$\dot{m} \Delta Q_{water} = \dot{m} Cp_{water} \Delta T_{water} = k' \Delta A [Cp_{air} (t'' - t_a) + L_{water} (w''_{sat} - w_{asat})]$$

$$\Delta Q_{water} = k' \Delta A [q'sat(t'') - q'sat(t_a)]$$

This equation can be modified and allows us to write, per unit of water mass flow:

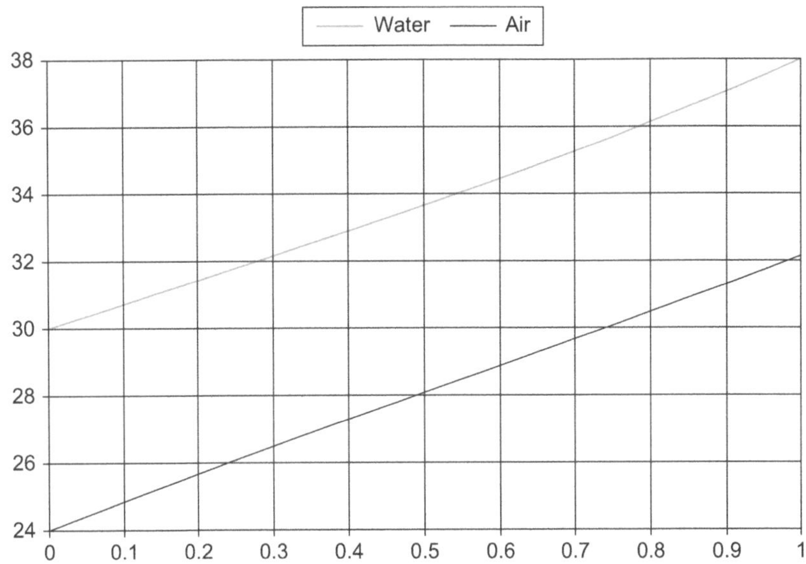

FIGURE 5.4.2 Profile of temperature in a direct contact cooling tower.

$$\frac{k'\Delta A}{\overset{\circ}{m}_{water}} = \frac{Cp_{water}\Delta T_{water}}{q'sat(t'') - q'sat(t_a)} \tag{5.4.2}$$

This equation can be numerically integrated along the cooling tower, leading to results in Figure 5.4.2, which shows the evolution of temperatures in the tower as a function of reduced distance between its inlet and outlet. The water temperature curve is of course located above that of air.

This provides a performance indicator of the cooling tower, which appears in good agreement with experimental data.

Let us now recall a characteristic of heat exchangers, the number of transfer units (NTU) introduced in Chapter 4.

We have shown that:

$$NTU = \frac{\Delta T_{max}}{\Delta T_{ml}}$$

NTU can be physically interpreted as the ratio of the maximum variation in temperature in one of the fluids to the log-mean temperature difference between them. This is indeed an indicator of the quality of the device: the more efficient the heat exchanger, the more it allows for heating or cooling a fluid with a low temperature difference with the other.

The concept of NTU can be generalized by considering not the ratio of the temperature variations, but that of the enthalpy differences. One may in particular put in the denominator the average logarithmic difference of specific enthalpy of saturated air taken at the temperatures of the two fluids, and in the numerator the variation of enthalpy δh of one of the two fluids, the two being equal.

$$NTU = \frac{\Delta h}{\Delta h_{ml}}$$

or

$$NTU = \frac{\Delta h}{\Delta q'_{sat\ ml}}$$

Under these conditions, equation (5.4.2) of the tower elementary heat balance that we integrated is like an elementary NTU and integration provides the NTU of the tower. To differentiate it from normal NTU in exchangers, we will note it with index m (Merkel).

$$\Delta NTU_m = \frac{k'\Delta A}{\dot{m}_{water}} = \frac{Cp_{water}\Delta T_{water}}{q'sat(t'') - q'sat(t_a)}$$

$$NTU_m = \int \frac{Cp_{water}\Delta T_{water}}{q'sat(t'') - q'sat(t_a)}$$

It is indeed an indicator of how difficult it is to realize the heat transfer in the cooling tower. Incorporating complex coupled phenomena of energy and mass exchange and dependent on technological implementation, it is not possible to give a simple expression.

We call the "approach" the pinch of the tower, i.e., the difference between the outlet temperature of the water and the wet bulb temperature of ambient air (4°C in the example in Figure 5.4.2) and the "range" the value of ΔT_{water} (8°C in this example).

NTU_m is usually given as a function of four variables: $\frac{\dot{m}_{water}}{\dot{m}_{air}}$ and the wet bulb air inlet temperature T'_{air} as well as either the two water temperatures T_{wi} (inlet) and T_{wo} (outlet), or the values of range $(T_{wi} - T_{wo})$ and approach $T_{wo} - T'_{air}$.

If they operate in the same environment with the same water inlet temperature and fluid mass flow ratios, two different cooling towers will cool water more or less depending on the quality of their design, with the better one resulting in the lowest value of T, and thus having a greater NTU_m.

5.4.3 Behavior Models

Many authors have investigated the modeling of cooling towers and sought expressions of NTU_m that are both easy to calculate and in agreement with experimental results. Since this is primarily an indicator of performance, it is not in itself unacceptable that there is a degree of arbitrariness in its definition, provided however that conventions are well specified.

5.4.3.1 Merkel Model

Experience shows that when we modify the four input variables of a given cooling tower, NTU_m takes different values that follow behavior similar to the following, where n and m are negative values between −1.1 and −0.35, and C is between approximately 1 and 2.5. C, m, and n are called the tower coefficients.

$$NTU_m = C\left[\frac{\dot{m}_{water}}{\dot{m}_{air}}\right]^n$$

or

$$NTU_m = C\left[\dot{m}_{water}\right]^n\left[\dot{m}_{air}\right]^{-m} \quad (5.4.3)$$

Note that one of the difficulties encountered when attempting to identify parameters C, n, and m is that available data are often incomplete. Manufacturers provide the values of inlet air wet bulb temperature, range, and approach, as well as water flow, but not the value of the air inlet humidity. They are often content to give the power consumed by the air fan, or at best its volume flow, so that identification of the thermodynamic state of air in the tower is inaccurate. When conducting tests, it is also often very difficult to measure the actual airflow that passes through the machine.

Figure 5.4.3 shows very good correspondence between the values of NTU_m calculated for the same model of cooling tower (Marley F1221) with four different fan engines (40, 50, 60, and 70 hp). This figure was obtained from values provided in the manufacturer's catalog and volume flow rates communicated to us. A linear

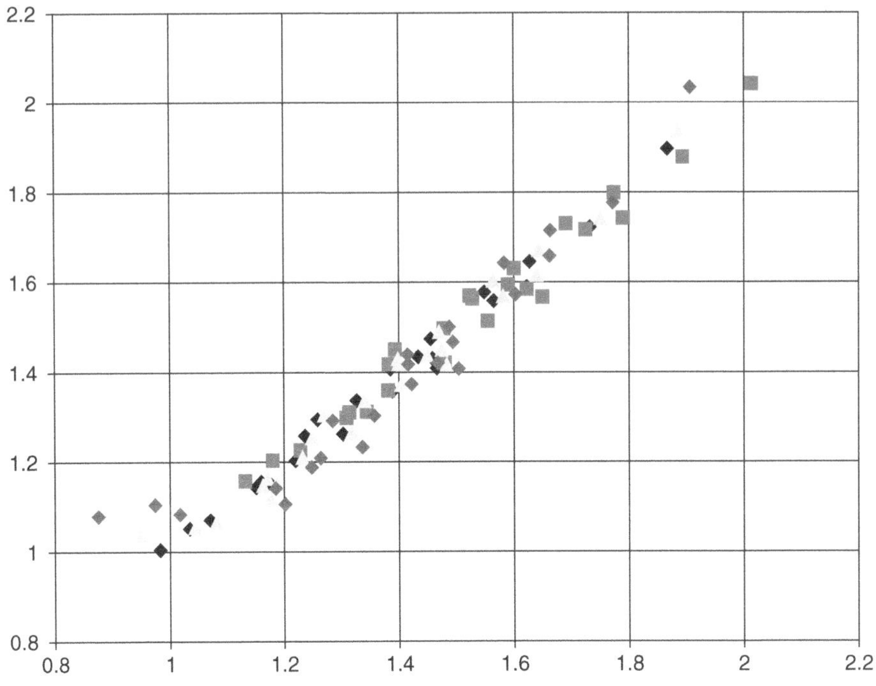

FIGURE 5.4.3 Mapping of manufacturer data and NTU_m correlation.

TABLE 5.4.1
Values of Coefficients

	F1221 40 hp	F1221 50 hp	F1221 60 hp	F1221 70 hp
C	2.19	1.87	1.93	2.06
n Merkel	−0.860	−0.968	−1.009	−1.171
r^2	0.970	0.982	0.963	0.853

regression allows us to identify, for correlation (5.4.3), the values of the coefficients given in Table 5.4.1, with correlation coefficients r^2.

Relation (5.4.3) also shows that for a given water flow, the higher the air flow, the smaller the tower can be. However, pressure drop and thus fan consumption are increasing functions of the airflow (air velocities are generally between 1.5 and 4 m/s).

For indirect contact towers, the fact that the fluid to be cooled is contained within the internal circuitry increases thermal resistance. We obtained good correlations by slightly increasing the approach, by values between 0.7 and 1.6 °C. This is easily explained when one considers that this device leads to lower NTU_m values, which is consistent with what should happen physically.

5.4.3.2 Cooling Technology Institute Calculator

The Cooling Technology Institute has developed a tool called Toolkit (http://www.cti.org/), which integrates the differential equation giving NTU_m. A comparison with our results yields a very small difference, less than 2%, presumably due to a difference in fluid property modeling.

With the mass flow of water noted L and that of air G, the expression of NTU_m in American literature is generally:

$$\Delta NTU_m = \frac{K_a V}{L} = \int \frac{Cp_{water} \Delta T_{water}}{q'sat(t'') - q'sat(t_a)}$$

In English units, Cp_{water} being equal to 1, it may disappear from the expression.

The cooling tower characteristics obtained have the appearance given in Figure 5.4.4, in bi-logarithmic coordinates. The abscissa is the ratio of the mass flow of water to that of air $\frac{L}{G} = \frac{\dot{m}_{water}}{\dot{m}_{air}}$, and the ordinate is NTU_m. The design value is displayed in orange, as well as a curve representing the behavior model law (5.4.3). Yellow curves are iso-approach lines. By selecting a rectangle on the chart, zoom is shown (double-click restores the original scale).

This chart can be used to characterize a given tower:

- The value of NTU_m for design conditions is calculated.
- The line of slope n and constant C is plotted and adjusted on the design point.

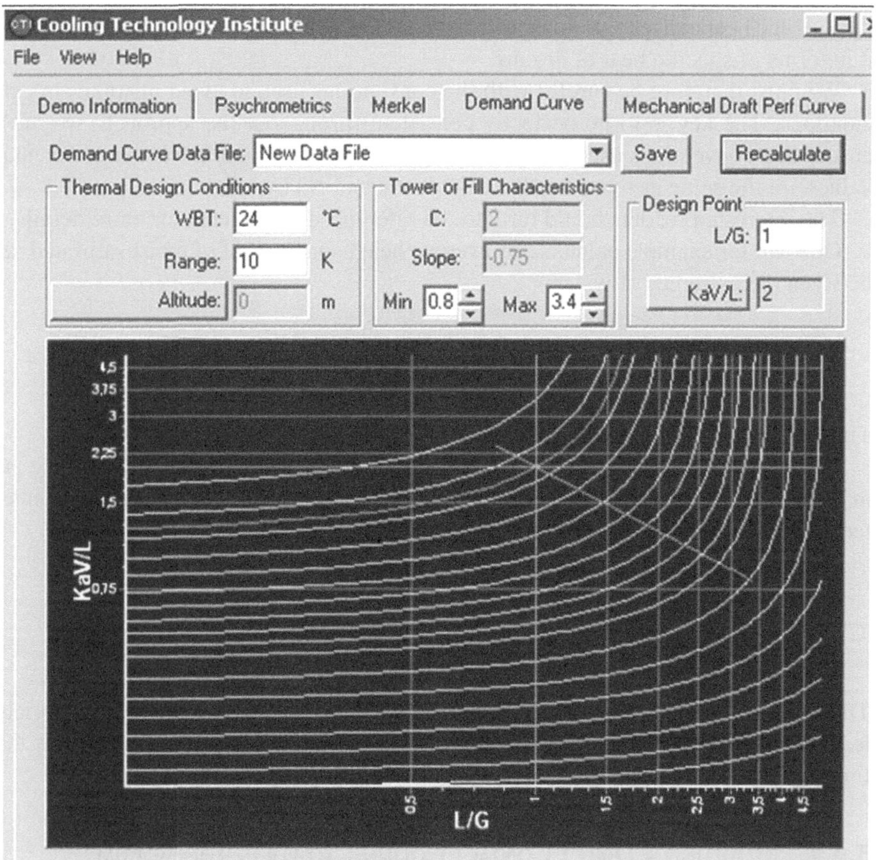

FIGURE 5.4.4 Cooling tower characteristics.

- For other operating conditions, determined by the fluid flows, the abacus allows us to find, at abscissa L/G and on the tower characteristic line, the new values of NTU_m and approach, which determine the outlet water temperature and therefore the capacity. The air outlet temperature is deduced. For different cooling conditions or other air wet bulb temperature, it is possible to recalculate another abacus. Knowing the values of variables (NTU_m, L/G) from the original chart, we obtain the new approach.

For example, the design conditions of Figure 5.4.4 correspond to an approach of 3.55°C (obtained by right-clicking the point in question).

5.4.3.3 Analogies with Heat Exchangers

Since it is the change in the enthalpy of saturated moist air that best characterizes the air thermodynamic behavior, a number of authors consider that it is as if water exchanged with a fictitious fluid whose specific heat at constant pressure is equal to the average heat capacity of moist saturated air Cp_a^{sat}, but it is also possible to express it in terms of specific heat of dry air.

We must therefore be careful with the conventions used to avoid mistakes in calculations. The key is to be perfectly consistent throughout the approach: we must ensure that conventions used for identifying the values of C and n from experimental values are the same as those used to calculate ε and NTU.

The interested reader should refer to the referenced documents for more details.

One can for example calculate as average heat capacity that of moist saturated air between the extremes of water:

$$Cp_a^{sat} = \frac{h_a^{sat}(T_a^{min}) - h_a^{sat}(T_a^{max})}{T_a^{min} - T_a^{max}} \tag{5.4.5}$$

The flux transmitted to the air can be written $\varphi = \varepsilon(q'sat(T_{wi}) - q'(T_a))$.

It is ε times the maximum flux, equal to the product of the mass flow of dry air $\dot{m}_{da}$ (index da) by the difference between the specific enthalpy of the air at water inlet temperature and its specific enthalpy at the entrance the tower.

$$\text{Futrhermore: } R = \frac{Cp_{water}}{Cp_a^{sat}}, \quad \text{and} \quad NTU_m = C\left[\frac{\overset{\circ}{m}_{water}}{\overset{\circ}{m}_{air}}\right]^n$$

The values of C and n are identified from experimental data and the appropriate (ε, NTU) relation (see Chapter 4). Values of these parameters are proposed in the literature (Kreith, 1999).

5.4.4 Modeling a Direct Contact Cooling Tower in Thermoptim

As an example, let us look at how a cooling tower can be modeled using external classes.

A direct contact cooling tower is crossed by two separate streams: air and water, which exchange matter and energy through an interface. It behaves like a quadrupole with two fluid streams at the inlet, and two others at the outlet.

This poses a slight difficulty in building a Thermoptim model, since the only components available are either processes or nodes. The solution is to form the quadrupole by combining an inlet mixer and an outlet divider, the two being connected by a process-point playing a passive role.

For the model to be coherent, the calculations made by both nodes are synchronized. More precisely, the divider takes control of the mixer, whose role is limited to performing an update of the coupling variables associated with the input streams. The model structure is given in Figure 5.4.5.

Under these conditions, we can reason at the overall enthalpy level knowing the inlet and outlet air side conditions and the inlet ones on the water side: model results are consistent with experimental values and those supplied by manufacturers.

The flow rates of both inlet streams are determined by the upstream conditions of the component and are not recalculated. If the water flow is insufficient to cool the

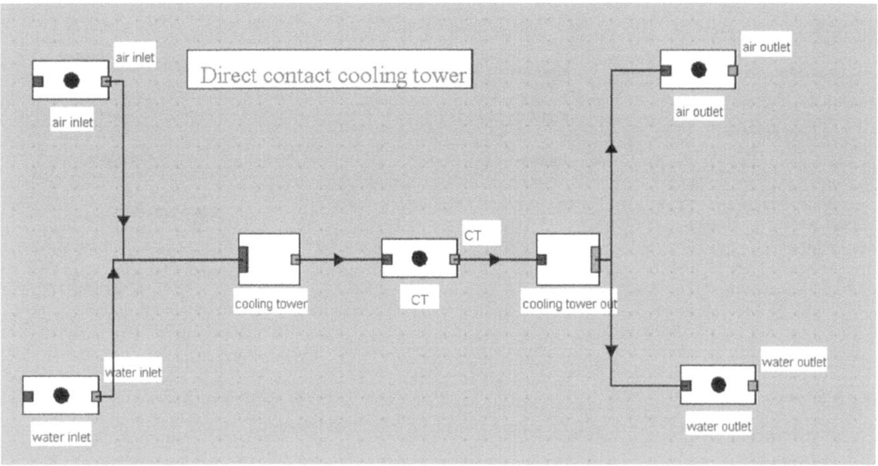

FIGURE 5.4.5 Diagram of the direct contact cooling tower component.

air (down to the wet bulb temperature of the incoming air) and thus provide it with the required enthalpy, a warning message is displayed to the user.

Recall that, as usual in calculations of moist functions, values are referred to dry gas, whose composition is invariant, while other calculations in Thermoptim are relative to the actual gas composition, i.e., are referred to the moist gas. It is therefore necessary to make the corresponding conversions.

The model that we can choose, when we know the outlet air temperature, is the following:

1. We begin by calculating inlet moist air properties and determining the dry gas mass flow from that of moist gas; inlet relative humidity ε is displayed on the screen.
2. The outlet relative humidity is set equal to 1, and the outlet moist air properties are calculated, which gives the specific and total enthalpy to be brought to air.
3. The flow of water carried by air is determined and the outlet moist air composition is changed.
4. The water enthalpy balance provides its outlet temperature, which must exceed the moist bulb temperature of air entering.
5. Values downstream of the node are updated.
6. NTU is calculated by integrating the Merkel equation (NTU integrated) established above and on the basis of the mean logarithmic enthalpies; effectiveness can be deduced from the classical relationship for counter-flow heat exchangers.

The tower is, as we said, represented by an external mixer connected to an external divider, with calculations being made by the latter. Classes are called DirectCoolingTowerInlet and DirectCoolingTower.

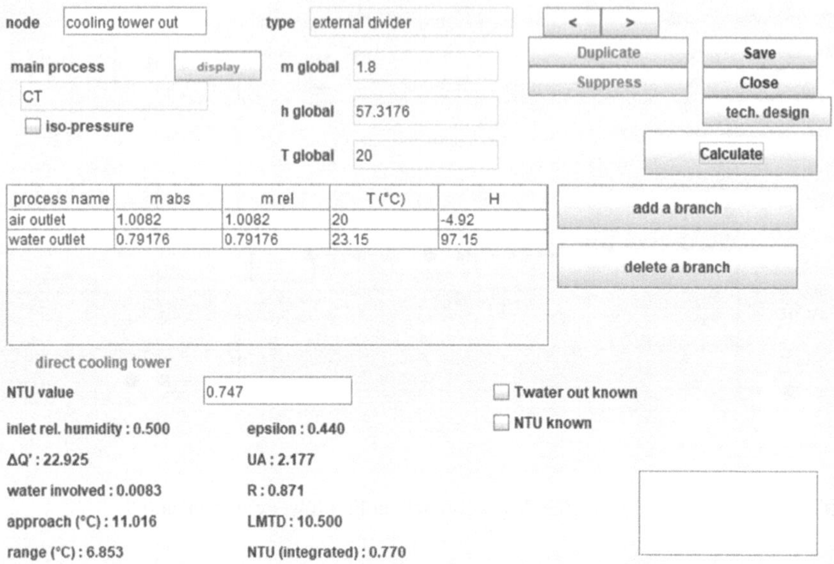

FIGURE 5.4.6 Cooling tower component screen.

The cooling tower component screen is given in Figure 5.4.6. We wish here to cool 1 kg/s of water from 30°C to 25°C. With a rate of 0.9 kg/s of air at 18°C and relative humidity equal to 0.5, the exhaust air temperature is 20.1°C and 7.5 g/s of water is evaporated. The tower capacity is 20.9 kW.

On this screen, we see two values of NTU. The one in the editable field is calculated by the approximate formula based on the logarithmic mean differences of enthalpies of saturated air between the tower inlet and outlet, while the one displayed at the bottom is the result of the integration of the Merkel differential equation for a counter-flow tower. It is in principle more accurate, but we can see that their values are close.

5.5 EXTERNAL CONTROLLERS

External classes also allow controlling Thermoptim either from another application to guide a user (intelligent tutoring system) or to control code execution (control or regulation, access to thermodynamic libraries). It is especially in Chapter 10 that external controller examples will be given because they allow finding solutions in off-design calculations.

5.6 EXTERNAL CLASS DEVELOPMENT ENVIRONMENT

To develop the external classes, if you do not already have a Java development environment, you can use an open-source tool called Eclipse, that you can download for free from the Internet.

Eclipse allows you to write your classes in Java, compile them, and directly test their integration into Thermoptim.

A page of the Thermoptim-UNIT portal explains how to install and configure this environment.

5.7 CONCLUSION

This chapter presented the extension mechanisms via external classes that enable Thermoptim to handle modeling problems beyond its core capabilities.

External classes allow for the creation of customized components in Java to represent specific substances and control mechanisms.

Detailed examples illustrated practical implementation with thermodynamic property servers, effectiveness-based models, and coupled heat and mass transfers. This extension mechanism, distributed following open-source principles, significantly broadens Thermoptim's applicability while maintaining consistency with its fundamental thermodynamic framework.

As indicated in the introduction, more detailed explanations are available through links on this book's website.[2]

NOTES

1 Available at: https://server.s4e2.com/crc/amtc/.
2 Available at: https://server.s4e2.com/crc/amtc/chapters/chapter5/_resources/index.html.

BIBLIOGRAPHY

ASHRAE (2000). *Handbook—HVAC systems and equipment*. American Society of Heating, Refrigerating and Air-Conditioning Engineers, Atlanta, GA. ISBN 0-910110-72-1.

Baker, D.R. & Shryock, H.A. (1961). "A comprehensive approach to the analysis of cooling tower performance." *ASME Transactions, Journal of Heat Transfer*, 50, 339–349.

Gicquel, R. (2022). *Thermoptim reference manual*, Vol. 3, English ed. Mines ParisTech. Available online: https://direns.minesparis.psl.eu/Sites/Thopt/en/res/refThopt3US.pdf

6 Component Sizing and Off-Design Operation

INTRODUCTION

This chapter presents a comprehensive treatment of component sizing and off-design simulation problems through a concrete example: a simple refrigeration cycle. In subsequent chapters, models for key components of the Thermoptim core will be explored in depth. The final chapter (Chapter 10) will focus on several case studies.

This chapter begins by explaining the challenges posed by sizing the main components used in a refrigerator (heat exchangers, compressor), and then presents the calculation principles for off-design operation, which will be examined in detail in Section 6.3.

6.1 INTRODUCTION

To address component sizing and off-design simulation, **two levels of models** have been distinguished:

- **Phenomenological models**, implemented in Thermoptim versions 2.5 and 2.6, provide access to thermodynamic cycle calculations independent of particular component technology choices.
- **Component sizing and off-design simulation models**, available in versions 2.7 and 2.8, not only provide the same results as phenomenological models but also enable users to geometrically size various components and, once sizing is complete, study system behavior under operating conditions different from the design point.

Such models are necessary when evaluating the performance of existing facilities operated under conditions different from those for which they were designed. This audit capability (particularly for proposing improvements) is of growing interest to many industrial and other organizations.

Off-design analysis of energy systems poses challenges far more difficult than those encountered when studying thermodynamic cycles purely phenomenologically at the rated point, independent of technology choices.

Off-design analysis, as defined here, corresponds to the stabilized operation of equipment under operating conditions other than those for which it was designed; it does not address fast transients caused by control actuators.

This work does not focus on dynamic changes at very short timescales, for which these models are not suitable. The assumption is that time constants of regulators controlling the systems studied are much shorter than those of the components

DOI: 10.1201/9781003606017-7

themselves. For example, the characteristic time of a refrigerator expansion valve (approximately 1 minute) is considered small compared to that of the condenser or evaporator (approximately 10 minutes). This assumption allows transient behavior to be neglected during off-design analysis of these two exchangers.

Consider, for example, a simple refrigeration cycle, easily modeled with the Thermoptim phenomenological version (Figure 6.1.1). To establish such a model, the user selects values for flow rate, evaporation and condensation pressures, and compressor isentropic efficiency, but need not specify the heat exchange surfaces involved or the type of compressor used (piston, scroll, centrifugal, etc.). Thermoptim calculates the thermodynamic state at all cycle points and the energies involved.

For the model to enable actual component sizing, the type and physical dimensions of various components must be specified as precisely as possible. For a refrigeration machine, a compressor type must be selected, for example from a list of devices available on the market. For a displacement machine, at minimum the swept volume and volumetric and isentropic efficiency laws as functions of compression ratio and rotation speed must be provided. For heat exchangers, the flow pattern must first be selected (plate, shell and tube, etc.), then exchange surfaces calculated, which involves determining internal heat transfer coefficients and therefore the dimensionless numbers governing thermal transfer between fluids and walls (Reynolds, Prandtl, and Nusselt numbers).

Many values that were set exogenously in the phenomenological model then depend on the initial design and machine adaptation to operating conditions.

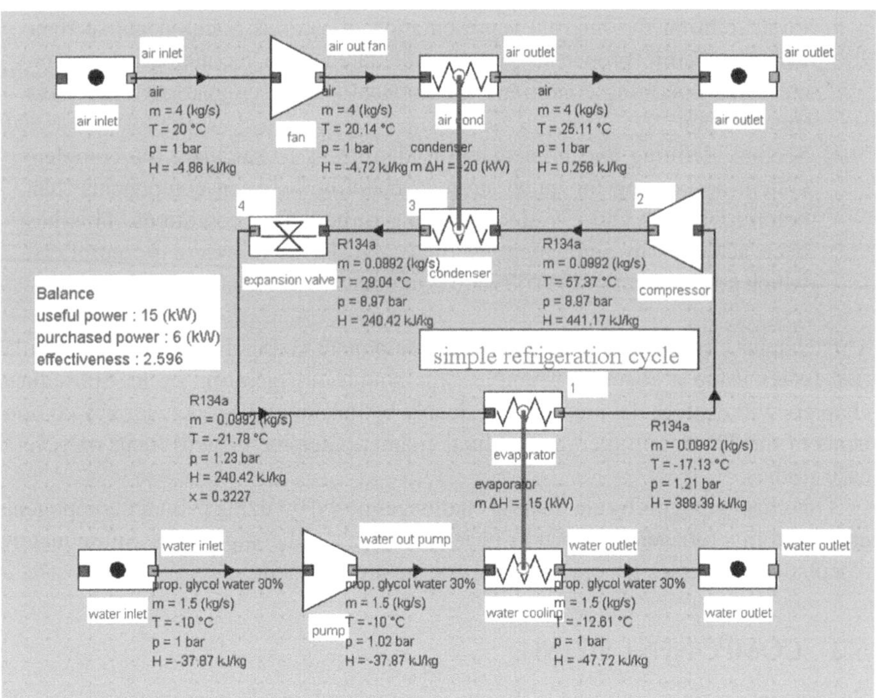

FIGURE 6.1.1 Refrigeration cycle model.

By making these technology choices, the user indirectly determines parameters that could previously be adjusted freely. For example, with a displacement compressor, the flow rate is not necessarily exactly as desired: it depends on the swept volume, compression ratio, suction specific volume, volumetric efficiency, rotation speed, etc.

For **off-design mode simulation** to be realistically feasible, not only must each component be sized, but the interactions between components due to their adaptation to new operating conditions must also be characterized.

For example, the condenser and desuperheater of a refrigerating machine actually form a single exchanger whose total area is known. Adapting the exchanger to changes in boundary conditions therefore requires coupled calculations far more complex than for the two separate parts.

Moreover, new systemic constraints emerge: in the example above, pressure levels are set by the thermal equilibrium of two phase-change heat exchangers—the evaporator and condenser—which determine saturation temperatures. The compressor sets the refrigerant volume flow (depending on the rotation speed and compression ratio, which determine volumetric efficiency), and therefore the mass flow rate (depending on refrigerant specific volume at suction). The main refrigeration machine parameters—pressure levels and flow rate—are determined by several strongly coupled components whose calculations cannot be performed independently.

Phenomenological models implemented in Thermoptim versions 2.5 and 2.6 are inadequate for addressing these two issues. To do so requires:

- First, refining the internal representation of various components so their thermodynamic properties can be calculated from technological parameters and operating conditions. Component sizing screens serve precisely this function.
- Second, defining and implementing algorithms to calculate the complete system accounting for much stronger coupling between components than before, with the entire system adapting to boundary conditions. This has been achieved by implementing these algorithms in external controllers, which take control of the software's automatic recalculation mechanism.

This chapter comprehensively addresses component sizing and off-design simulation issues using a concrete example—the simple refrigeration cycle. Subsequent chapters will explore the matter more deeply by introducing models for key components of the Thermoptim core. The final chapter (Chapter 10) will focus on several case studies.

This chapter begins by elucidating challenges posed by sizing primary components employed in a refrigerator, such as heat exchangers and compressors. Subsequently, calculation principles for off-design operation are outlined.

6.2 COMPONENT SIZING

Component sizing requires refining phenomenological models used in the Thermoptim core, supplemented to reflect mechanisms operating in off-design conditions, if any.

To perform component sizing, new screens have been introduced, complementary to those performing phenomenological modeling. They can be activated via the "tech. design" button in the standard screens of different components.

These new screens allow definition of geometric characteristics representative of different technologies used and parameters necessary for their calculation. For a given component, these obviously depend on the technology type used.

This section briefly introduces models enabling component sizing for a refrigeration cycle, and then demonstrates how such a cycle can be sized in practice.

Note that implementation of these new screens was done outside the Thermoptim core, enabling users who wish to modify or supplement them with other parameters to do so, with all corresponding classes distributed according to open source principles.

6.2.1 Heat Exchangers

The most significant changes relate to heat exchangers, as earlier Thermoptim versions only determine the UA value—the product of overall heat transfer coefficient U and exchanger surface A—without evaluating the two terms separately. To design a heat exchanger, i.e., calculate its surface, a geometric configuration must first be selected, and second, U must be computed, which depends on that configuration, fluid thermophysical properties, and operating conditions.

Recall that although the approach adopted in Thermoptim is unconventional, it has proved quite fruitful for studying complex systems: a heat exchanger ensures coupling between two "exchange" processes, one representing the hot fluid being cooled, and the other the cold fluid being heated. Consequently, definition of exchanger flow patterns and geometry is made at the process level, not globally.

6.2.1.1 Calculation Method

The method for calculating heat exchangers in off-design operation is based on the NTU method. When the two inlet temperatures and flow rates are known, an exchanger can be represented by a quadrupole whose generalized matrix formulation can be expressed in various ways, depending on which temperatures are known (see Chapter 4).

Formulation (4.2.20) corresponds to the case where inlet temperatures of both fluids are known, and equation (4.2.22) where the inlet and outlet temperatures of the cold fluid are known (if these are for the hot fluid, simply invert matrix C).

$$\begin{pmatrix} T_{ho} \\ T_{co} \end{pmatrix} = \begin{pmatrix} 1-\varepsilon_h & \varepsilon_h \\ \varepsilon_c & 1-\varepsilon_c \end{pmatrix} \begin{pmatrix} T_{hi} \\ T_{ci} \end{pmatrix} = A \begin{pmatrix} T_{hi} \\ T_{ci} \end{pmatrix} \tag{4.2.20}$$

$$\begin{pmatrix} T_{ho} \\ T_{hi} \end{pmatrix} = \begin{pmatrix} \dfrac{1-\varepsilon_h}{\varepsilon_c} & 1-\dfrac{1-\varepsilon_h}{\varepsilon_c} \\ \dfrac{1}{\varepsilon_c} & 1-\dfrac{1}{\varepsilon_c} \end{pmatrix} \begin{pmatrix} T_{co} \\ T_{ci} \end{pmatrix} = C \begin{pmatrix} T_{co} \\ T_{ci} \end{pmatrix} \tag{4.2.22}$$

The notations used in these equations are:

$$\varepsilon_h = \inf\left(\varepsilon, \frac{(\dot{m}c_p)_c}{(\dot{m}c_p)_h}\varepsilon\right) \quad \text{and} \quad \varepsilon_c = \inf\left(\varepsilon, \frac{(\dot{m}c_p)_h}{(\dot{m}c_p)_c}\varepsilon\right)$$

Furthermore:

$$NUT = \frac{UA}{(\dot{m} \cdot c_p)_{min}}$$

$$R = \frac{(\dot{m}c_p)_{min}}{(\dot{m}c_p)_{max}} \leq 1$$

$$\varepsilon = f(\text{ NUT, R, flow pattern })$$

In design mode, heat exchanger calculation proceeds in three steps:

- Knowing inlet and outlet temperatures and flows of both fluids, effectiveness ε is first determined, and UA is deduced by the NTU method.
- U is then estimated by correlations depending on the exchanger flow pattern and geometry.
- Surface area A calculation follows immediately: A = UA/U.

In off-design mode, knowing inlet temperatures and flow rates of both fluids, exchanger area A and its geometry (flow patterns and technological parameters), calculation proceeds in three steps:

- Determining U by correlations depending on the exchanger flow pattern and geometry (see Chapter 7).
- Calculating UA, the product of U and A, and then NTU.
- Determining exchanger effectiveness ε by the NTU method, and calculating hot and cold fluid outlet temperatures, with equations (4.2.20) and (4.2.22) determining how the heat exchanger adapts to operating conditions.

Knowing T_{hi}, T_{ci}, m_h, m_c, and U, equation (4.2.20) enables calculating R and NTU to deduce ε and determine outlet temperatures T_{co} and T_{ho}.

Knowing T_{ci}, T_{co}, m_h, m_c, and U, equation (4.2.22) enables calculating R and NTU to deduce ε, and determine temperatures T_{ho} and T_{hi}.

6.2.1.2 Multizone Heat Exchangers

Multizone heat exchanger calculation is more complex than for simple ones but relies on the same principle. Consider the refrigeration machine condenser whose temperature/enthalpy diagram is shown in Figure 6.2.1.

The refrigerant leaving the compressor is first desuperheated (T_{hi}_T_{hv}), then condensed at constant temperature, and then slightly subcooled (T_{hl}_T_{ho}). In this case,

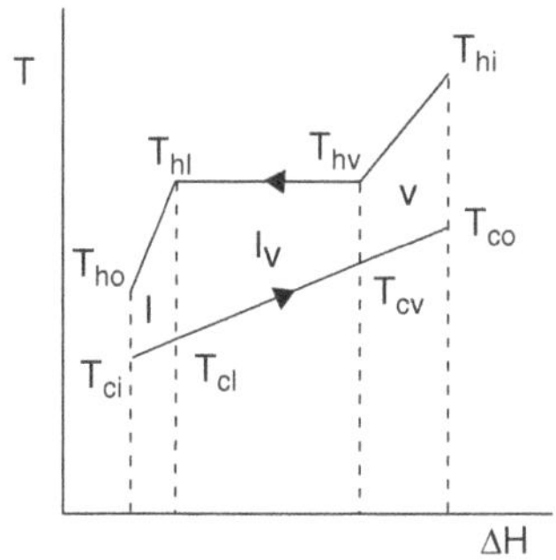

FIGURE 6.2.1 Temperature/enthalpy diagram.

calculation must be performed for each zone, with the heat exchanger surface being the sum of the three zone areas. Heat transfer coefficients obviously differ significantly across these three areas.

In off-design mode, total area remains constant, but its distribution among the three areas varies. Precise calculation is therefore required to find the solution corresponding to the same total area while meeting all other thermal and thermodynamic constraints.

Refer to Section 7.2.5 for detailed operational explanations.

6.2.1.3 Geometric Setting

Heat exchanger geometric setting is a complex subject. This section introduces the quantities used without discussing rationale for their selection, which will be addressed in Chapter 7.

To calculate heat transfer coefficients and pressure drops, two geometric quantities are particularly important besides exchanger surface: cross-sectional flow area $\mathbf{A_c}$ and hydraulic diameter $\mathbf{d_h}$. When fluid heat transfer coefficients differ significantly, various devices such as fins can compensate for the difference between their values. These are called extended surfaces, which can be characterized by a surface factor **f** and fin effectiveness $\boldsymbol{\eta}_0$.

These four parameters have been selected in Thermoptim to characterize heat transfer in each fluid. Exchanger length is also used for certain calculations such as pressure drops.

In the heat exchanger sizing screen (Figure 6.2.2), the following conventions are adopted:

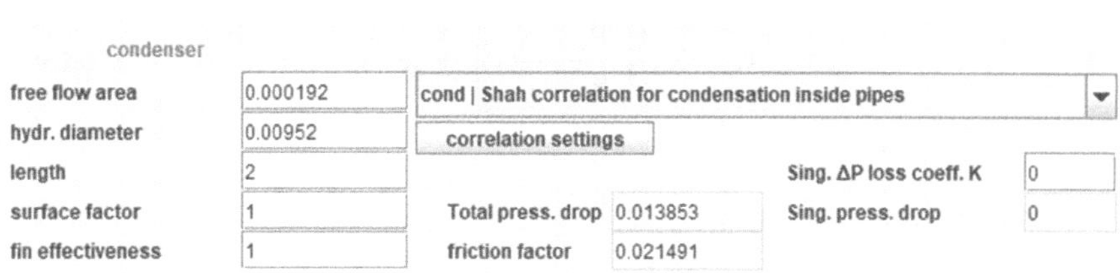

FIGURE 6.2.2 Condenser sizing screen detail.

- Type configuration (here, "condlShah") is selected from a combo box.
- "free flow area" represents the cross-sectional flow area A_c.
- "hydr. diameter" is the standard hydraulic diameter d_h.
- "length" is the exchanger length (used for calculating the boiling number Bo and pressure drops).
- "surface factor" is the surface factor f for extended surfaces.
- "fin effectiveness" is the fin effectiveness η_0.

6.2.1.4 Correlations for Calculating Heat Transfer Coefficients

A major difficulty is calculating the overall transfer coefficient U, which depends on both fluid heat transfer coefficients h, through general formula (6.2.1), which is simply a variant of equation (4.1.7). Many correlations have been proposed in the literature, and selecting the most appropriate is not always straightforward.

$$\frac{1}{U_c A_c} = \frac{1}{A_c \eta_{0,c} h_c} + \frac{e}{A\lambda} + \frac{1}{A_f \eta_{0,f} h_f} \qquad (6.2.1)$$

Chapter 7 explains options available in Thermoptim and presents the main configurations proposed.

6.2.2 Displacement Compressors

This section presentation is limited to displacement compressors, with dynamic compressors presented in Chapter 9.

A displacement compressor is geometrically defined by its swept volume, and its technological parameters enable calculation of volumetric and isentropic efficiencies according to rotation speed and suction and discharge conditions. Refer to Chapters 3 and 8 for a detailed presentation.

Models implemented in Thermoptim assume that volumetric compressor behavior can be represented with reasonable accuracy by two parameters: volumetric efficiency, which characterizes actual swept volume (3.1.1), and classical isentropic efficiency η (3.1.2).

As we have already mentioned, if these do not suit you, it is entirely possible for you to define other models by creating external classes that allow them to be calculated.

$$\lambda = \alpha_0 - \alpha_1 \frac{P_{ref}}{P_{asp}} \qquad (3.1.1)$$

$$\eta_s = K_1 + K_2 \cdot \left(\frac{P_{ref}}{P_{asp}} - R_1 \right)^2 + \frac{K_3}{\frac{P_{ref}}{P_{asp}} - R_2} \qquad (3.1.2)$$

Compressor calculation proceeds as follows:

- Isentropic and volumetric efficiencies are calculated from equations (3.1.1) and (3.1.2).

- Knowing rotation speed, swept volume, and volumetric efficiency, volumetric flow can be calculated as:

$$\dot{V} = \frac{\lambda N V_s}{60} \quad (6.2.2)$$

- Knowing suction volume v, mass flow is deduced:

$$\dot{m} = \frac{\dot{V}}{v} = \frac{\lambda N V_s}{60V}$$

In design mode, the rotation speed or swept volume required to provide the desired flow is determined. Calculation accounts for inlet and outlet pressures and the flow value entered in the compressor flow field.

In off-design mode, the compression ratio enables determining λ and η_s, which sets compressor flow and outlet temperature. The calculation sequence is as follows:

1. Knowing inlet and outlet conditions, the compression ratio P_{ref}/P_{asp} and suction volume v are updated.
2. Volumetric and isentropic efficiencies λ and η_s and actual flow are calculated knowing rotation speed N.
3. Flow volume $\dot{V}$ is deduced by (6.2.2).
4. The compressor sets $\dot{M} = v\dot{V}$ and propagates this value to connected components.
5. Useful work can then be calculated knowing isentropic efficiency η_s.

6.2.3 Expansion Valves

Expansion valve sizing is addressed in a simplified manner because their dynamics are considered much faster than other cycle components, and therefore they act as regulators. The only parameter introduced at this level is the evaporator superheat value.

6.2.4 Practical Example: Design of a Cycle

Suppose a refrigeration cycle must be sized to provide 15 kW cooling capacity at an outside temperature of 20°C, with brine temperature equal to −10°C at the evaporator inlet.

For this cycle (Figure 6.1.1), the cold fluid is 30% by volume propylene glycol, available among external substances. Evaporation temperature must naturally be lower than brine temperature, approximately −22°C, i.e., a pressure of 1.22 bar for R134a.

Compressor isentropic efficiency is assumed to be approximately 0.8.

Evaporation superheat ΔT_{surch} is 5 K, which for a coolant flow of 0.1 kg/s sets cooling capacity at approximately 15 kW. With a brine flow of 1.5 kg/s, this produces cooling of 2.64°C.

For the condenser, air temperature is 20°C and flow rate is 4 kg/s. Condensing temperature is estimated at 36°C, with subcooling ΔT_{ssrefr} of 6.5 K, corresponding to a condensation pressure of approximately 9 bar. The machine Coefficient of Performance (COP) under these conditions is 2.6.

Design proceeds in two separate steps: first is the classic cycle setup in Thermoptim, while sizing is performed from component sizing screens.

For simplicity, assume the Thermoptim cycle model has already been constructed. Otherwise, it would first be built, and then the method would be the same.

Given the relative complexity of the problem, which involves off-design modeling of three coupled components—the evaporator, compressor, and condenser—creating an external controller is preferable to using the available generic controller. The case study on heat exchanger parameterization will demonstrate implementation of the latter.

6.2.4.1 Refrigeration Machine External Controller

Chapter 10 examples will demonstrate how to create the external controller to be used.

Figure 6.1.1 shows a synoptic view of the refrigerator model to be controlled.

To perform off-design regime calculations, a sizing screen must be created and set up for each component. In this case, creation is done through the external controller.

6.2.4.2 Sizing Components

With sizing screens created, component setup and sizing can proceed. This step requires careful attention, involving a series of choices about internal configurations, geometric dimensions, etc. Some clarification is provided in this section, but refer to Chapter 10 and the third volume of the Thermoptim reference manual for more details.

Consider the evaporator to be shell and tube type. Begin by choosing configuration type: "evap|Gungor Winterton …" for evaporation inside the tubes for the "evaporator" (refrigerant), and "ext tube|Colburn correlation…" for brine, with screen settings in Figure 6.2.3 (on the water side, cross-sectional flow area of 0.008 m^2 and hydraulic diameter of 1 cm were considered, and on the refrigerant side, flow area of 0.0005 m^2 and hydraulic diameter of 2 cm, with a tube length of 3 m).

For the condenser, first choose configuration type: "cond|Shah…" for condensation inside the tubes for the "condenser" (refrigerant), and "airCoil_wcc|Wang, Chi & Chang…" for air, with screen settings in Figure 6.2.4 (on the air side, a fluid free flow area of 0.3 m^2 and hydraulic diameter of 3.84 mm with a length of 0.2 m were considered, and on the refrigerant side, a free flow area of 1.92 cm^2 and hydraulic diameter of 9.52 mm, with a tube length of 2 m).

Additional parameters were set for the Wang, Chi, and Chang correlation to define fin configuration.

The compressor is now sized from its sizing screen (Figure 6.2.5). For now, the setup is not detailed, as it will be explained in Chapter 10. Simply note that swept

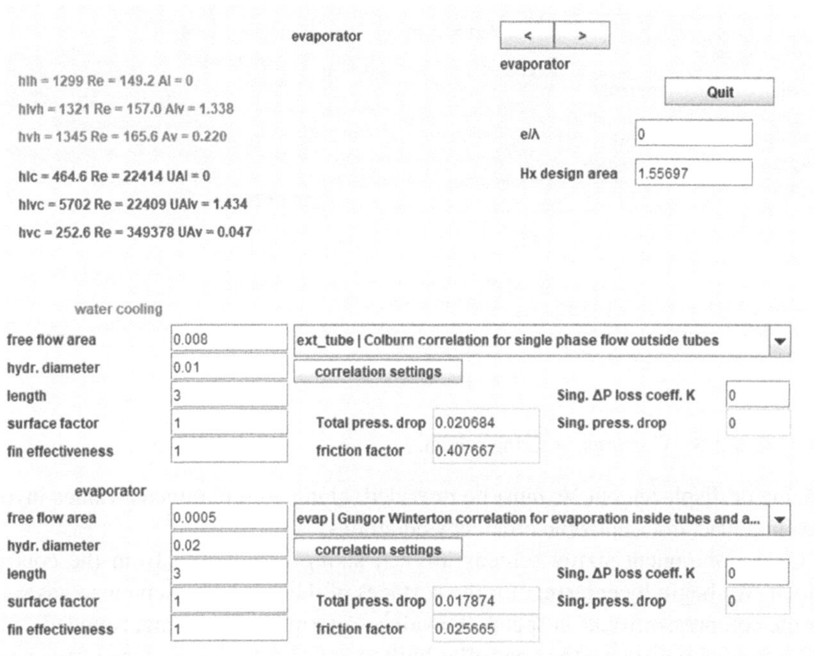

FIGURE 6.2.3 Evaporator sizing screen.

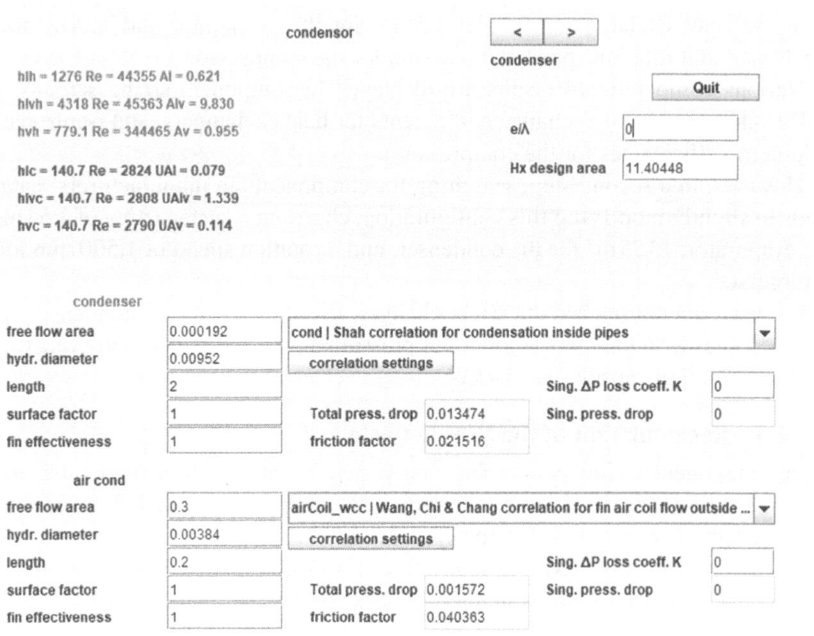

FIGURE 6.2.4 Condenser sizing screen.

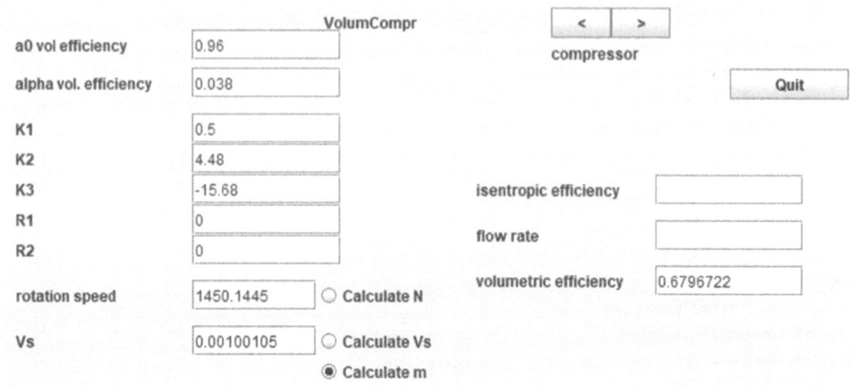

FIGURE 6.2.5 Compressor sizing screen.

volume or displacement Vs must be provided, along with parameter values involved in volumetric and isentropic efficiency equations.

Once component sizing screens are set, sizing is achieved from the controller screen. We begin by entering, for the surfaces of the two heat exchangers, as well as for the compressor rotation speed, the values determined by sizing, rounded.

Figure 6.2.6 displays the controller built to coordinate calculations of the various components. It allows changing input variables such as compressor rotation speed or outside air temperature.

To finalize component setup, click "Initial settings." Results are displayed in sizing screens: exchange surface of 1.55 m^2 for the evaporator and 11.4 m^2 for the condenser, and rotation speed of 1,450 rpm for the compressor.

Various computational results are displayed in component sizing screens, such as Re values and local exchange coefficients for heat exchangers, and isentropic and volumetric efficiencies for the compressor.

However, in a second step, searching for components in manufacturers' catalogs leads to slightly modifying this configuration, choosing a surface area of 1.54 m^2 for the evaporator, 13.75 m^2 for the condenser, and a rotation speed of 1,500 rpm for the compressor.

Cycle recalculation leads to slight changes: flow rate increases, condensing pressure becomes 8.55 bar, and evaporation pressure 1.2 bar. Refrigeration cycle COP is 2.96, and cooling capacity is 15.4 kW.

6.2.4.3 Recalculation of the Whole Cycle

Once component sizing is complete, only an estimate of the actual refrigerating machine state is obtained, and the entire cycle must be recalculated to more accurately reflect interactions within the system.

By proceeding this way, characteristics of a machine capable of meeting specified requirements have been defined. Verification with manufacturers is needed to confirm they have such components: heat exchangers will be selected from the manufacturer's range so their surface is as close as possible to the desired value, but probably slightly different. Compressor rotation speed will not necessarily match

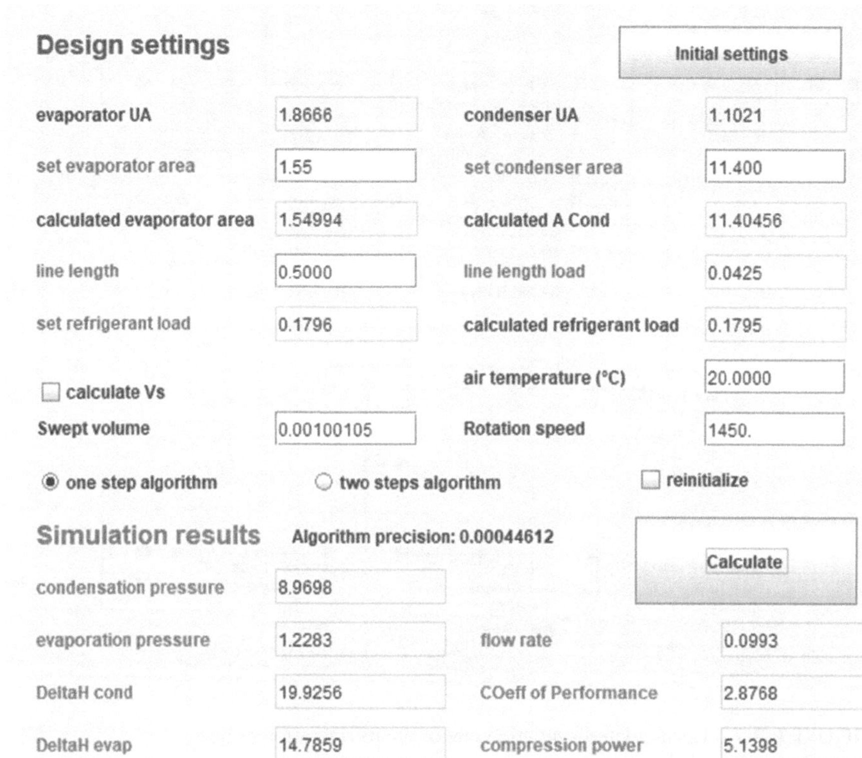

FIGURE 6.2.6 Controller screen after initial sizing.

the calculated value: this depends on the available speed control system. Generally, devices with similar but not identical performance can be found in their ranges.

Therefore, actual technological values must be entered and the machine performance recalculated in off-design mode, as explained in the next section, for selected settings to assess deviation from specifications. The cycle is then slightly modified. If deviation is acceptable, the solution can be retained.

For example, suppose compressor rotation speed is chosen as 1,200 rpm. Cycle recalculation leads to slight changes: flow rate decreases, condensing pressure becomes 8.15 bar, evaporation pressure 1.26 bar, and isentropic efficiency increases slightly. Refrigeration cycle COP is 3.16, and cooling capacity is 13.6 kW. Figure 6.2.7 shows the result.

Note that the COP value displayed on the controller screen is that of the refrigeration cycle alone, whereas the one shown in the diagram editor is lower because it takes into account the consumption of auxiliary equipment.

6.3 OFF-DESIGN CALCULATIONS

This section presents principles for calculating off-design operation of a refrigerator.

Cases treated have revealed two major challenges:

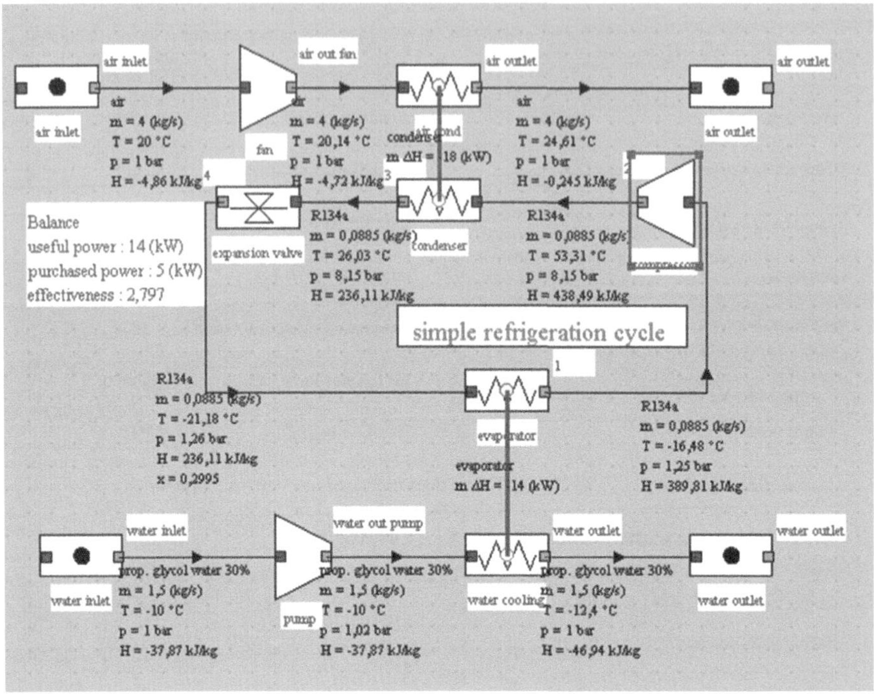

FIGURE 6.2.7 Recalculated synoptic view of the refrigeration cycle.

- The first concerns development of physical phenomenon equations. The behavior of different components, which may be far from simple, must be carefully analyzed and correctly modeled. This book and examples posted on the Thermoptim-Unit portal propose numerous models, but they are not the only possible ones, and not all interesting cases are covered. Choosing these equations is the user's responsibility. Several are proposed, but users must decide which are appropriate for the problem of interest, even if they must develop their own external classes to implement them.
- The second corresponds to solving model equations, usually numerically, which requires finding a suitable algorithm and properly initializing it.

The primary focus being applied thermodynamics, concentration has mainly been on the first difficulty. Numerical tools offered use libraries either included in Thermoptim for performing simple dichotomy searches, or generic ones like minPack, which is a set of classes enabling users to perform nonlinear optimization, particularly to seek solutions to nonlinear equation systems.

Following a brief reminder of how the Thermoptim core manages recalculations, the approach is illustrated by showing how the refrigeration cycle case can be solved in off-design mode, with results provided for the refrigerator sized in the previous section.

Recall that off-design regime refers to steady-state facility operation under operating conditions other than those for which it was designed, not fast transients such as control actuators. Dynamic changes at very short timescales, for which these models are unsuitable, are not addressed.

The solution is to implement main algorithms in **external controllers** using component sizing screens built themselves as external classes.

After recalling how Thermoptim calculates various system components, specific off-design problems are introduced.

6.3.1 Principle of Computing Coupled Systems in Thermoptim

The Thermoptim simulator calculates project elements step by step. This sequential calculation method differs from other modeling environments (matrix-based) where all problem equations are solved simultaneously. This procedure has the advantage that calculating elements one by one successively is much easier than solving the entire system at once. However, it induces two problems: first, iterating calculations several times may be necessary to find the correct solution, especially if the system is coupled, and second, for somewhat complex projects, the question arises of the order in which calculations should be performed. To solve this latter problem, a set of algorithms was developed. Called the Thermoptim **automatic recalculation engine**, it is a key package element. A particular screen can be displayed if desired to follow recalculation steps, thus ensuring modeling relevance.

The automatic recalculation engine, however, can only calculate phenomenological models, where couplings between components are much simpler than when studying the system in off-design mode. It was therefore necessary to supplement it with appropriate algorithms, with **external controllers** appearing to be the best solution.

Calculating the off-design operation of an energy system composed of several coupled components requires solving a typically highly nonlinear set of equations.

Since each component may involve up to several dozen equations, a complete project usually contains several hundred. Powerful numerical solvers can be employed, but this approach assumes that an explicit formulation of all equations is available. In practice, the main challenge lies in ensuring the consistency of the coupling equations between components. Without an appropriate modeling environment, this task can become extremely complex and difficult to make reliable.

Thermoptim offers a major advantage by facilitating the description of the analyzed technology through its diagram editor, which automatically ensures consistency between coupled components. The software can generate the equations corresponding to its models, although solving them may prove difficult in the general case.

For off-design simulations to be realistically feasible, each component must first be sized, and the interactions between components—arising from their adaptation to new operating conditions—must also be taken into account.

To illustrate this point, let us consider the equations governing the behavior of a refrigerator, presented in Section 6.2.

6.3.2 Off-Design Equations of the Refrigerator

Evaporation temperature is set by evaporator heat balance, which depends mainly on coolant temperature T_f and flow rate (brine in this example), and, second, on refrigerant flow.

$$Q_e = \dot{m}\left(h\left(T_e + \Delta T_{surch}, P_e\right) - h\left(T_c - \Delta T_{ssrefr}, P_c\right)\right) = U_e A_e \Delta T_{ml\_ef} \qquad (6.3.1)$$

Note that exchange coefficient U_e also depends on many parameters, including temperature and fluid flow through the exchanger, so solving this equation is particularly complex.

Similarly, condensation temperature is determined by condenser thermal balance, which depends mainly on cooling air temperature and flow rate, on refrigerant flow, and finally on condenser outlet temperature.

The latter depends on the compression ratio and compressor isentropic efficiency, itself also depending on this ratio.

$$Q_c = \tau + Q_e = U_c A_c \Delta T_{ml\_ac} \qquad (6.3.2)$$

Compressor balance expresses that compression work equals the product of mass flow by isentropic compression work, divided by isentropic efficiency. It also depends on several variables and is given by:

$$\tau = \dot{m} \frac{\Delta h_s\left(T_e, P_e, P_c\right)}{\eta_s\left(P_c / P_e\right)} \qquad (6.3.3)$$

These three variables are connected by (6.3.4), which is simply the First Law expression.

$$Q_c = \tau + Q_e \qquad (6.3.4)$$

Searching for a triplet (coolant flow, evaporation temperature, condensation temperature) corresponds to solving a relatively complex nonlinear equation set.

The adopted solution is coding an external controller that provides resolution of this equation system and updates Thermoptim once it has determined that triplet. A solution using the minPack library will be presented in Chapter 10.

6.3.3 After Treatment of Simulation Results

When performing off-design simulations, each setting produces a new result, which can be saved in a specific project file. This generates a file set with identical structure that must then be exploited.

An Excel spreadsheet containing a macro has been specifically developed for this. Usage instructions are provided in Volume 1 of the software reference manual as well as on this page. The principle is as follows:

- Save each new simulation in a separate project file.
- Once all desired states are obtained, open the Excel spreadsheet and remove all sheets except "macro" and "graph."
- Open Windows Explorer in the project directory, select all desired project files, and drag them into the spreadsheet window. Excel opens them all in different windows.
- Go to the "macro" sheet of the spreadsheet and click "Load Project," which automatically loads files into new sheets in the workbook. However, there is no control over the order in which sheets will be placed, and manual reordering may be needed.
- Back in the "macro" sheet, update the value table to be extracted from result sheets, changing labels in column A to match your needs, and entering in column B the cell number corresponding to each value. Finally, enter in B4 the number of values to be extracted.
- Click "extract." The macro updates the "graph" sheet by extracting from result files the values you wish to monitor. If applicable, remove from this sheet values that may still exist from another study.
- Use spreadsheet functions to calculate additional values if desired, and to create desired graphics.

A second option is using an application called DataExtractor, which operates similarly.

- Save each new simulation in a separate project file.
- Once all desired states are obtained, run DataExtractor and set the parameter file with names and references of variables of interest.
- Select all desired project files and copy them to a specific directory.
- Load the parameter file, edit fields if necessary, and launch extraction.
- Open the extraction file and copy results into a spreadsheet or another application.
- Use spreadsheet functions to calculate additional values if desired, and to create desired graphics.

6.3.4 Effect of the Variation of the Ambient Temperature

Figure 6.3.1 shows the evolution of different indicators when outside temperature changes. The setup is slightly different from the one that has just been presented, but this does not change the overall behavior of the variations of the different quantities in any way.

U evolution is far from negligible, particularly for the condenser, fully justifying the effort to account for heat transfer coefficient influence in off-design models.

Figure 6.3.2 shows results obtained when compressor speed varies.

Comparison between these two cases shows several behavioral differences, such as between refrigerant flow and COP. The latter decreases in both cases, while flow decreases in the first and increases in the second, thus demonstrating the difficulty in predicting evolution laws for even a cycle as simple as this.

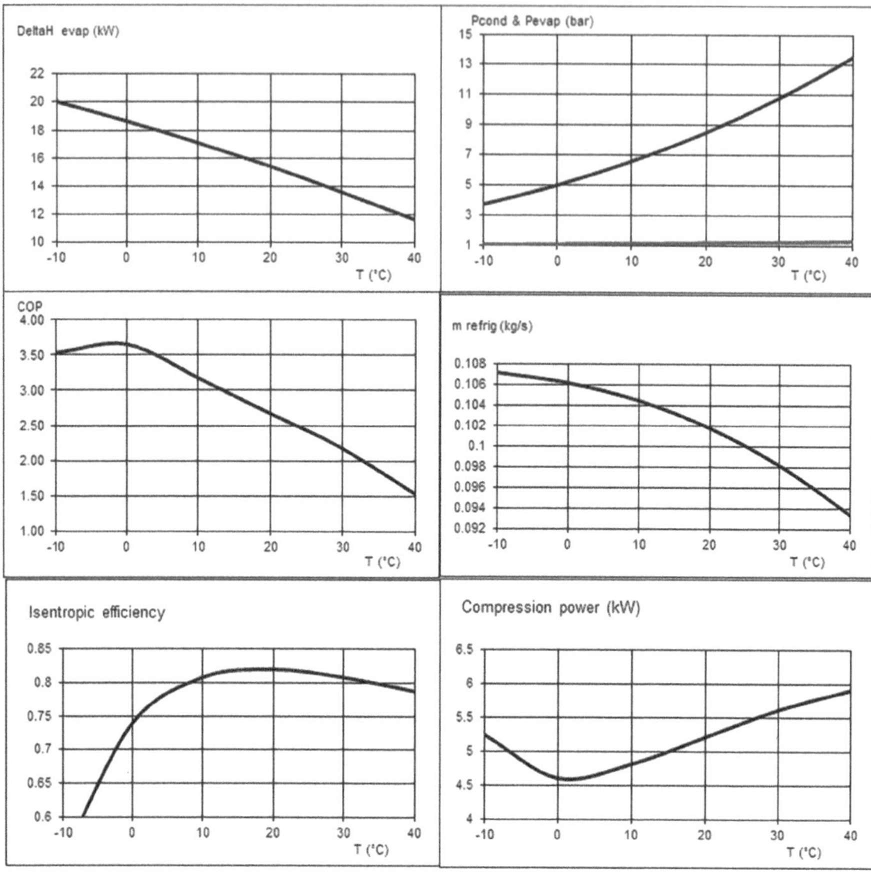

FIGURE 6.3.1 Effect of the variation of the ambient temperature.

Also note that the isentropic efficiency law as a function of compression ratio is highly nonlinear, with a very clear maximum at Rp = 7. Consequently, when temperature drops below 0°C, compression work increases when it would be expected to decrease…

Results obtained show that, particularly for refrigerators involving multizone exchangers, off-design evolution laws are not simple and do not reduce to a power function of flow. This arises because total area available for heat exchange is divided into two-phase and vapor zones, where heat transfer coefficients differ significantly.

6.4 METHODOLOGICAL DIFFICULTIES

When attempting to perform off-design calculations, one faces numerous methodological difficulties. These include:

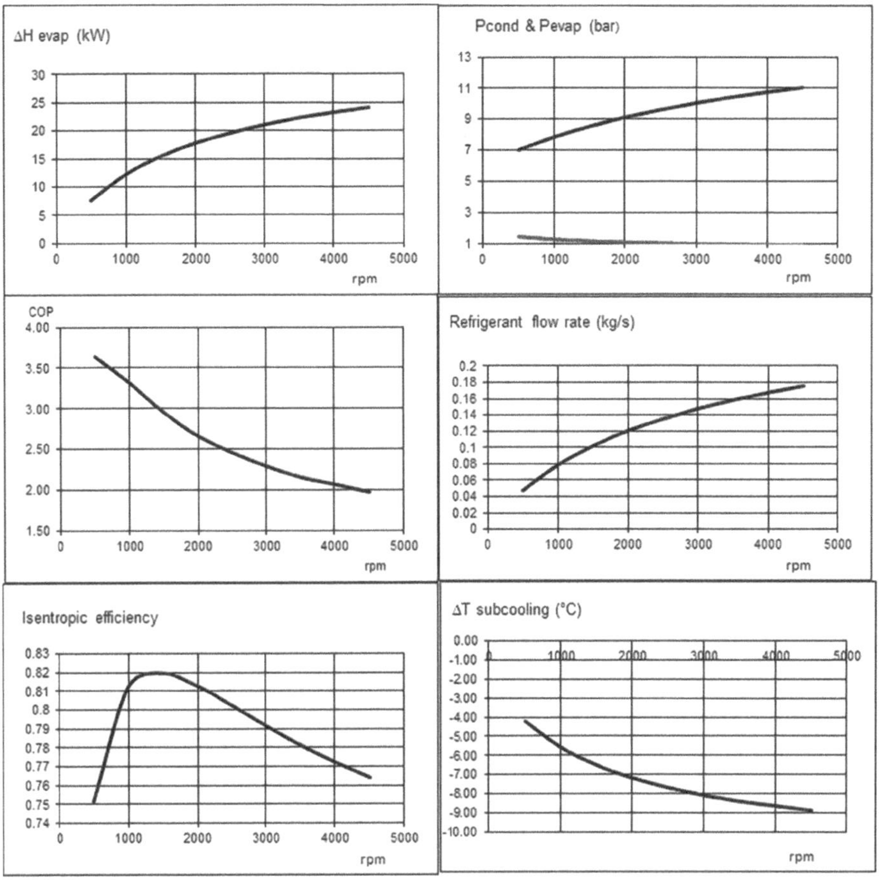

FIGURE 6.3.2 Effect of rotation speed.

- Having relevant technological models, both at the component level and for their interaction. For this, two main approaches can be adopted: either purely numerical fitting, if a satisfactory one exists, or maintaining as physical an approach as possible. As appropriate, one choice or the other will be made.
- Obtaining usable experimental data. Experience shows this can be a very difficult task.
- Setting models of selected components, i.e., identifying model parameters. This is probably one of the main difficulties overall.
- Having a sufficiently powerful reference modeling environment capable of simulating different configurations encountered in practice. As seen, Thermoptim responds to some extent to these specifications.
- Whenever possible, developing simplified models from reference tools (data analysis, model reduction, etc.).

Subsequent chapters will elaborate on these difficulties, first for various components, and then for examples to be presented.

It has just been mentioned that two of the main methodological difficulties are obtaining exploitable experimental data and model parameterization of selected components. A necessary condition for solving this is having reliable test benches, which represents a considerable investment, both materially and in human resources.

6.5 CONCLUSION

This chapter demonstrated how to size components and simulate the off-design operation of a simple refrigeration cycle.

Two levels of models were distinguished: phenomenological models for thermodynamic cycle calculations independent of technological choices, and sizing/off-design simulation models for geometric sizing and performance evaluation under off-design conditions.

Sizing requires refining internal representations to calculate thermodynamic properties from technological parameters. For heat exchangers, the NTU method enables both sizing and off-design calculation via matrix formulations. Positive displacement compressors are characterized by their swept volume and isentropic efficiency laws.

Off-design simulation requires external controllers to manage the strong coupling between components and adaptation to boundary conditions. The methodology was illustrated using the example of a refrigeration cycle, showing how the evaporation and condensation temperatures are determined by the coupled thermal balances.

Additional examples illustrating how the concepts presented in this chapter can be applied using Thermoptim are available on this book's website.[1]

NOTE

1 Available at: https://server.s4e2.com/crc/amtc/chapters/chapter6/_resources/index.html.

BIBLIOGRAPHY

Boyce, M.P. (2012). *Gas turbine engineering handbook*, 4th ed. Elsevier, San Diego, CA.

Gicquel, R. (2022). *Thermoptim reference manual,* Vol. 4, English ed. Mines ParisTech. Available online: https://direns.minesparis.psl.eu/Sites/Thopt/en/co/manuel-reference-tome-4.html

Gordon, J.M. & Ng, K.C. (2000). *Cool thermodynamics: The engineering and physics of predictive, diagnostic and optimization methods for cooling systems.* Cambridge International Science Publishing, Cambridge.

Martelli, E., Alobaid, F. & Elsido, C. (2021). "Design optimization and dynamic simulation of steam cycle power plants: A review." *Frontiers in Energy Research*, 9, 676969.

Pérez-Segarra, C.D., Oliva, A. & others (2004). "Modelling and numerical simulation of combined cycle power plants." *Applied Thermal Engineering*, 24(11–12), 1583–1595.

7 Sizing and Off-Design Behavior of Heat Exchangers

INTRODUCTION

The purpose of this chapter is to explain how to model and configure heat exchangers so that they can be sized and calculated in off-design mode. It complements Chapter 4, which provides the foundation for simple heat exchanger calculations in energy systems.

Although not particularly complex, this chapter employs non-trivial thermal engineering concepts, which are reviewed here. However, it assumes that the reader has basic knowledge of the discipline. If this is not the case, it is advisable to refer to a comprehensive thermal engineering textbook. The notations used are those of Chapter 4.

7.1 HEAT EXCHANGER SIZING AND OFF-DESIGN CALCULATIONS

Recall that the NTU method, presented in Chapter 4 and used in the Thermoptim phenomenological version, provides only the product UA of the overall exchange coefficient and the exchanger surface area, without evaluating the two terms separately.

As explained in Section 6.2.1, to size the heat exchanger—that is, to calculate its surface area—one must first determine its geometry and then calculate the overall exchange coefficient U, which depends on the configuration and the thermophysical properties of the fluids.

As stated in Chapter 6, there is actually a significant qualitative difference between the basic calculations for balancing a heat exchanger and what we call its sizing.

The NTU method is presented in Section 4.2.1. We recommend that you refer to it if you are not very familiar with it.

To **size a heat exchanger**, we start from a known thermodynamic state and seek to identify a set of characteristic parameters for the component.

In **off-design operation**, these values are not known, and we attempt to estimate them realistically. Based on experimental data that we have analyzed, we propose some approaches, though they are only partial.

In design mode, if we know the flow rates of both fluids, their inlet temperatures, and the heat flux transferred, the procedure is as follows:

- We first determine the outlet temperatures of the fluids from equation (4.2.4).
- We deduce the fluid heat capacity rates $\dot{m}cp$ and their ratio R.

DOI: 10.1201/9781003606017-8

- The effectiveness ε is calculated from equation (4.2.5).
- The value of NTU is determined from the appropriate (NTU, ε) relationship.
- UA is calculated from equation (4.2.1).

In off-design mode, if UA is known, equation (4.2.1) gives the value of NTU, and ε can be determined from the appropriate (NTU, ε) relationship. Equation (4.2.5) allows calculation of the heat transferred.

In both cases, the enthalpy balance provides the outlet temperatures.

In practice, we do not know UA but rather A. However, we can often consider that U varies only to second order and seek an approximate solution by assuming U constant, then recalculate its value for the new operating conditions, and iterate until reasonable accuracy is obtained. This approach is particularly necessary when the exchanger is multizone, because only the total area is known, not its distribution among the different zones.

For multizone exchangers, we deduce NTU and UA for each sub-exchanger from the relationships NTU = f(ε, R, flow pattern) and $NTU = \frac{UA}{(\dot{m} \cdot c_p)_{min}}$. If U is known, we obtain A.

During sizing, we determine the three surface areas Al, Alv, and Av.

Knowing all temperatures and heat capacity rates, we can estimate the values of ε and R for each zone, deduce the UA_j, and hence A_j. A equals their sum.

In off-design mode, the general algorithm for finding the solution is implemented as a controller and depends on the specific problem.

The calculation sequence can then be:

- For an evaporator, the superheat value ΔT_0 is usually assumed to be known. T_{co}, T_{ho}, and T_{hv} are recalculated, which allows estimation of the values of ε and R for each zone, from which the UA_j and hence A_j are deduced.
- For a condenser, the surface area A_l corresponding to liquid subcooling is usually assumed to be known. We begin by recalculating T_{ho} for subcooling, and then recalculate T_{co}, T_{cl}, and T_{cv}, which allows estimation of the values of ε and R for each zone, from which the UA_j and hence A_j are deduced.

The issues addressed below correspond to a refrigerator but could easily be applied to other devices.

For superheat, we usually assume a constant value set by the expansion valve. This assumption requires substantiation, but it has the advantage of simplicity.

For subcooling, one hypothesis is to consider that the wetted surface of the condenser is constant. For greater precision, if experimental data are available, we can attempt to identify the subcooling value using a linear or quadratic relationship with respect to the control variables used (evaporation and condensation temperatures, compressor rotation speed, compression ratio, etc.).

One can also directly estimate the wetted surface of the condenser, which is directly compatible with the Thermoptim calculation method.

7.2 PRESSURE DROP CALCULATION

7.2.1 Gas or Liquid State Pressure Drop

With C being the fluid velocity and v the specific volume, the so-called regular or linear pressure drop is generally expressed as:

$$\Delta P = f\frac{C^2}{2}\frac{L}{vd_h}$$

where f is the friction coefficient, given by correlations depending on the geometry (as for heat exchange coefficients) and as a function of the relative roughness RR. For an initial rapid assessment, f can be taken as approximately 0.02 for a smooth tube.

Expressed in terms of mass flow rate, for a total cross-sectional flow area equal to Ac, the pressure drop equals:

$$\Delta P = f\frac{(\dot{m}v)^2}{2Ac^2}\frac{L}{vd_h} = fv\frac{\dot{m}^2}{2Ac^2}\frac{L}{d_h}$$

The calculation of pressure drop in turbulent flow can be performed using a relationship of the form $\Delta P = \frac{4fG^2Lv_m}{2d_h}$ where f is the friction coefficient, given by correlations based on Re.

In laminar flow, pressure drop is proportional to flow rate and fluid viscosity: f = 64/Re.

For smooth tubes, f is given by the Blasius formula ($f = 0.316\,Re^{-0.25}$) if Re < 30,000, or by the formula ($0.0032 + 0.221\,Re^{-0.237}$) for higher values of Re.

For rough tubes in turbulent flow (Re > 2,100), the friction coefficient is given by the Colebrook equation:

$$\frac{1}{\sqrt{f}} = -0.868589\ln\left[\frac{RR}{3.71} + \frac{2.51}{Re\sqrt{f}}\right]$$

$$RR = \frac{\varepsilon}{D} \quad \varepsilon = \text{where absolute roughness of the tube}$$

It can be shown that this implicit equation for f can be approximated by:

$$\frac{1}{\sqrt{f}} = -0.78173\ln\left[\left[\frac{RR}{3.71}\right]^{1.1} + \frac{6.9}{Re}\right]$$

7.2.2 Two-Phase Pressure Drop

In two-phase systems, pressure drop calculation can be very complex, not least because the lengths of the various exchanger zones must first be determined.

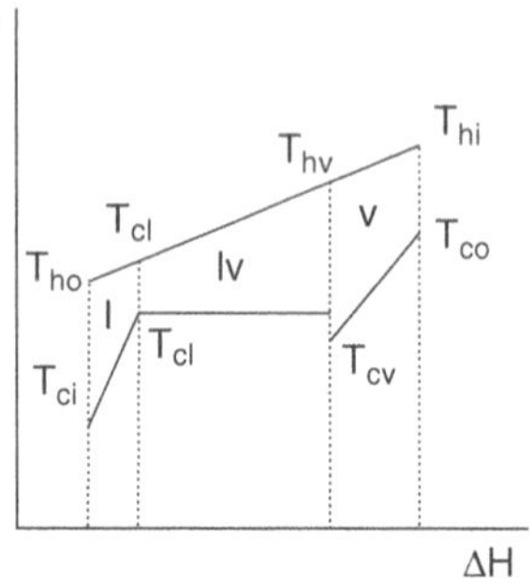

FIGURE 7.3.1 Evaporator with pressure drop.

In the liquid-vapor equilibrium zone, we use the approximate formula for nuclear steam generators proposed by Gallori and Herre, adjusted to account for nonzero quality at the inlet (Techniques de l'Ingénieur, BN 3050).

$$\Delta P_{diff} = \varphi \Delta P_{liq}$$

$$\varphi = \left[1 + |x_o - x_i|\left(\frac{\rho_l}{\rho_v} - 1\right)\right]\left[1 + |x_o - x_i|\left(\frac{\mu_l}{\mu_v} - 1\right)\right]^{-0.2}$$

The pressure drop is thus calculated from the saturated liquid state using this enhancement factor φ.

Accounting for pressure drop in multizone evaporators requires some care: if we assume that pressure drop occurs during evaporation, the temperature at the end of evaporation is lower than that at the exchanger inlet, resulting in a negative heat capacity rate and leading to a miscalculation of the exchanger by the NTU method. The calculation is performed assuming that the pressure drop occurs after the outlet, as shown in Figure 7.3.1. Superheat is calculated using the outlet pressure, and vaporization uses the pressure at the exchanger inlet.

We will provide some examples of pressure drop calculations in the section on nucleate boiling.

7.3 MODELING OF HEAT TRANSFER

The overall heat transfer coefficient U depends on the distribution of thermal resistances in the exchanger. If the exchange surfaces are equal for both fluids:

$$\frac{1}{U} = \frac{1}{h_h} + \frac{e}{\lambda} + \frac{1}{h_c} \tag{7.3.1}$$

The values of convection coefficients h_h and h_c are based on fluid thermophysical properties and exchange configurations, with convective regimes being strongly dependent on flow velocity.

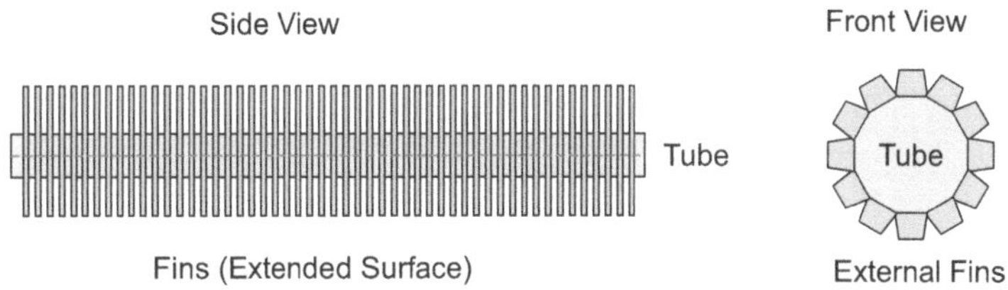

FIGURE 7.3.2 Tube with external fins.

They can be obtained from correlations that give the value of the Nusselt number $Nu = \frac{hd_h}{\lambda}$, depending on Reynolds $Re = \frac{\rho C d_h}{\mu}$ and Prandtl $Pr = \frac{\mu c_p}{\lambda}$ numbers.

In the implementations proposed in Thermoptim, we account for the thermal resistance of the wall (as well as that of the fins if they exist). Calling S_h and S_c the total exchange areas of the hot and cold sides, respectively, we relate these two surfaces to a common reference (or primary) surface S or A, called the exchanger surface. For this, we introduce two "surface factors" f_h and f_c, such that $S_h = f_h A$, and $S_c = f_c A$. If there are no extended surfaces, $f_h = f_c = 1$.

The calculation of U is ultimately performed by:

$$\frac{1}{U} = \frac{d_{hh}}{\lambda_h Nu_h} + \frac{e}{\lambda} + \frac{d_{hc}}{\lambda_c Nu_c}$$

7.3.1 Extended Surfaces

If extended surfaces are present, we take as reference the surface without extended surfaces and calculate a surface factor greater than 1 for the other side (Figure 7.3.2). Accounting for the existence of fins, $\eta_{0,h}$ and $\eta_{0,c}$ being the overall fin effectiveness on the hot and cold sides, respectively, we have:

$$\frac{1}{U_h A_h} = \frac{1}{A_h \eta_{0,h} h_h} + \frac{e}{A_w \lambda} + \frac{1}{A_c \eta_{0,c} h_c} \tag{7.3.2}$$

where A_h and A_c are the total exchange areas on the hot and cold sides, respectively, A_w is the exchanger wall surface, and $\eta_{0,h}$ and $\eta_{0,c}$ are the overall fin effectiveness on the hot and cold sides, respectively.

There is also a cold-side overall exchange coefficient U_c, with:

$$U_c A_c = U_h A_h$$

7.3.2 Calculation of Reynolds and Prandtl Numbers

$Re = \frac{\rho C d_h}{\mu}$ can be expressed directly in terms of fluid mass flow, fluid free flow area A_c, and mass velocity $G = \rho C$.

Since $\dot{m} = GA_c$

$$Re = \frac{Gd_h}{\mu} = \frac{\dot{m}d_h}{\mu A_c} \quad (7.3.3)$$

Knowing the mass flow, one must specify the cross-sectional flow area A_c and the hydraulic diameter d_h.

7.3.3 Calculation of the Nusselt Number

The Nusselt number is almost always given by the following equation:

$$Nu = C_1 Re^a Pr^b \left(\mu/\mu_p\right)c \quad (7.3.4)$$

where C_1, a, b, and *c* are constants. Coefficients C_1, a, and b vary depending on the configurations encountered, and the additional term accounting for viscosity variations is often neglected.

The viscosity correction is made at the wall temperature. However, it is not calculated in Thermoptim. Initially, the solution is to estimate it as equal to the average temperature of both fluids, but this is only valid as a first approximation, and in the case of a multizone heat exchanger, this can lead to discrepancies (Figure 7.3.3).

In general, the Reynolds number exponent is between 0.5 and 0.8, and that of the Prandtl number is between 0.33 and 0.4. Values of correlations for very specific finned geometries can be found in the literature.

The flow regime has a significant influence on the value of Nu. When Re is less than 2,000, the regime is laminar, and if Re is greater than 5,000, it is turbulent, with a transition zone between these boundaries, which are not perfectly stable.

For fully developed laminar flow, Nu = Const, with the value being between approximately 4 and 8, depending on the tube shape and heat exchange mode (constant temperature or constant flux).

Calculating the Nusselt number can be challenging for some single-phase flows, but it becomes even more so when one of the fluids changes phase, as in evaporators or condensers.

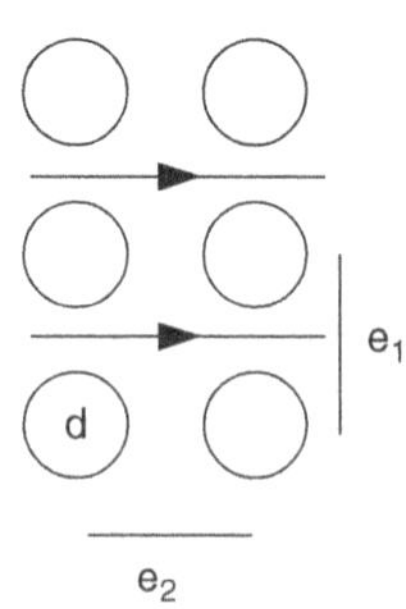

FIGURE 7.3.3 Arrangement in line.

In the following sections, we provide some guidance on how calculations can be performed, without claiming to be exhaustive.

7.3.3.1 Inside Tubes

Inside tubes, the most commonly used formula is that of Mac Adams and Dittus-Boelter:

$$Nu = 0.023\,Re^{0.8}\,Pr^{0.4} \tag{7.3.5}$$

7.3.3.2 Flow Perpendicular to Tubes

For flows perpendicular to tubes, the tube arrangement influences the exchange coefficients, and several formulas exist, such as Colburn:

$$Nu = 0.33 Kp\,Kr\,Re^{0.6}\,Pr^{0.33} \tag{2.3.6}$$

Coefficients Kp and Kr depend on the arrangement and relative values of e_1, e_2, and d (Kp on the number of rows, and Kr on the network geometry and Re) (Figure 7.3.4). Initially, to avoid overcomplicating matters, we have neglected the influence of these two terms in the external classes provided with Thermoptim, taking both equal to 1.

7.3.3.3 Finned Coils

For finned coils, predictive correlations are relatively complex. However, in the end, matters are simplified, and it can be shown that simple relationships such as (7.3.7) can be used.

$$h = K_1\,Re^{0.77} = K_2 G^{0.77} = K_3 V^{0.77} \tag{7.3.7}$$

One needs only identify coefficient K based on the chosen parameter to obtain a simple relationship characterizing the air-side heat exchange.

For finned coils, we have selected four correlations:

- Kim & Jacobi correlation for fin air coil flow outside tubes
- Manglik & Bergles correlation for fin air coil flow outside tubes
- Wang, Chi & Chang correlation for fin air coil flow outside tubes
- Morisot correlation for air coil flow outside tubes

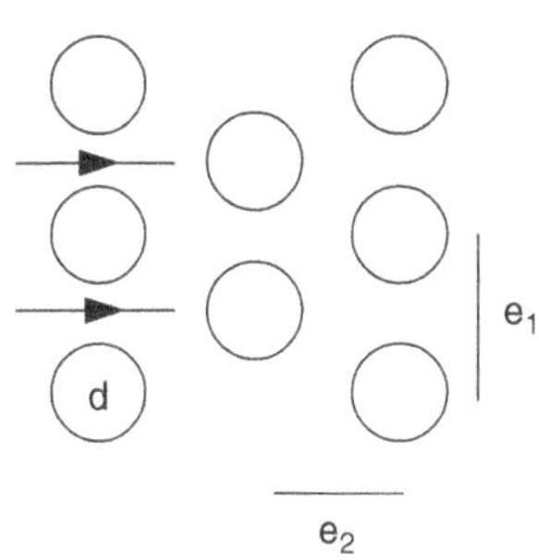

FIGURE 7.3.4 Staggered arrangement.

7.3.3.4 Plate Heat Exchangers

For plate heat exchangers, exponent b is almost always one-third, and c is 0.14 for Muley and Manglik (1999) and 0.17 for Kumar (1984).

For chevron angles ≤ 30°, Kumar (1984) gives 0.348 for *C1* and 0.663 for a. Alfa Laval provides *C1* = 0.142, a = 0.72, b = 0.37, and c = 0.108 for the A20 exchanger. Manglik and Muley (1999) propose a more complex formula for chevron angles $\beta = 90 - \alpha$ ranging from 30° to 60° and surface factors μ from 1 to 1.5 for Re > 103.

$$Nu = \left[0.2668 - 0.006967\beta + 7.244 \times 10^{-5}\beta^2\right] \times \left[20.78 - 50.84\mu + 41.16\mu^2 - 10.51\mu^3\right]$$

$$\times Re^{\left[0.728+0.0543\sin\left[(\pi\beta/45)+3.7\right]\right]} Pr^{1/3} \left(\eta/\eta_w\right)^{0.14}$$

Figure 7.3.5 shows the significant influence of the correlation used on the Nu value as a function of Re for Pr = 0.7.

The correlations used are listed below:

- Inside tubes (Mac Adams): $Nu = 0.023\,Re^{0.8}\,Pr^{0.4}$
- Outside tubes (Colburn): $Nu = 0.33\,Re^{0.6}\,Pr^{0.33}$
- Plate with viscosity correction: $Nu = 0.2\,Re^{0.674}\,Pr^{0.33}\left(\mu/\mu_p\right)^{0.14}$

Figure 7.3.6 shows the influence of the hydraulic diameter on the exchange coefficient.

As expected, we note that the correlation for plate heat exchangers leads to higher U values than those of Mac Adams and Colburn. However, it is not the only one proposed (Ayub, 2003).

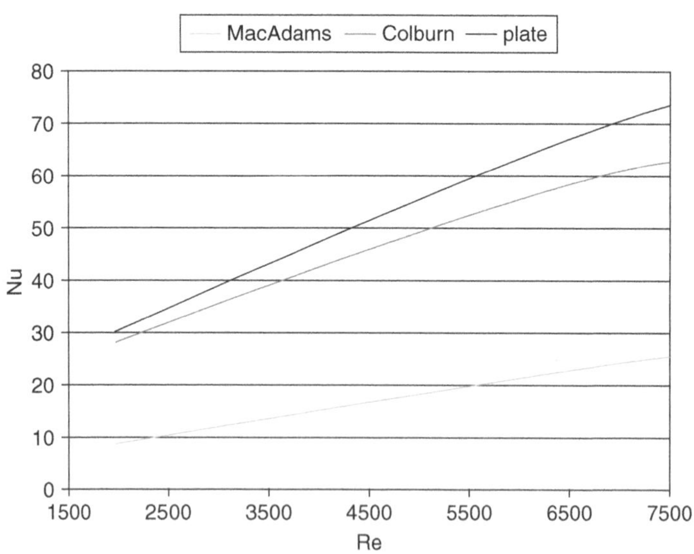

FIGURE 7.3.5 Influence of correlations used on Nu.

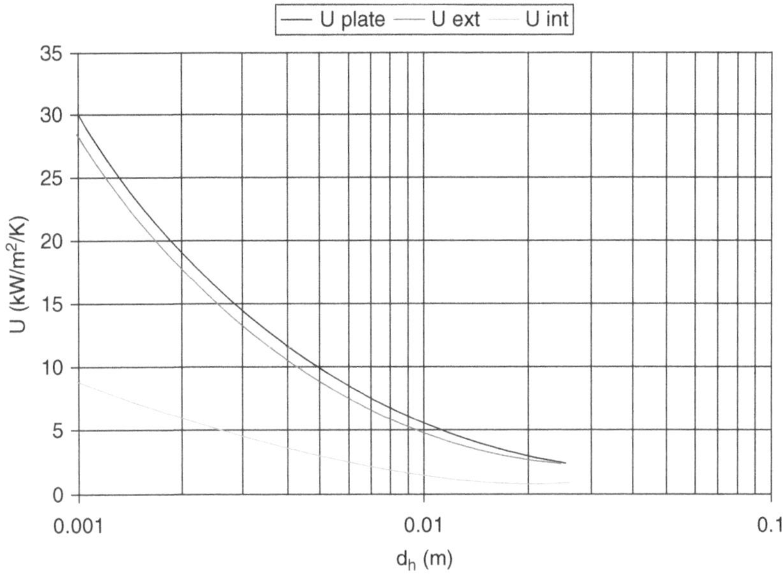

FIGURE 7.3.6 Influence of dh on U.

7.3.3.5 Two-Phase Exchange

For two-phase exchange, matters are much more complicated. The most accurate methods account for several zones depending on vapor quality and a related quantity called void fraction. Some correlations calculate h from an implicit relationship, while others avoid this complication.

In correlations used for two-phase exchange, the Nusselt number is often calculated as the product of a liquid Nusselt number by an amplification factor, with Nu_{liq} being calculated from the MacAdams correlation, valid inside a tube. It is a legitimate question whether this calculation remains valid for other types of heat exchangers including plate heat exchangers. Can we recalculate Nu_{liq} for the chosen configuration and retain the amplification factor? As discussed below, tests suggest otherwise.

7.3.3.5.1 Formulation for Condensation Outside Tubes

For condensation outside tubes, we have selected three correlations:

- The Nusselt correlation for condensation outside vertical tubes is one of the oldest, often regarded as a reference, with corrective terms added to it.
- Dhir-Lienhard correlation for condensation outside horizontal tubes.
- Levy correlation for condensation outside horizontal tubes.

7.3.3.5.2 Formulation for Condensation in Tubes

For condensation inside tubes, we have selected three correlations:

- Shah correlation for condensation inside tubes
- Akers correlation for condensation inside tubes
- Rabiee et al. correlation for condensation inside horizontal tubes

For instance, the Shah (1979) correlation, modified by Bivens and Yokozeki (1994), is implemented as follows:

$$\mathrm{Nu} = \mathrm{Nu}_{10}\mathrm{F} = 0.023\,\mathrm{Re}_{10}^{0.8}\,\mathrm{Pr}_{10}^{0.4}\left[(1-x)^{0.8} + \left(\frac{3.8x^{0.76}(1-x)^{0.04}}{\left(\frac{P}{P_c}\right)^{0.38}}\right)\right]$$

$$\mathrm{Nu} = \mathrm{Nu}_{\mathrm{Shah}}\left[0.78738 + 6{,}187.89G^{-2}\right]$$

In this expression, the calculation of Nu_{10} must be performed with the total mass flow $\dot{m}$.

This relationship is applied by averaging the F factor over the range of variation of x, which simply requires calculating average values for the integrated functions of x. The following equation was identified numerically to obtain a simple expression for the averaged F:

$$\int x^{0.76}(1-x)^{0.04}\,dx = -0.0155x^3 + 0.1669x^2 + 0.1083x - 0.0065$$

$$\int (1-x)^{0.8}\,dx = -0.5555556(1-x)^{1.8}$$

formulation for the evaporation in tubes

For evaporation inside tubes, we have selected three correlations:

- Borishanskij correlation for forced convective boiling
- Forster Greif correlation for forced convective boiling
- Gungor-Winterton correlation for evaporation inside tubes and annuli

For instance, the Gungor and Winterton (1987) correlation is implemented as follows:

$$\mathrm{Nu} = \mathrm{Nu}_{\mathrm{liq}}\mathrm{F}$$

$$\mathrm{Nu} = 0.023\,\mathrm{Re}_{\mathrm{lip}}^{0.8}\,\mathrm{Pr}_{\mathrm{lip}}^{0.4}\left[\left(1 + 3{,}000\mathrm{Bo}^{0.86} + 1.12\left(\frac{x}{1-x}\right)^{0.75}\left(\frac{v_v}{v_l}\right)^{0.41}\right)\right]$$

In this expression, the calculation of Nu_{liq} must be performed with the liquid phase flow alone: $(1-x)\dot{m}$.

This relationship is applied by averaging the F factor over the range of variation of x, accounting for the liquid phase flow alone. The following equation was identified numerically to obtain a simple expression for the averaged F:

$$\int \left(\frac{x}{1-x}\right)^{0.75} (1-x)^{0.8} dx = -0.0079x^4 - 0.131x^3 + 0.6128x^2 + 0.0651x - 0.0054$$

$$\int (1-x)^{0.8} dx = -0.5555556(1-x)^{1.8}$$

The expression for Bo, the boiling number, is given below in general form, then for a tube, where q_{th} is the heat flux density, A_{ext} is the outer surface, and G is the mass velocity:

$Bo = \frac{q_{th}}{GL_v}$, which can be approximated as follows for a tube:

$$Bo = \frac{mL_v}{A_{ext}} \frac{1}{GL_v} = \frac{GA_cL_v}{A_{ext}} \frac{1}{GL_v} = \frac{A_c}{A_{ext}} = \frac{d_h}{4L}$$

7.3.4 Taking into Account Nucleate Boiling in Steam Generators

The consideration of nucleate boiling in steam generators poses specific problems that have led to the creation of a dedicated class which, although based on the correlations just presented, performs calculations in a slightly different manner.

The TechnoEvaporator class is specific to evaporators designed as two- or three-zone exchangers calculated globally.

It enables calculation of the triple exchanger by distinguishing the zones corresponding to the economizer, evaporator, and superheater, but has several limitations. Although it allows pressure drop to be calculated, it assumes uniform pressure in each of the fluids. It also does not account for possible nucleate boiling in the economizer section.

The TechnoSteamGenerator class was developed to address these limitations.

To do this, it implements:

- A specific calculation of pressure drops that modifies pressures inside the steam generator, with Nusselt number calculations being performed in the usual way via the FlowConfig.
- A generic method accounting for nucleate boiling.

7.3.4.1 Calculation of Pressure Drops

The calculation of pressure drops is performed in the classical manner for the single-phase case and for singular losses. In the two-phase case, they are estimated from various correlations. A well-documented synthesis can be found in (Thome & Cioncolini, 2015).

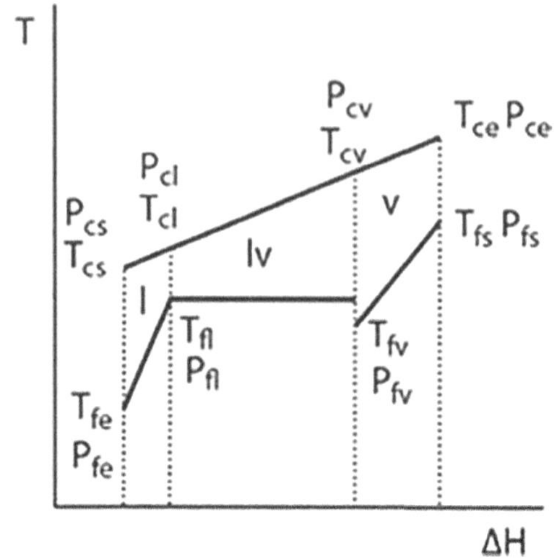

FIGURE 7.3.7 Evaporator with pressure drop.

The Lockhart-Martinelli (1949) correlation is one of the most widely used, even if it is not among the most accurate according to some authors: many publications state that it significantly overestimates pressure drops. It appears that those of Müller-Steinhagen and Heck (1986) and Sun and Mishima (2009) or those of Friedel (1979) are much more accurate. All three have been implemented in the TechnoSteamGenerator class.

The notations used in the class are those of Figure 7.3.7, which schematically represents the cooling of the hot fluid and the heating of water to the state of superheated steam.

With the exception of singular pressure losses at the cold fluid entrance, the most significant losses occur during vaporization. However, if we assume that pressure decreases during evaporation, the temperature at the end of evaporation is lower than that at the exchanger inlet, resulting in a negative heat capacity rate, which leads to an erroneous calculation of the exchanger by the NTU method. The calculation is therefore performed assuming that the pressure drop occurs after the end of vaporization, as shown in the figure. Superheat is calculated with the outlet pressure.

7.3.4.2 Calculation of Saturated Boiling

The TechnoEvaporator class uses the Gungor-Winterton correlation calculated by discretizing the evaporation zone into 100 elements.

Five correlations for estimating exchange coefficients during boiling are also implemented, those of Shah (1982), Borishanskiy et al. (1971), Kim and Mudawar (2013), Kandlikar (2017), and Saitoh et al. (2007).

One difficulty is that they lead to significant discrepancies in estimating h_{lvf}. It is therefore difficult to know which one to choose if data on the studied steam generator are not available.

7.3.4.3 Calculation of Nucleate Boiling

The calculation of nucleate boiling is much more delicate than that of saturated boiling. This is a relatively new area of investigation, and many publications exist without truly comprehensive summaries being available.

Figure 7.3.8 shows the definitions of the different boiling phases. The curves displayed allow visualization of the phenomena at stake.

On the ordinate is the temperature of the fluid T_b or wall T_w, and on the abscissa is the quality x, which is considered negative in the nucleate boiling zone.

As the wall temperature T_w increases, three thresholds appear:

- When T_w reaches T_{sat}, bubbles may begin to appear.
- But they only appear with some delay, when $T_w = T_{w,onb}$ (Onset of Nucleate Boiling ONB). A complementary flux to the purely convective flux then appears $q_{\_fc}$. It equals $q_{\_nb}$ for nucleate boiling.
- Boiling becomes fully developed when $T_w > T_{w,fdb}$ (Fully Developed Boiling FDB), which corresponds to a quality x greater than or equal to 0.

In the TechnoSteamGenerator class, the ONB is detected by a relationship such as that of Thom et al. (1965), and the FDB by a relationship such as that of Bowring (1962).

It is considered that the exchange coefficient $h_{\_fnb}$ equals that calculated for saturated boiling, depending on the chosen correlation.

For the transition between the ONB and the FDB, a ramp is made between the local h_{fl} and h_{lvf} reached during the FDB. More sophisticated methods are proposed in the literature, but they are complex to implement and their robustness is not guaranteed.

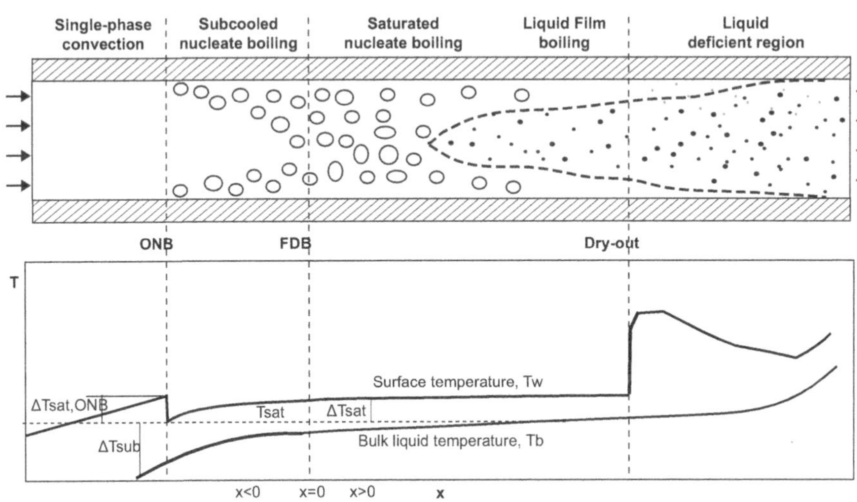

FIGURE 7.3.8 Temperature profiles in an evaporator. (Derived from Ohrby, 2014.)

7.3.4.4 Calculation Process

The general principle of calculations is as follows:

- In sizing mode, once the geometric parameters have been set and the device operating conditions have been chosen, a calculation determines the total surface area of the exchanger and its possible distribution among different zones if it is a steam generator or condenser. For example, for a steam generator, these are the zones corresponding to the economizer, evaporator, and superheater.
- In off-design mode, with the inlet conditions known, the controller seeks a solution such that the total area equals that determined during sizing, but the internal distribution of surfaces can generally be different. The outlet conditions of the hot and cold fluids are then recalculated.

After loading the values of the upstream points of the two fluids, it begins by performing pressure drop calculations. The internal pressures defined in Figure 7.3.7 are then updated, as are the outlet states of the exchanger, which is recalculated several times.

The different zones are then re-evaluated and recalculated as in the TechnoEvaporator class, except that pressure is no longer uniform within the exchanger.

As is the case in water-cooled nuclear reactors, when steam exits the generator in a two-phase state, the accuracy of the heat exchanger calculation using the NTU method is not as good as when there is significant superheat. This results in small errors potentially existing in the steam's exit state.

7.3.4.4.1 Nucleate Boiling

The exchanger is divided into 100 parts, and the different thermal equilibria are calculated. The wall temperature is determined for each interval, assuming that the exchange coefficient on the hot side remains constant.

Different ONB detection correlations have been implemented, and the one that leads to the highest value is retained. This can be replaced with another if desired.

Similarly, different correlations for FDB detection (or NVG) have been implemented, and the one that leads to the highest value is retained.

Some examples are given in Wang et al. (2011) and Ghione (2017).

$h_{\_fdb}$ is estimated to equal that of saturated boiling.

Figure 7.3.9 shows the types of results obtained. It gives the Tw-Tsat difference in degrees and the value of the exchange coefficient in kW/m^2/K as a function of the reduced length of the economizer. The single-phase exchange coefficient decreases slowly due to the increase in cold fluid temperature; then, from the ONB, the two-phase exchange coefficient takes over at an abscissa of approximately 0.7. In this case, the FDB is not reached when the quality becomes equal to 0.

7.3.4.4.2 Saturated Boiling

The six available correlations are calculated, with the chosen one providing the two-phase Nusselt value.

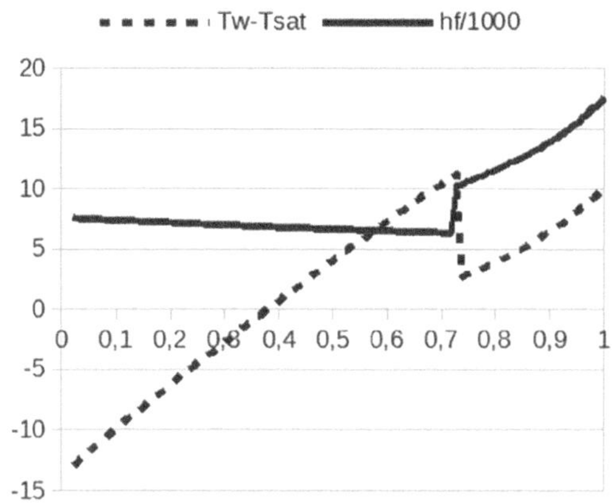

FIGURE 7.3.9 Changes in the economizer.

7.3.5 Calculation of Multizone Exchangers

The calculation of U by $\frac{1}{U} = \frac{d_{hh}}{\lambda_h Nu_h} + \frac{e}{\lambda} + \frac{d_{hc}}{\lambda_c Nu_c}$ is only valid for a simple exchanger through which two fluids with constant thermophysical properties flow.

For multizone exchangers such as evaporators and condensers for refrigeration machines or boilers and steam generators, the calculation of U is more difficult because the overall exchanger is actually a series or series–parallel heat exchanger due to the linkages among the three sub-exchangers—liquid (l), liquid-vapor (lv), and vapor (v)—which share the total area of the exchanger. Since only the total area is known, the three parts of the exchanger must be computed simultaneously to find the solution.

The additivity rules that apply are as follows: the total area A is the sum of the areas of the different zones, and the total flux exchanged φ is the sum of fluxes exchanged in the different zones.

For multizone exchangers, the calculation must be conducted in several iterations since UA and NTU must be determined for each zone. Convergence problems may occur.

In this section, we present the equations that need to be solved based on the exchange configurations encountered. In Section 7.4.3, we will outline the correlations implemented in Thermoptim, and we will detail the calculations to be performed to account for nucleate boiling in steam generators.

Several cases must be considered:

- The hot fluid is two-phase and the cold fluid is sensible heat.
- The cold fluid is two-phase and the hot fluid is sensible heat.
- Both cold and hot fluids are two-phase.

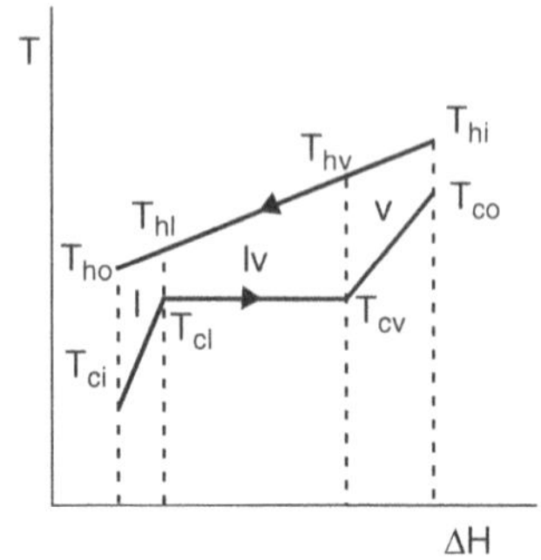

FIGURE 7.3.10 Evaporator temperature/enthalpy diagram.

7.3.5.1 Equations for Evaporators

7.3.5.1.1 Two-Phase Cold Fluid and Hot Fluid Sensible Heat

The equations are as follows (Figure 7.3.10):

$$m_h Cp_h (T_{hi} - T_{hv}) = m_c Cp_{cv} (T_{co} - T_{cv})$$

$$m_h Cp_h (T_{hv} - T_{hl}) = m_c Cp_{clv} (T_{cv} - T_{cl}) = m_c L_c$$

$$m_h Cp_h (T_{hl} - T_{ho}) = m_c Cp_{cl} (T_{cl} - T_c i)$$

The relationships giving epsilon and R are then:

$$\varepsilon_v = \frac{T_{hV} - T_{hl}}{T_{hV} - T_{ci}} \qquad R_{lv} = \frac{m_h C_{Ph}}{m_c C_{pclv}}$$

$$\varepsilon_{lv} = \frac{T_{hv} - T_{hl}}{T_{hv} - T_{ci}} \qquad R_{lv} = \frac{m_h Cp_h}{m_c Cp_{clv}}$$

$$\varepsilon_l = \frac{T_{cl} - T_{ci}}{T_{hl} - T_c i} \qquad R_l = \frac{m_c Cp_{cl}}{m_h Cp_h}$$

If the fluid enters the evaporator in a two-phase state, the equations are slightly different.

7.3.5.2 Equations for Condensers

7.3.5.2.1 Two-Phase Hot Fluid and Cold Fluid Sensible Heat

The equations are as follows (Figure 7.3.11):

$$m_h C_{hv} (T_{hi} - T_{hv}) = m_c Cp_c (T_{co} - T_{cv})$$

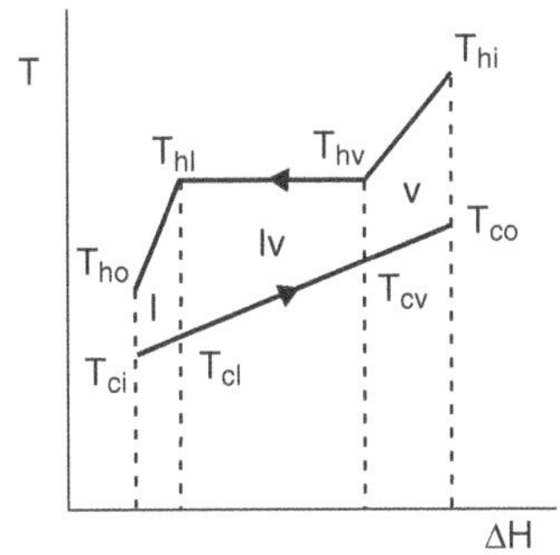

FIGURE 7.3.11 Condenser temperature/enthalpy diagram.

$$m_h Cp_{hlv}(T_{hv} - T_{hl}) = m_c Cp_c (T_{cv} - T_{cl}) = m_h L_h$$

$$m_h Cp_{hl}(T_{hl} - T_{ho}) = m_c Cp_c (T_{cl} - T_{ci})$$

The relationships giving epsilon and R are then:

$$\varepsilon_v = \frac{T_{hi} - T_{hv}}{T_{hi} - T_{cv}} \quad R_v = \frac{m_h Cp_{hv}}{m_c Cp_c}$$

$$\varepsilon_{lv} = \frac{T_{cv} - T_{cl}}{T_{hv} - T_{cl}} \quad R_v = \frac{m_h Cp_{hv}}{m_c Cp_c}$$

$$\varepsilon_l = \frac{T_{hv} - T_{ho}}{T_{hv} - T_{ci}} \quad R_l = \frac{m_h Cp_{hl}}{m_c Cp_c}$$

7.3.5.3 Equations for Two-Phase Inlet Condensers

7.3.5.3.1 Two-Phase Hot Fluid and Cold Fluid Sensible Heat

The equations are as follows (Figure 7.3.12):

$$m_h Cp_{hlv}(T_{hi} - T_{hl}) = m_c Cp_c (T_{co} - T_{cl}) = m_h L_h x_{hi}$$

$$m_h Cp_{hl}(T_{hl} - T_{ho}) = m_c Cp_c (T_{cl} - T_{ci})$$

The relationships giving epsilon and R are then:

$$\varepsilon_{lv} = \frac{T_{co} - T_{cl}}{T_{hi} - T_{cl}} \quad R_{lv} = \frac{m_c Cp_c}{m_h Cp_{hlv}}$$

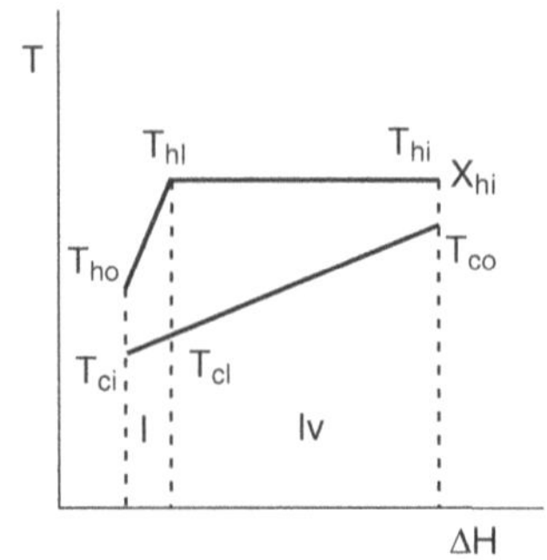

FIGURE 7.3.12 Two-Phase Inlet Condenser temperature/enthalpy diagram.

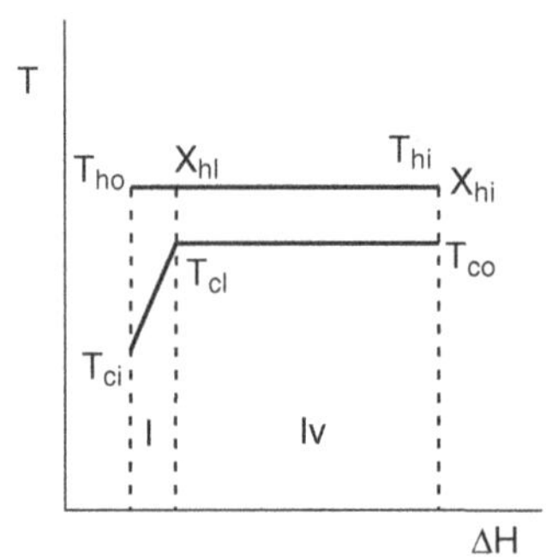

FIGURE 7.3.13 Evapo-condenser temperature/enthalpy diagram.

$$\varepsilon_l = \frac{T_{hi} - T_{ho}}{T_{hi} - T_{ci}} \quad R_l = \frac{m_h Cp_{hl}}{m_c Cp_c}$$

7.3.5.4 Equations for Condensers-Two-Phase Inlet Evaporators

7.3.5.4.1 Two-Phase Hot Fluid and Cold Fluid Sensible Heat

Assuming that the cold fluid exits in the saturated vapor state, the equations are as follows (Figure 7.3.13):

$$m_h L_h (x_{hi} - x_{hl}) = m_c L_c$$

$$m_h L_h x_{hl} = m_c Cp_c (T_{cv} - T_{ci})$$

For the lv part, the notions of epsilon and R lose their meaning, and we have:

$$m_c L_c = U_{lv} A_{lv} (T_{h_v} - T_{c_v})$$

which provides A_{lv} in design mode. Note that if the liquid zone is small compared to the lv zone, this relationship very simply gives the difference in saturation temperatures of both fluids.

For part l, the relationships giving epsilon and R are:

$$\varepsilon_l = \frac{T_{cv} - T_{ci}}{T_{hv} - T_{ci}} \quad R_l = \frac{m_c Cp_{cl}}{m_h Cp_{hlv}}$$

with $m_h C_{p_{hlv}} (T_{hi} - T_{ho}) = m_h L_h$

These equations allow determination of the two saturation temperatures that meet both the constraint on equal fluid state change plateaus and the thermal constraint related to the exchange surface.

7.4 HEAT EXCHANGER SIZING

7.4.1 Heat Exchanger Sizing Screen

The heat exchanger sizing screen (Figure 6.2.6) has been presented in Sections 6.2.1.3 and 6.2.4.1.

Recall that, to calculate heat transfer coefficients and pressure drops, two geometrical quantities are particularly important besides the exchanger surface area: the cross-sectional flow area $\mathbf{A_c}$ and the hydraulic diameter $\mathbf{d_h}$. When the fluid heat exchange coefficients are very different, various devices such as fins can be used to compensate for the difference between their values. These are called extended surfaces, which can be characterized by a surface factor **f** and a fin effectiveness $\boldsymbol{\eta}_0$.

These four parameters are those chosen in Thermoptim to characterize heat transfer in each fluid. The length of the exchanger is also used for certain calculations such as pressure drops.

When calculations are performed, the values of h and Re are displayed in the upper left for each phase of each fluid, and the exchange surface area appears beneath the field dedicated to thermal resistance. Values are displayed in red for the hot fluid and blue for the cold fluid.

7.4.2 Correlations Implemented in Thermoptim

The external classes provided with Thermoptim offer several exchange configurations for each fluid. Parameter values are provided by default but can be modified using the screens shown below. It is also very easy to add new configurations by subclassing those proposed.

As indicated above, approximately 15 correlations are available in the Thermoptim FlowConfigs, and additional ones have been implemented in the TechnoSteamGenerator class for calculating nucleate boiling.

Changes to Nusselt correlation parameters must be made very carefully; otherwise, results may be erroneous. These are numerical fitting parameters whose physical meaning is far from clear. They cannot be estimated based on simple geometric considerations alone.

A screen accessible via a small button labeled "correlation settings" placed in the exchanger sizing screen allows the user, for each fluid, to determine or change the values of correlations used for the calculation (Figure 7.4.1).

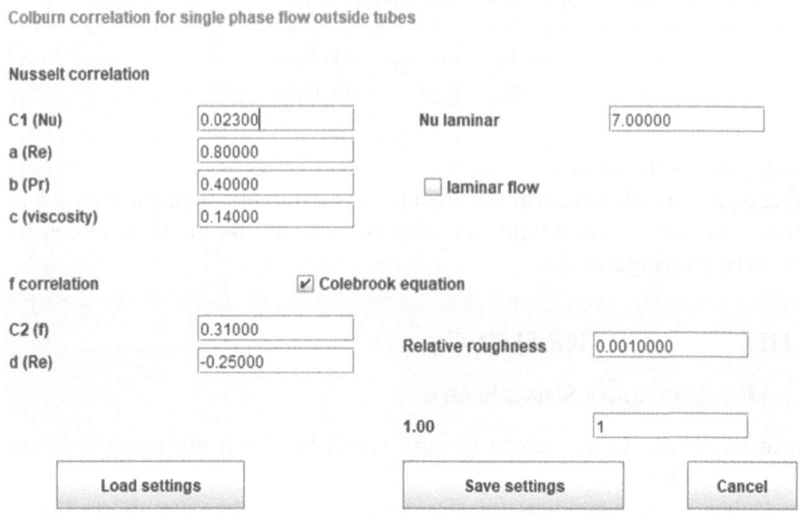

FIGURE 7.4.1 Correlation setting screen.

By clicking “Load Settings,” you load the current values, and by clicking “Save settings,” you apply the values you entered on the screen.

7.5 GEOMETRIC PARAMETER ESTIMATION

The model used being, as stated, a physical model and not purely numerical except for correlations giving Nu, its parameters can be directly estimated from available geometric data or identified from experimental data.

Note that in practice, we rarely know enough about the internal geometry of the exchanger to accurately estimate the parameters just presented. Moreover, given the multiplicity of proposed correlations for heat exchangers, including plates, it is difficult to rely on the literature to know which to choose and how to set the values of geometric parameters characteristic of the exchanger.

The parameters that need to be defined are as follows: always the hydraulic diameter, the cross-sectional flow area, the length of the exchange zone, and optionally the surface factor and fin efficiency if applicable, as well as the wall resistance e/λ.

7.5.1 Direct Setting from Geometric Data

The exchanger geometric description depends on its configuration, which can be quite complex.

By definition, the hydraulic diameter d_h equals the ratio of four times the fluid flow area to the wetted perimeter p: $d_h = 4\ Ac/p$ (if there are insulated walls, the heat exchange perimeter must be considered).

Caution: in the following, expressions are sometimes given for a single mesh and sometimes for the entire section.

7.5.1.1 Flow Outside Tubes

The following example relating to flow outside tubes, parallel to tubes, illustrates this feature (Figure 7.5.1). For a mesh, calling e_1 and e_2 the vertical and horizontal pitches:

$$A_c = e_1 e_2 - \pi d^2 / 4$$

To calculate the wetted perimeter, do not include straight portions that do not participate in exchange:

$$p = 4(\pi d / 4) = \pi d$$

$$d_h = 4A_c / p = (4e_1 e_2 / \pi d) - d$$

For the entire section, these results remain valid if the number of tubes is large enough. Otherwise, corrections may be necessary to account for edge effects.

7.5.1.2 Rectangular Section

Assuming that on one fluid side the flow section can be approximated as a rectangle of dimensions l and h (Figure 7.5.2), we obtain:

$$A = L \cdot l \qquad A_c = hl$$

$$p = 2(l + h)$$

$$d_h = 4A_c / p = 2A_c / (l + h) = 2hl / (l + h)$$

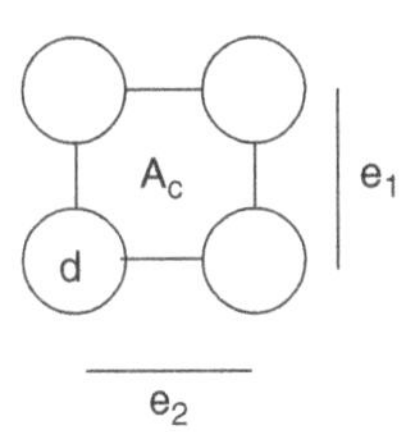

FIGURE 7.5.1 Flow outside of tubes.

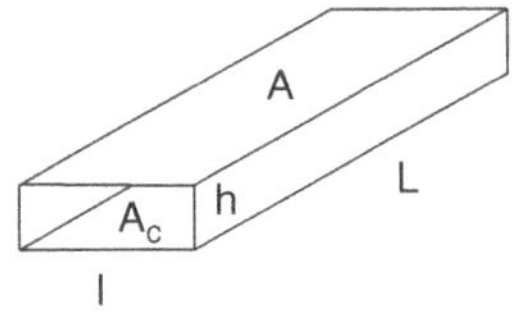

FIGURE 7.5.2 Rectangular section.

In the general case, matters are much more complex, but knowing A, A_c, and d_h is sufficient in practice to characterize the exchange for a fluid.

7.5.1.3 Case of Plate Heat Exchangers

In plate heat exchangers (Figure 7.5.3), the geometry of each face can sometimes be approximated in first approximation as a rectangular section, and the above formulas apply. If the internal details of the corrugations are known, a more precise calculation can be performed. The profiles are generally symmetric, but this is not always the case.

Estimation techniques for A_c and d_h can be found in the literature for some specific configurations, including tube bundles.

Depending on circumstances, the reference length is L_p or L_A (see Figure 7.5.3). The average gap between plates, b, equals the pitch, p, minus the plate thickness, t.

$$b = p - t$$

The free flow area for both fluids, A_c, is:

$$A_c = bw / 2$$

Corrugation surface amplification factor

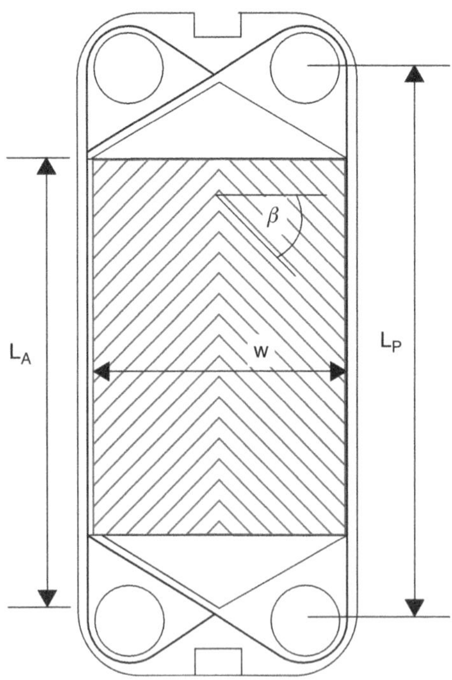

FIGURE 7.5.3 Plate heat exchanger.

The equivalent diameter of a channel d_e, used for calculating the Reynolds number, must be based on actual flow area and not projected section, which leads to defining an amplification factor, μ, the ratio of developed length to projected length, which varies from 1.1 to 1.25 with a typical average of 1.17 (Kumar, 1984).

Knowing that the area is generally developed in a single dimension, the ratio of developed and projected surfaces is the same as that of their lengths.

If corrugations are sinusoidal, calling a the half-height of the sinusoid, μ is given by:

$$\frac{\int_0^{2\pi} \sqrt{1+a^2 \cdot (\sin(t))^2}\, dt}{2\pi}$$

With h/l being the ratio of corrugation height to its pitch (4/7.8 in Figure 7.5.4), this relationship can be replaced by the approximate formula (Figure 7.5.4):

$$\mu = 0.9075e^{(0.9576h/l)}$$

Channel length

If the corrugation pitch p_c, the plate width c, and the chevron angle α (complementary to angle β) are known (Figure 7.5.5), we can determine the number of channels per plate:

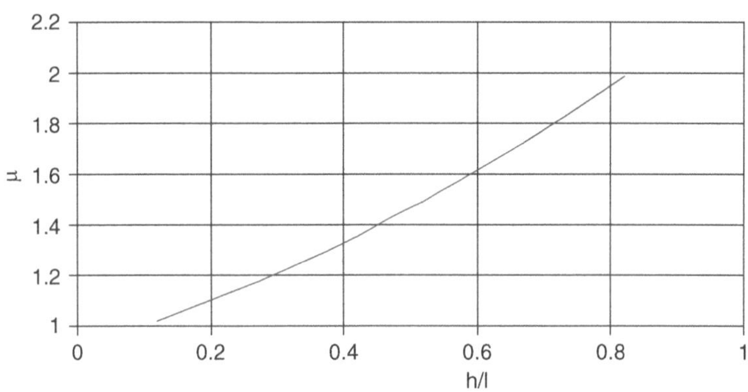

FIGURE 7.5.4 Corrugation surface amplification factor.

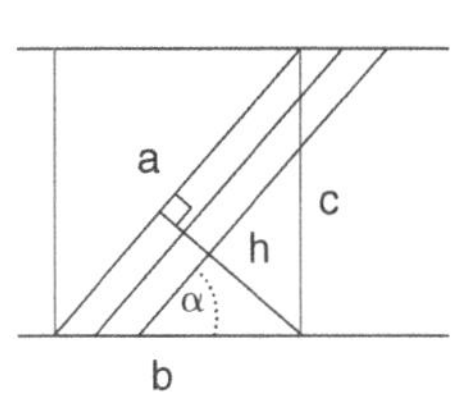

FIGURE 7.5.5 Plate exchanger.

$$c = a \sin\alpha$$

$$b = a \cos\alpha$$

$$b = c \cot\alpha$$

$$h = b \sin\alpha = c \cos\alpha$$

The number of channels per plate is $N_c = h / p_c = c\cos\alpha / p_c$.

If the number of plates N_p and the number of channels per plate N_c are known, we can determine the average channel length L by the relationship $A = N_p N_c p / 2L$, or by making the simplifying assumption $p = \pi d_h : A = N_p N_c \pi d_h / 2L$.

Factor 2 appears because each plate has two sides and the exchange area to consider is that of a single face.

Moreover, the exchange surface equals the product of the number of plates by their width c and useful length L_u:

$$A = N_p c L_u$$

We therefore have $N_c \pi d_h / 2L = c L_u$.

Hydraulic diameter

The hydraulic diameter is given by $d_h = 4A_c / P_w$.

where $P_w = 2(b + w)$

which gives: $d_h = \dfrac{4bw}{2(b+w)}$

As b is very small compared to w, $d_h = 2b / \mu$

It is common not to account for the amplification factor μ. In this case, dh = 2b.

7.5.1.4 Case of Shell and Tube Exchangers

For standard shell and tube configurations, the fluid free flow is approximately one-quarter the shell diameter Dc (Figure 7.5.6).

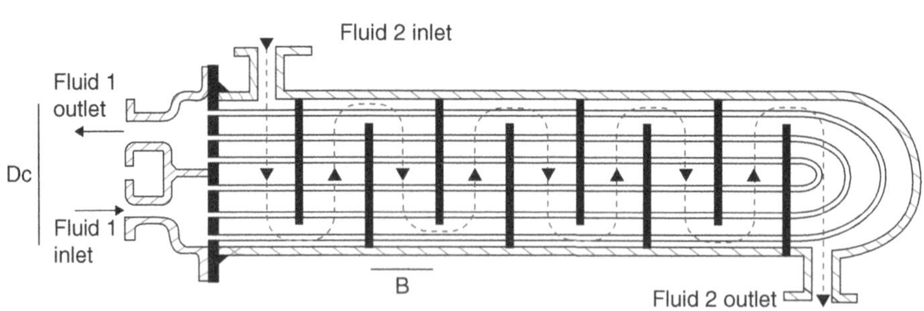

FIGURE 7.5.6 Cross section of a shell and tube exchanger.

Calling B the baffle spacing, e_l the pitch perpendicular to flow, and w the smallest space between two tubes $(w = e_l - d)$, the free flow area between two tubes equals wB, and the total section depends on the number of tubes to be considered. At the center of the shell:

$$Ac = wBDc/e_l$$

The wetted perimeter between two tubes is $p = 2(w + B)$, and the hydraulic diameter $d_h = B2w / (w + B)$.

7.5.2 Identification of Exchanger Parameters

A second approach is to identify the exchanger parameters from experimental data. To facilitate this, specific features have been implemented in Thermoptim to help conduct a series of sizing calculations of area A by varying A_c and d_h for a set of project files corresponding to test points.

If well-validated experimental data are available, the first challenge is to ensure their quality, and the second is to succeed in configuring the selected components.

To facilitate parameter identification, Thermoptim has been equipped with special screens accessible from the "Component sizing" menu of the simulator screen. These screens allow you to:

- Import experimental data into project files.
- Automate simulations on a series of project files.
- Perform sensitivity analysis on the influence of parameters involved in the correlations used to size heat exchangers.

7.6 CONCLUSION

This chapter explained how to model and configure heat exchangers for both sizing and off-design calculations, complementing the fundamentals introduced in Chapter 4.

While the NTU method provides the UA product, sizing requires the separate evaluation of both these terms through the selection of geometric configuration and the calculation of heat transfer coefficients.

Following the NTU fundamentals, pressure drop calculations were addressed for both single-phase flows (friction factors and correlations) and two-phase flows (Lockhart-Martinelli, Müller-Steinhagen & Heck, and Friedel correlations). The modeling of heat transfer covers extended surfaces, Reynolds and Prandtl numbers, and Nusselt correlations for various configurations (tubes, perpendicular flow, fins, plates).

Two-phase exchange received in-depth treatment with correlations for condensation and evaporation.

Equations for multizone exchangers were detailed, and geometric estimation methods range from direct calculation to identification from experimental data via Thermoptim's specialized screens.

BIBLIOGRAPHY

Ayub, Z.H. (2003). "Plate heat exchanger literature survey and new heat transfer and pressure drop correlations for refrigerant evaporators." *Heat Transfer Engineering,* 24(5), 3–16.

Bivens, D.B. & Yokozeki, A. (1994). "Heat transfer coefficients and transport properties for alternative refrigerants." *International Refrigeration Conference Proceedings,* Purdue University, 299–304.

Borishanskiy, B.M., Andreevskij, A.A., Fromzel, V.N., Fokin, B.S., Cistgakov, V.A., Danilowa, G.N. & Bikov, G.S. (1971). "Heat transfer during two-phase flows" (in Russian). *Teploenergetika,* 11, 68–69.

Bowring, R.W. (1962). *Physical model based on bubble detachment and calculation of steam voidage in subcooled region of a heated channel.* Institute for Atomenergi, Halden, Norway, Report No. HPR-10.

Friedel, L. (1979). "Improved friction pressure drop correlation for horizontal and vertical two-phase pipe flow." *European Two-Phase Flow Group Meeting,* Paper E2, Ispra, Italy.

Garcia-Valadares, O. (2003). "Review of in-tube condensation heat transfer correlations for smooth and microfin tubes." *Heat Transfer Engineering,* 24(4), 6–24.

Ghione, A. (2017). *Assessment and improvements of thermal-hydraulic correlations and methods for the analysis of the Jules Horowitz Reactor.* Doctoral thesis, Chalmers University of Technology, Gothenburg, Sweden.

Gungor, K.E. & Winterton, R.H.S. (1986). "A general correlation for flow boiling in tubes and in annuli." *International Journal of Heat and Mass Transfer,* 29, 351–358.

Gungor, K.E. & Winterton, R.H.S. (1987). "Simplified general correlation for saturated flow boiling and comparisons of correlations with data." *Chemical Engineering Research and Design,* 65, 76–84.

Herer, C. & Gallori, D. (1998). *Thermohydraulique des réacteurs à eau sous pression. Techniques de l'Ingénieur,* BN 3050.

Incropera, F.P., DeWitt, D.P. & Bergman, T.L. (2007). *Fundamentals of heat and mass transfer*, 6th ed. Wiley, Hoboken, NJ.

Kakac, S. & Liu, H. (2002). *Heat exchangers: Selection, rating, and thermal design*, 2nd ed. CRC Press, Boca Raton, FL. ISBN 0-8493-0902-6.

Kandlikar, S.G. (2017). "Boiling." In Crowe, C.T., Michaelides, E. & Schwarzkopf, J.D. (eds.), *Multiphase flow handbook,* 2nd ed. CRC Press, Boca Raton, FL, pp. 559–604.

Kays, W.M. & London, A.L. (1984). *Compact heat exchangers.* (3rd ed.). McGraw-Hill, New York.

Kim, S.M. & Mudawar, I. (2013). "Universal approach to predicting saturated flow boiling heat transfer in mini/micro-channels. Part II: Two-phase heat transfer coefficient." *International Journal of Heat and Mass Transfer,* 64, 1239–1256.

Kumar, H. (1984). "The plate heat exchanger: Construction and design." *First UK National Conference on Heat Transfer,* I Chem E Symposium Series, No. 86, 1275–1288.

Levy, W. (1980). *Condenseurs par surface dans les centrales thermiques.* Techniques de l'Ingénieur, B1540.

Lockhart, R.W. & Martinelli, R.C. (1949). "Proposed correlation of data for isothermal two-phase, two-component flow in pipes." *Chemical Engineering Progress,* 45, 39–48.

Muley, A. & Manglik, R.M. (1999). "Experimental study of turbulent flow heat transfer and pressure drop in a plate heat exchanger with chevron plates." *Transactions of the ASME, Journal of Heat Transfer,* 121, 110–117.

Müller-Steinhagen, H. & Heck, K. (1986). "A simple friction pressure drop correlation for two-phase flow in pipes." *Chemical Engineering and Processing,* 20, 297–308.

Ohrby, F. (2014). *Numerical modeling of subcooled nucleate engine cooling systems.* Master's thesis, Chalmers University of Technology, Gothenburg, Sweden.

Saitoh, S., Daiguji, H. & Hihara, E. (2007). "Correlation for boiling heat transfer of R134a in horizontal tubes including effect of tube diameter." *International Journal of Heat and Mass Transfer,* 50, 5215–5225.

Saunders, E.A.D. (1988). *Heat exchangers: Selection, design and construction.* Longman Scientific and Technical, Harlow, Essex.

Shah, M.M. (1979). "A general correlation for heat transfer during film condensation inside pipes." *International Journal of Heat and Mass Transfer,* 22, 547–556.

Shah, M.M. (1982). "Chart correlation for saturated boiling heat transfer: Equations and further study." *ASHRAE Transactions,* 88(1), 185–196.

Shah, R.K. & Bell, K.J. (2000). "Heat exchangers." In Kreith, J.F. (ed.), *The CRC handbook of thermal engineering.* CRC Press, Boca Raton, FL, pp. 4-118–4-170. ISBN 0-8493-9581-X.

Sun, L. & Mishima, K. (2009). "Evaluation analysis of prediction methods for two-phase flow pressure drop in mini-channels." *International Journal of Multiphase Flow,* 35, 47–54.

Thom, J.R.S., Walker, W.M., Fallon, T.A. & Reising, G.F.S. (1965). "Boiling in subcooled water during annuli." *Proceedings of Symposium on Boiling Heat Transfer in Steam Generation Units and Heat Exchangers.* Manchester, United Kingdom.

Thome, J.R. & Cioncolini, A. (2015). *Encyclopedia of two-phase heat transfer and flow I: Fundamentals and methods, Vol. 3: Flow boiling in macro and microchannels.* World Scientific, Singapore.

Wang, J., Huang, Y. & Wang, Y. (2011). "Visualized study on specific points on demand curves and flow patterns in a single-side heated narrow rectangular channel." *International Journal of Heat and Fluid Flow,* 32, 982–992.

8 Modeling and Setting of Displacement Compressors

INTRODUCTION

In this chapter, we focus on off-design modeling of displacement compressors. The study of dynamic compressors, pumps, and fans has been merged with that of turbines in Chapter 9.

The displacement compressor models we propose are based on the assumption that their behavior can be represented with reasonable accuracy by their volumetric efficiency and classical isentropic efficiency. Typical laws allowing representation of these two parameters are given, and we discuss problems related to their identification, which can cause many difficulties due to the scarcity of experimental data available in practice and uncertainties about them.

8.1 BEHAVIOR MODELS

Models implemented in Thermoptim are extensions of those described in Section 3.2. As we have seen, two main approaches can be adopted: either a purely numerical fit, if a satisfactory one exists, or maintaining an approach as physical as possible. The first choice was made here.

Displacement compressors are classified into two main types: reciprocating piston compressors and RPDCs.

Displacement compressors models implemented in Thermoptim are based on the assumption that their behavior can be represented with reasonable accuracy by two parameters: volumetric efficiency λ, which characterizes the actual swept volume, and classical isentropic efficiency η_s. As we have already mentioned, if these do not suit you, it is entirely possible for you to define other models by creating external classes that allow them to be calculated.

Let us recall that **volumetric efficiency** is not an efficiency in the classic sense: it appears in the numerator of the work per cycle expression and not in the denominator, as it reflects the fact that, since the actual swept volume is only a fraction of the total displacement, the work is lower than that theoretically corresponding to the displacement.

Analysis and experimentation on piston compressors have shown that λ can be written as follows (Conan, 1988), with the two largest losses being first β_3, then β_1:

$$\lambda = 1 - (\beta_1 + \beta_2 + \beta_3 + \beta_4 + \beta_5)$$

DOI: 10.1201/9781003606017-9

- $\beta_1 = \varepsilon\left(\left(\frac{P_2}{P_1}\right)^{1/\gamma} - 1\right)$ is the loss due to dead space.
- β_2: loss due to pressure drop in the intake and discharge manifolds.
- β_3: losses due to wall effects (thermal shunt).
- β_4: losses by leakage (at the piston rings and valves).
- β_5: loss due to pressure drop in the inlet and outlet valves.

There is a limit compression ratio r_{max} beyond which the compressor no longer provides any fluid flow. Physically, this means that at the end of expansion, the mass contained in the dead space at pressure P_1 occupies the entire volume of the cylinder, so that the compressor can no longer draw in the upstream gas.

Equation $\lambda > 0$ allows us to express this condition. Limited to its first term, it gives:

$$r_{max} = \frac{P_2}{P_1} < \left(1 + \frac{1}{\varepsilon}\right)^{\gamma}$$

The combination of all losses yields volumetric efficiency laws of variation with respect to the compression ratio and rotation speed that appear as shown in Figures 8.1.1 and 8.1.2.

At a constant compression ratio, they show an optimum at fairly high speed. Indeed, at low speeds, wall effects (β_3) and leakage (β_4) are relatively large and decrease as speed increases. However, at high speed, pressure drops in collectors (β_2) and especially valves (β_5) play an increasingly important role.

Three sources of losses linked to the previous ones mainly affect **isentropic efficiency**, which can be written:

$$\eta = 1 - (\gamma_1 + \gamma_2 + \gamma_3 + \gamma_4 + \gamma_5)$$

- Wall effects (γ_3) are very disadvantageous because they greatly increase the compression work by deforming the isentropic.
- Losses due to valves (γ_5) come next, followed by leakage losses (γ_4).

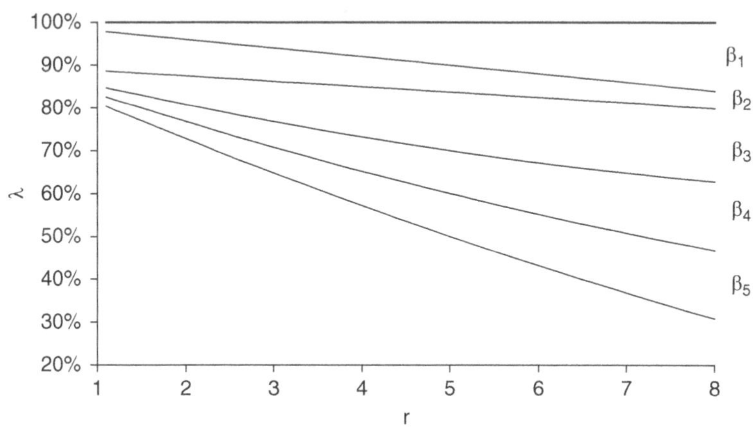

FIGURE 8.1.1 Volumetric efficiency function of compression ratio.

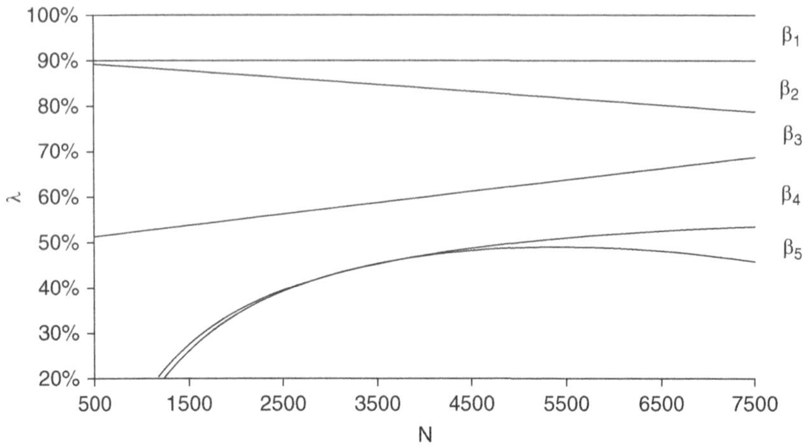

FIGURE 8.1.2 Volumetric efficiency function of rotation speed.

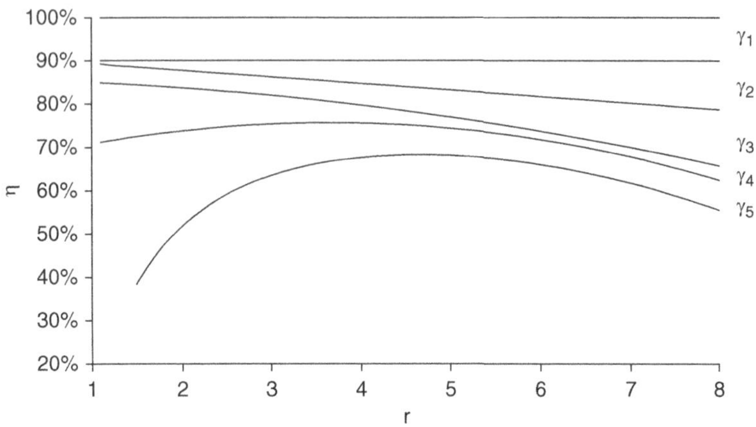

FIGURE 8.1.3 Isentropic efficiency function of compression ratio.

At a given speed, the shape of the compression efficiency law as a function of the compression ratio is given in Figure 8.1.3. It shows the existence of a maximum at relatively low compression ratios (3–5). At a given compression ratio, when speed varies, compression efficiency passes through a maximum close to that of volumetric efficiency.

All of these phenomena are extremely complex to model, so we usually simply adjust numerical models to represent the behavior of displacement compressors.

8.1.1 Operation at Rated Speed and Full Load

At rated speed, the equations we use are extensions of those proposed in Section 3.2 (3.2.6 and 3.2.7) for adiabatic displacement compressors. We show later how to adapt these models to the case of cooled compressors. Depending on the compression ratio, these two characteristics are given by equations (8.1.1) and (8.1.2 and 2bis):

$$\lambda = \alpha_0 - \alpha_1 \frac{P_{ref}}{P_{asp}} \tag{8.1.1}$$

Typically, for reciprocating compressors, α_0 is close to 0.95 and α_1 to 0.038, and for rotary compressors, α_0 is close to 0.9 and α_1 to 0.008.

For isentropic efficiency, we propose two equations, roughly equivalent: the first uses five parameters, and the second three.

$$\eta_s = K_1 + K_2 \cdot \left(\frac{P_{ref}}{P_{asp}} - R_1 \right)^2 + \frac{K_3}{\frac{P_{ref}}{P_{asp}} - R_2} \tag{8.1.2}$$

For the first, typical parameter values are, for both piston and rotary compressors: K_1 close to 0.8, K_2 to 0.0037, K_3 to −0.16, R_1 to 7, and R_2 to 1.2.

Note that, although positive values of K_2 are acceptable, they lead to an increase in η_s for high values of compression ratio, which is physically questionable.

The second equation, of type $y = a + bx + cx^2$, is very easy to identify.

$$\eta_s = K_1 + K_2 \cdot \frac{P_{asp}}{P_{ref}} + K_3 \cdot \left(\frac{P_{asp}}{P_{ref}} \right)^2 \tag{8.1.2a}$$

Note that it is expressed linearly as a function of the inverse of compression ratio and its square. Its parameters have the advantage of having a physical meaning, since it can be rewritten as follows (8.1.4):

$$\eta_s = \eta_{lim} + (\eta_{max} - \eta_{lim}) \cdot \left(2 \left[\tau_{max} \frac{P_{asp}}{P_{ref}} \right] - \left[\tau_{max} \frac{P_{asp}}{P_{ref}} \right]^2 \right)$$

η_{lim} is the asymptotic value of isentropic efficiency for high compression ratios, and η_{max} is the maximum efficiency obtained for a compression ratio equal to τ_{max}.

Knowing K_1, K_2, and K_3, we obtain these values very easily:

$$\eta_{lim} = K_1$$

$$\eta_{max} = K_1 - \frac{K_2^2}{4K_3}$$

$$\tau_{max} = -2 \frac{K_3}{K_2}$$

Figure 8.1.4 shows $\eta_{lim} = 0.6$, $\eta_{max} = 0.8$, $\tau_{max} = 3$.

The estimation of K_i from physical parameters is straightforward:

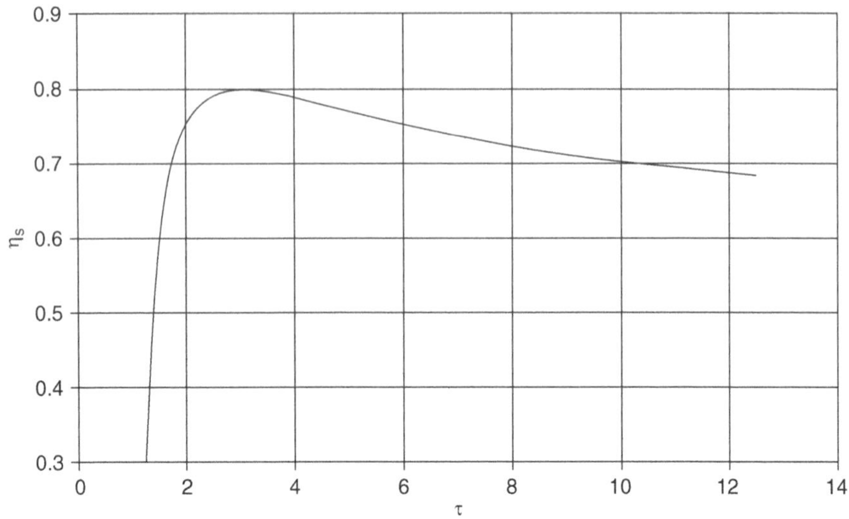

FIGURE 8.1.4 Three-parameter equation.

$$K_1 = \eta_{lim}$$

$$K_2 = \frac{\tau_{max}}{2}\left(\eta_{max} - \eta_{lim}\right)$$

$$K_3 = -\frac{\tau_{max}}{4}\left(\eta_{max} - \eta_{lim}\right)$$

Unlike (8.1.2), (8.1.2a) cannot increase when the compression ratio tends to infinity. However, a blind identification on a relatively small number of grouped points can lead to negative parameter values, which of course has no physical meaning. Therefore, if the number of points is limited, it may be desirable to manually adjust the equation in the form (8.1.2).

Although both models are much less detailed than others, comparisons with existing reciprocating compressors usually give satisfactory results. If this is not the case, it is also possible to use a polynomial model. The model choice depends primarily on available data, the range of compression ratio or rotation speed values we seek to cover, and desired accuracy.

With N being the rotation speed (rpm), and $V_{s,fl}$ the volume swept by the cylinders at full load, the volumetric flow rate at full load is given by (8.1.3):

$$\dot{V} = \lambda \frac{N}{60} V_{s,fl} \qquad (8.1.3)$$

8.1.2 Operating at Partial Load and Speed

Partial load operation of a displacement compressor can be modeled simply by introducing into the calculation of volumetric and isentropic efficiencies a multiplier that is more or less a simple function of partial load factor X_{pl}, itself defined as the ratio of the volume swept at partial load to the volume swept at full load:

$$\lambda_{pl} = X_{pl}\lambda_f \tag{8.1.4}$$

$$\eta_{s,pl} = \eta_{s,fl}\left[1-\left(0.682\cdot e^{-3.96X_{pl}}\right)\right] \tag{8.1.5}$$

The influence of the rotation speed can in turn be taken into account by a simple law of type (8.1.6).

$$\eta_{s,rpm2} = \eta_{s,rpm1}\left[\frac{rpm_1}{rpm_2}\right]^m \tag{8.1.6}$$

The exponent m value depends on compressors: it may, as appropriate, be positive or negative. Everything depends on the (implicit or explicit) manufacturer's choice: whether to favor the compressor efficiency at rated or reduced speed. A positive value of m indicates better efficiency at reduced speed.

We draw the reader's attention to the fact that there may be ambiguity in the definition of the partial load factor, depending on whether one considers the influence of speed or that of a cylinder bypass. For our part, we reserve X_{pl} for the second case and are content to speak of variable speed for the first.

The indications given in this section are tentative: they have not been sufficiently validated on experimental data to be considered final. They have not yet been implemented in external classes on technological design and off-design operation provided with Thermoptim.

8.2 PRACTICAL MODELING PROBLEMS

8.2.1 Technological Screen of Displacement Compressors

The screen in Figure 8.2.1 corresponds to the five-parameter isentropic efficiency model. For other models, screens are slightly different: fields R_1 and R_2 are absent for the three-parameter model.

The fields on the left side of the compressor technological screen allow you to enter the compressor characteristics: the two volumetric efficiency parameters α_0 and α_1, the five isentropic efficiency parameters K_1, K_2, K_3, R_1, and R_2, the reference speed (for which the five parameters were determined), the swept volume at full load $V_s\left(m^3\right)$, and finally a coefficient to estimate possible cooling power.

The fields on the right side of the screen show calculated values of isentropic efficiency, mass flow, and volumetric efficiency.

For rotary compressors, the speed is set and cannot be changed.

In this screen, the compressor is assumed to be adiabatic. If it is cooled, which is sometimes necessary to limit the fluid discharge temperature, we must take into account the thermal power extracted, which can be done as follows. This approach

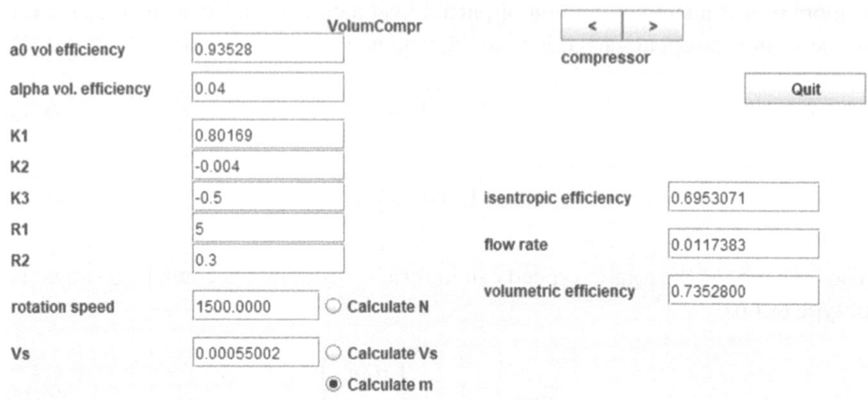

FIGURE 8.2.1 Compressor sizing screen.

is somewhat schematic but corresponds to a case we have treated. The external class can easily be subclassed to implement another algorithm.

The state of fluid leaving the compressor is first calculated as if the machine were adiabatic, and the heat load is applied at constant output pressure, which shifts the temperature value.

The thermal load is considered equal to the product of a coefficient $(K_{NonAdiab})$ by the difference between the isentropic outlet temperature and 290 K:

$$Q = K_{NonAdab}\,(Ts - 290) \tag{8.2.1}$$

Two possibilities exist depending on whether the heat load is known or not:

- If yes, the value of the thermal load is entered in a field of the screen.
- Otherwise, heat load is estimated by (8.2.1).

If $Q_{NonAdiab}$ or $K_{NonAdiab}$ is negative, the compressor is cooled; otherwise, it is heated (which of course has no technological or thermodynamic interest).

8.2.2 Identification of Compressor Parameters

The identification of compressor parameters can cause many difficulties due to the scarcity of experimental data available in practice and uncertainties about them.

The problem is further complicated when the compressor is cooled, because you must either measure the cooling power Q or estimate it.

The identification methodology comprises a number of stages: we first identify λ, then Q, then η at rated speed, and conclude with η at partial speed or load. It is quite possible to apply this methodology to data from catalogs or software published by manufacturers.

8.2.3 Calculation in Design Mode

In design mode, the rotation speed or the swept volume required to provide the desired flow is determined. Calculation is performed taking into account the inlet and outlet pressures and the flow value entered in the compressor flow field.

8.2.4 Calculation in Off-Design Mode

In off-design mode, the compression ratio allows determination of λ and η_s, which sets the compressor flow and outlet temperature. The sequence of calculations is as follows:

1. Knowing the inlet and outlet conditions, compression ratio P_{ref}/P_{asp} and suction volume v_{su} are updated.
2. Volumetric and isentropic efficiencies λ and η_s and actual flow are calculated knowing rotation speed N.
3. Flow volume $\dot{V}$ is deduced by (8.1.3).
4. The compressor sets $\dot{M} = v_{su}\dot{V}$ and propagates this value to connected components.
5. Useful work can then be calculated knowing the isentropic efficiency η_s.

8.2.5 Fixed V_i Rotary Compressors

As explained in Section 3.2.2, a specific characteristic of fixed V_i RPDCs is the possibility of under-compression or over-compression if the internal pressure ratio π differs from the constructive π_i.

The inadequacy of the compressor pressure results in losses as shown in Figure 8.2.2. If the constructive pressure P_i is known, calculation can be performed as follows, τ_{is} being the isentropic work between P_1 and P_i, and τ_{ref} the reference work for an internal isentropic compression:

- if $P_2 > P_i$ $\tau_{ref} = \tau_{is} + (P_2 - P_i)v_1 / V$
- if $P_2 < P_i$ $\tau_{ref} = \tau_{is} - (P_i - P_2)v_1 / V_i$

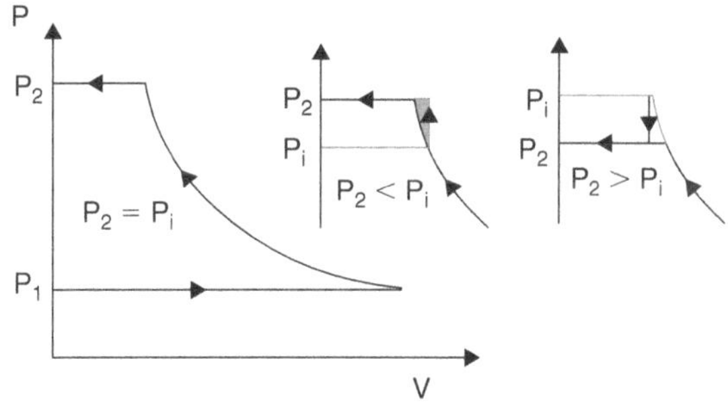

FIGURE 8.2.2 Adaptation of the pressure for fixed V_i RDPCs compressors.

In all cases, the "isentropic" compression work is equal to the real isentropic work between P_1 and P_i, plus the transfer work between P_i and P_2:

$$\tau_{ref} = \tau_{is} + (P_2 - P_i) v_1 / V_i \qquad (8.2.1)$$

The actual useful work is then given by: $\tau = \tau_{ref} / \eta_s$ (8.2.2)

To calculate P_i, it suffices to determine the intersection of isentrope $S = S_1$ and isovolume $v = v_1 / V_i$.

8.3 CONCLUSION

This chapter focused on the off-design modeling of positive displacement compressors. The models assume that their behavior can be represented with reasonable accuracy by two parameters: the volumetric efficiency λ, characterizing the effective swept volume, and the classical isentropic η_s.

The analysis of loss mechanisms includes the effects of clearance volume (dead space), pressure drop in manifolds and valves, thermal wall effects, and leakages.

Volumetric efficiency laws show variation with the compression ratio and rotational speed, with an optimum occurring at relatively high speeds. For isentropic efficiency, two formulations were proposed: a five-parameter model and a simplified three-parameter model with a clear physical interpretation.

Special consideration was given to rotary compressors with a fixed internal volume ratio V_i, where under-compression or over-compression occurs when the operating pressure ratio differs from the built-in ratio.

Practical implementation was demonstrated via the Thermoptim screens, accommodating both adiabatic and cooled compressors.

BIBLIOGRAPHY

Bernier, M. & Bourret, B. (1999). "Pumping energy and variable frequency drives." *ASHRAE Journal,* 41(12), 37–40. December.

Cohen, R., Groll, E., Harden, W.H., Hickman, K.E., Mistry, D.K. & Muir, E. (2000). "Compressors." In Kreith, J.F. (ed.), *The CRC handbook of thermal engineering.* CRC Press, Boca Raton, FL, pp. 4-37–4-58. ISBN 0-8493-9581-X.

Conan, J.G. (1988). *Réfrigération industrielle.* Eyrolles, Paris.

Giampaolo, T. (2010). *Compressor handbook: Principles and practice.* The Fairmont Press, Lilburn/CRC Press, Boca Raton, FL. ISBN 0-88173-615-5.

Sauls, J.R. (1982). "Performance characteristics of fixed volume ratio, positive displacement compressors." *International Compressor Engineering Conference,* West Lafayette: Purdue University. Available at: https://docs.lib.purdue.edu/cgi/viewcontent.cgi?article=1393&context=icec

Trebilcock, F., Lebrun, J. & Winandy, E. (2002). *Reciprocating and screw compressors simulation through polynomial expressions.* Belgium: University of Liège, October.

Pronk, O.I. (2025). "Compressor performance analysis under off-design and surge conditions." *Journal of Fluids Engineering,* 147(10), Article 101209. https://doi.org/10.1115/1.4069229

9 Modeling and Setting of Dynamic Compressors and Turbines

INTRODUCTION

In this chapter, we present different models that can be used to represent the off-design behavior of turbomachinery, mainly turbines and dynamic compressors, but also pumps and fans.

It is important to note that this is a very difficult subject, on which there are few books or reference publications, and therefore one on which there is no real consensus approach in the scientific community and engineering practice.

We present different models that seem relevant, and that we have at least partially implemented in the examples given either in Chapter 6 or in the Thermoptim-UNIT portal. We therefore make no claim to be exhaustive, and we limit our ambition to guiding the reader in exploring this issue, so that they can, if desired, begin building their own models.

The off-design study of turbomachinery is in fact much more complex than that of positive displacement compressors, given the flows that take place in these components.

Two main approaches are generally used for this:

- either consider how the velocity triangle is deformed due to changes in the operating conditions of the machine;
- or use similarity laws and experimental performance maps of the machine to find its new operating point.

So far, for the sake of simplicity, we have limited developments relating to the thermodynamics of turbomachinery. We will now deepen our analysis by making some additions to what was presented in Section 3.3.

9.1 SUPPLEMENTS ON TURBOMACHINERY

9.1.1 Analysis of the Velocity Triangle

We presented in Section 3.3.2.4 the Euler relationship, which links the enthalpy change and thus the useful work to the velocity profile in turbomachinery. In this section, we will complete this presentation with a more detailed analysis of equations for flows in turbomachinery, especially those corresponding to the velocity triangle.

DOI: 10.1201/9781003606017-10

This analysis is necessary if we are to understand the physical phenomena that govern turbomachinery off-design behavior.

In general, the Euler relationship is written:

$$\tau = \overrightarrow{U_2}\,\overrightarrow{C_{t2}} - \overrightarrow{U_1}\,\overrightarrow{C_{t1}}$$

where C_t is the tangential component of absolute velocity C: $C_t = C\cos\alpha$.

If the fluid enters the wheel radially, the Euler relation is simplified and written as:

$$\tau = \overrightarrow{U_2}\,\overrightarrow{C_{t2}}$$

or $\tau = UC_t$, dropping indices and vector notations for ease of writing.

Moreover, the velocity triangle gives in the case of a turbine (Figure 9.1.1):

$$C_t = w\cos\beta - U$$

Since $w = \dfrac{C_r}{\sin\beta}$, we have $C_t = Cr\,\mathrm{cotg}\,\beta - U$

For a compressor, we would have:

$$C_t = U + w\cos\beta$$

$$\text{And } C_t = U + Cr\,\mathrm{cotg}\,\beta$$

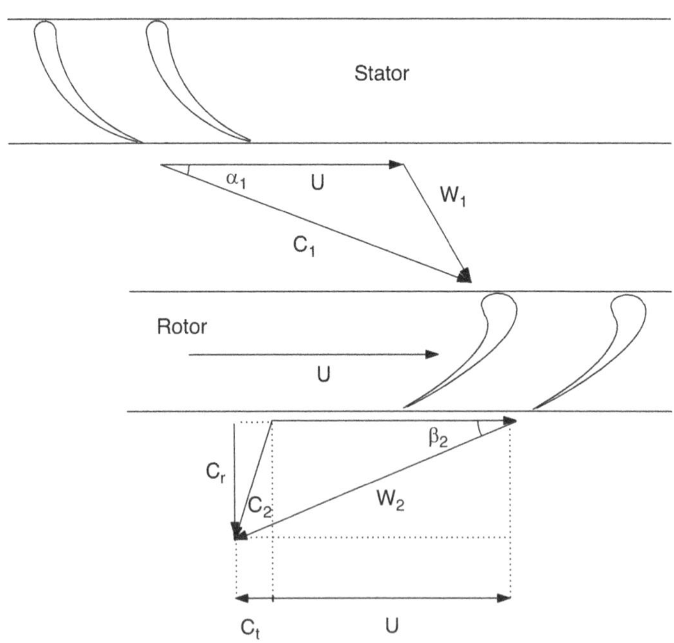

FIGURE 9.1.1 Turbine velocity triangle.

$$\text{or: } \frac{C_t}{U} = 1 + \frac{C_r}{U}\text{cotg}\beta \tag{9.1.1}$$

This relationship can also be written:

$$\frac{\tau}{U^2} = 1 + \frac{C_r}{U}\text{cotg}\,\beta \tag{9.1.2}$$

Yet $\frac{\tau}{U^2}$ is precisely the power factor μ defined in Section 3.1.4, if τ is positive (compressors), and its opposite if τ is negative (turbines). We assume temporarily, in order not to burden the presentation, that τ is positive.

In addition, $\frac{C_r}{U}$ can be expressed in terms of volume flow $\dot{V}$:

$$\frac{\pi De}{v}C_r = \dot{m} \quad \dot{V} = \dot{m}v = \pi DeC_r$$

With $\delta = \frac{\dot{V}}{UD^2}$, we obtain:

$$\mu = 1 + \delta\frac{D}{\pi e}\text{cotg}\,\beta \tag{9.1.3}$$

Relation (9.1.3) binding μ and δ is very general because it is derived immediately from the velocity triangle. It must be verified, as (9.1.1) or (9.1.2), for all types of turbomachinery.

μ, the power factor, represents the work that is put into play in a turbomachinery stage. δ is called the flow coefficient. It is representative, up to a factor, of the ratio between the flow velocity and the circumferential velocity. These two are Rateau coefficients introduced in Section 3.3.3, which are directly involved in the study of turbomachinery similarity.

For a wheel of given shape, the Rateau coefficients perfectly determine the flow in the wheel:

$$\text{cotg}\,\beta = \frac{\pi e(\mu - 1)}{\delta D} \quad \text{and} \quad C_t = \mu U$$

The shape of the velocity triangle is characterized by these coefficients and the geometric characteristics of the wheel. For example, consider the case $\mu = 1$.

We have $\beta = \frac{\pi}{2}$ and $C_t = U$

The outlet triangle is right-angled.

9.1.2 Degree of Reaction of One Stage

In Section 3.4.5, we defined the degree of reaction of a stage, ε, the fraction of enthalpy change that takes place in the rotor.

We can relate the **Rateau** coefficients to the degree of reaction:
By definition, we have:

$$\varepsilon = \frac{h_2 - h_1}{h_r - h_1} = \frac{h_2 - h_1}{\Delta h}$$

$$\varepsilon = 1 - \frac{h_r - h_2}{\Delta h} = 1 - \frac{h_r - h_2}{\tau}$$

If we neglect the kinetic energy at the diffuser outlet,

$$h_r - h_2 = \frac{1}{2}C_2^2 = \frac{1}{2}\left(C_{t2}^2 + C_{r2}^2\right)$$

By introducing the power factor μ, we obtain:

$$\varepsilon = 1 - \frac{h_r - h_2}{\mu U^2} = 1 - \frac{C_{t2}^2 + C_{r2}^2}{2\mu U^2} \tag{9.1.4}$$

Each of the velocity squares can be expressed in terms of the Rateau coefficients.
After completing all calculations, we obtain:

$$2(1-\varepsilon)\mu = \mu^2 + \frac{1}{\pi^2}\left(\frac{D}{e}\right)^2 \delta^2$$

Generally, the ratio $\frac{C_r}{U}$ of flow velocity to circumferential velocity is low, between 0.1 and 0.2. We can neglect its square, and (9.1.4) simplifies to:

$$\varepsilon = 1 - \frac{\mu}{2}$$

Thus, the radial exit vane (μ = 1) has a degree of reaction close to 0.5.

9.1.3 Theoretical Characteristics of Turbomachinery

Let us return to the general equation (9.1.3) and study its form for various turbomachinery types.

9.1.3.1 Centrifugal Compressors

In the case of centrifugal compressors, the velocity triangle can be arbitrary, as β_2 can take values above or below $\pi/2$. Three types of blades may exist (Figure 9.1.2):

$\beta_2 < \pi/2$ **blade lying forward** $\varepsilon < 0.5$ $\mu > 1$

The velocity at the impeller outlet is very large, which allows a high compression ratio, but only on condition that the diffuser is very effective (Figure 9.1.3). We must therefore use a vaned diffuser, whose efficiency drops when operating conditions

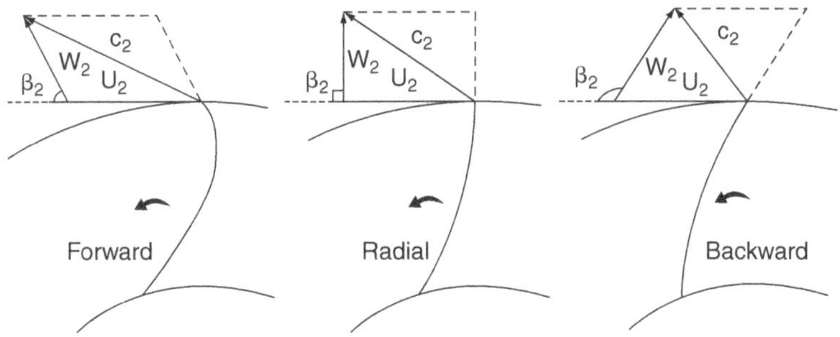

FIGURE 9.1.2 Velocity triangles of centrifugal compressors.

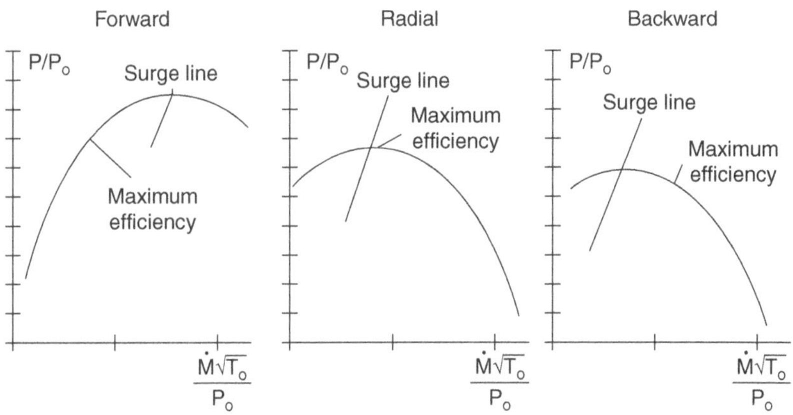

FIGURE 9.1.3 Influence of exit angle on surge.

deviate from the rated conditions. Moreover, the point of best efficiency of these compressors is in the surge zone, which makes them unstable.

$\beta_2 < \pi/2$ **radial exit vane** $\varepsilon = 0.5 \quad \mu = 1$

The velocity at the impeller outlet is still high, and it is therefore best to use a vaned diffuser. Moreover, the point of best efficiency is at the limit of the surge zone, so that the stability of radial vane compressors remains low.

$\beta_2 < \pi/2$ **blade lying back** $\varepsilon > 0.5 \quad \mu < 1$

The velocity at the impeller outlet is much lower than in the previous cases, so that a vaneless diffuser should be adequate. Moreover, the point of best efficiency is outside the surge zone, so that these lying-back blade compressors are much more stable than others. If we desire to have an operating range as wide as possible, these compressors are better suited than others.

Equation (9.1.3) results in a diagram (δ,μ) by a straight line whose slope is positive for $\beta_2 < \pi/2$, zero for $\beta_2 = \pi/2$, and negative for $\beta_2 > \pi/2$.

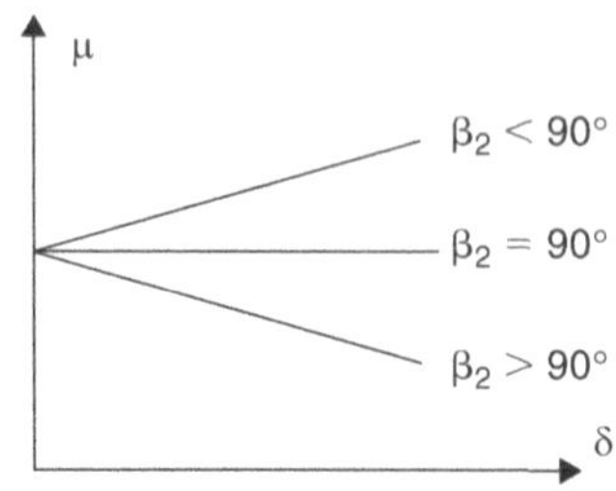

FIGURE 9.1.4 Centrifugal compressor.

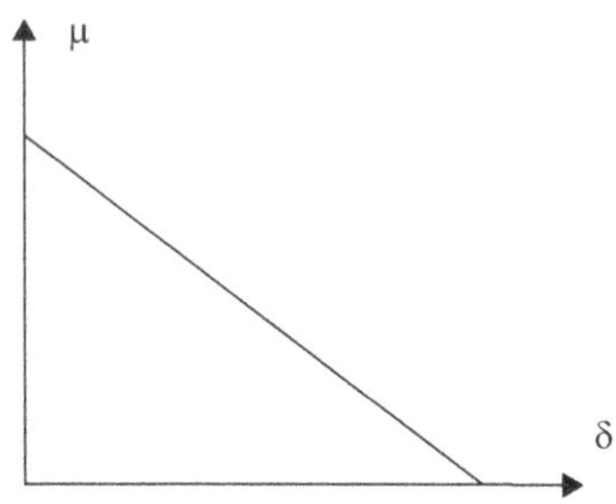

FIGURE 9.1.5 Axial compressor.

The intercept is 1, and its slope depends only on geometry (Figure 9.1.4).

9.1.3.2 Axial Compressors

In the case of axial compressors, the velocity triangle is such that $\beta_2 > \pi/2$, due to conservation of flow velocity and circumferential velocity. In practice, we do not use β, but $\pi - \beta$, and equation (9.1.3) becomes:

$$\mu = 1 - \delta \frac{D}{\pi e} \operatorname{cotg} \beta$$

The line always has a negative slope, with an intercept of 1. Its slope depends only on the geometry (Figure 9.1.5). μ is always below 1, meaning that the compression ratio achievable in an axial compressor is relatively low.

9.1.3.3 Turbines

In the case of turbines, the useful work τ is negative, so μ must be replaced by its opposite $-\mu$ in equation (9.1.3).

For the same reason as in the case of axial compressors, the velocity triangle is such that $\beta_2 > \pi/2$, β_2 is also replaced by $\pi - \beta_2$. Moreover, the assumption $U_1 C_{u1} = 0$ is no longer valid in turbines (Figure 9.1.1), and equation (9.1.3) becomes:

$$-\mu = 1 - \delta \frac{D}{\pi e} \left(\operatorname{cotg} \alpha_1 + \operatorname{cotg} \beta_2 \right)$$

or

$$\mu = \delta \frac{D}{\pi e} \left(\operatorname{cotg} \alpha_1 + \operatorname{cotg} \beta_2 \right) - 1 \tag{9.1.6}$$

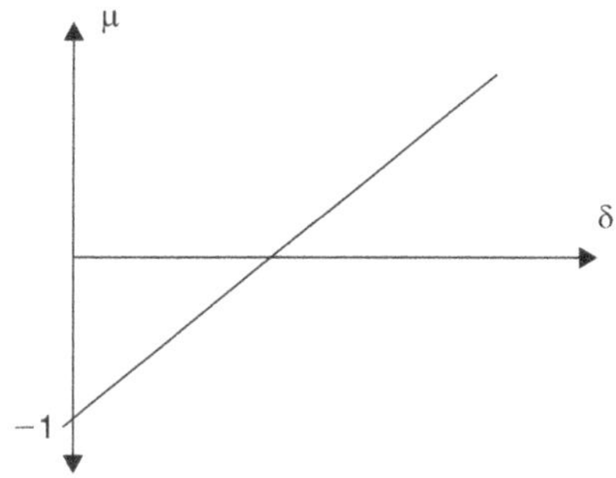

FIGURE 9.1.6 Turbine.

The line has a positive slope and an intercept equal to −1. Its slope depends only on geometry (Figure 9.1.6).

9.1.4 Real Characteristics of Turbomachinery

9.1.4.1 Form of Real Compressor Performance Map

The characteristics that were highlighted in the previous section are theoretical lines, valid for perfect flows.

As previously mentioned, thermodynamics alone is insufficient to theoretically account for the effects of irreversibilities; this requires the use of fluid mechanics.

This allows us to understand the nature of losses due to friction and shocks in the flow, and suggests the appearance of actual characteristics.

On the Mollier (h, s) chart, the theoretical compression results in a vertical line AS, and the actual process in a curve AR, which can often be approximated by a polytropic curve (Figure 9.1.7).

We therefore define a factor $\mu_s = \tau_s / U^2$ which is lower than μ, the ratio μ_s/μ being the internal isentropic efficiency η_{int}.

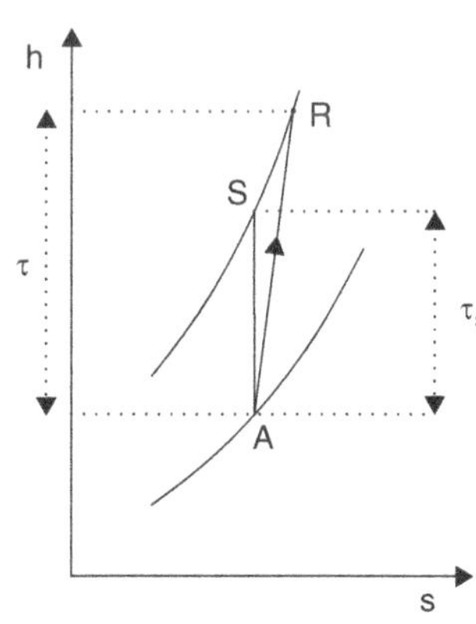

FIGURE 9.1.7 Mollier (h, s) chart.

For each flow-rate value, and thus δ, there is a value of μ corresponding to work expended, and a value of μ_s corresponding to the useful compression effect.

The curve $\mu_s = f(\delta)$ is called the performance curve of the compression stage.

9.1.4.2 Qualitative Analysis of Losses

9.1.4.2.1 Friction Losses

In turbomachinery stage channels, flow is always turbulent, due to the very high speeds reached by the fluid and discontinuities corresponding to the passage of the moving blades (Figure 9.1.8).

Experimental fluid mechanics tells us that in such flows, enthalpy losses are proportional to the square of velocity. In dimensionless variables, this results in a law like:

$$\Delta\mu_f = -k_f\delta^2$$

We must therefore subtract from the theoretical line of heights a parabolic quantity $k_f\delta^2$.

9.1.4.2.2 Shock Losses

The wheel blades are designed for a rated operating flow rate, for which fluid streamlines pass through the impeller with minimum possible impact. When operating conditions deviate from this regime, shock losses are inevitable, as shown by the fixed β_1 and β_2 velocity triangle. Fluid mechanics tells us here that shocks have the effect of creating losses proportional to the square of the difference between the actual flow and the design flow rate. In dimensionless variables, the corresponding loss is of the form (Figure 9.1.9):

$$\Delta\mu_{ch} = -k_{ch}\left(\delta - \delta_0\right)^2$$

If the diffuser is vaned, the same phenomenon occurs, whereas if it is vaneless shock, losses do not occur. It is therefore better, if machines are working in a wide flow range, to adopt vaneless diffusers, while vaned devices can contribute more effectively to the recovery of kinetic energy when the flow-rate variation range is narrow.

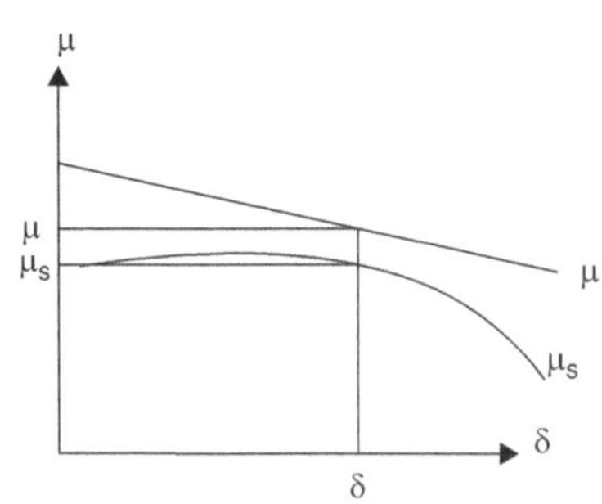

FIGURE 9.1.8 Real performance curve.

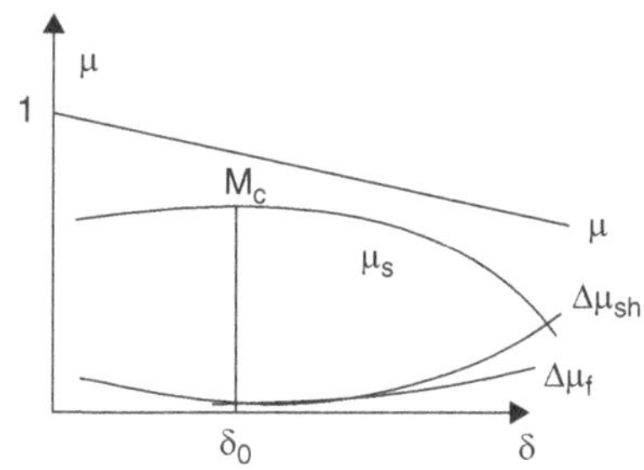

FIGURE 9.1.9 Losses due to friction and shock.

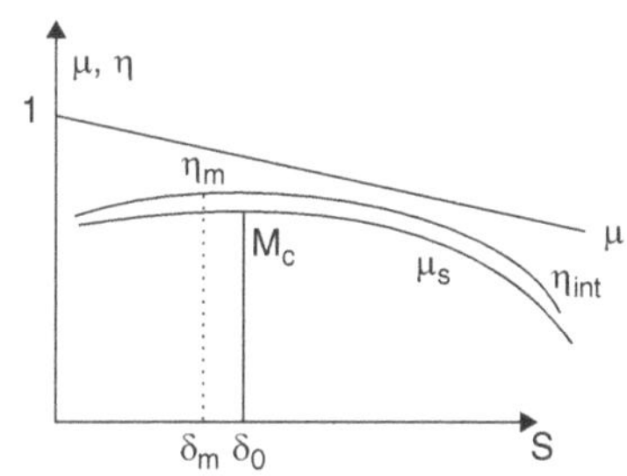

FIGURE 9.1.10 Real performance curve.

9.1.4.2.3 Summary

By bringing together all the above results on one graph in the plane (δ,μ), we can plot point by point the curve giving the power factor μ_s, by subtracting from the theoretical line the two parabolas representing losses. The resulting curve is the real performance curve (Figure 9.1.10).

We can also plot point by point the internal isentropic efficiency η_{int} curve, which is the ratio, for a value of δ, of the ordinate of the actual characteristic to that of the theoretical line.

The maximum efficiency is obtained for a value δ_m slightly less than rated δ_0, due to the shape of loss laws.

We can also plot the internal power P_{int} curve (i.e., the real power minus mechanical losses). We introduce a dimensionless variable for this:

$$\Pi = \frac{vP_{int}}{U^3D^2} \text{ We find : } \Pi = \mu\delta$$

The above analysis is however incomplete in that it neglects mechanical losses and internal leaks, which have the effect of recirculating some of the fluid.

For a given type of equipment, it is often preferable to reason about macroscopic quantities rather than reduced variables such as Rateau coefficients (Figure 9.1.11).

Therefore, instead of working in the (δ,μ) plane, we use the (μ,H) or (μ,r) plane, where H is the head, and r is the compression ratio. We then plot the performance map for different rotation speeds N.

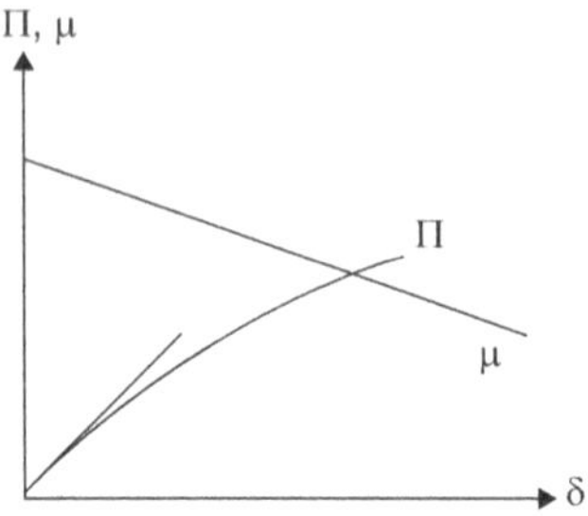

FIGURE 9.1.11 Internal power.

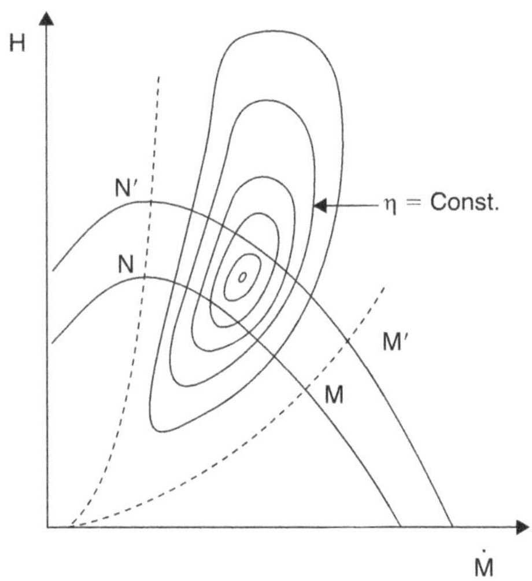

FIGURE 9.1.12 Performance map.

Two homologous points, i.e., points with the same (δ,μ), lie on a vertical axis parabola, since for fixed μ and δ, we have:

$$\frac{\dot{m}'}{\dot{m}} = \frac{U'}{U} = \frac{N'}{N}$$

$$\frac{H'}{H} = \frac{U'^2}{U^2} = \frac{N'^2}{N^2}$$

In particular, the peaks of the different characteristics are arranged on a parabola.

According to this analysis, iso-efficiency curves should be parabolas with vertex at the origin. In fact, given the interference phenomena at very low and very high speeds, as well as leakage and mechanical losses, they are closed curves, and the name efficiency island is sometimes given to the chart prepared by experiment (Figure 9.1.12).

The developments we have just made on velocity triangle analysis and characteristics analytically justify the appearance of turbomachinery performance maps, which we briefly presented in Section 3.3.3, and which we will examine more thoroughly in the rest of this chapter.

They also show the limits of off-design modeling based on the analysis of velocity triangle deformation, which cannot take into account losses by shock and friction. The approach based on similarity rules and machine performance maps allows us to do so and is therefore in our view more accurate.

9.1.5 Factors of Similarity

In Section 3.3.3, we also introduced, in addition to Rateau coefficients μ and δ, two dimensionless magnitudes widely used in the Anglo-Saxon literature, which are particularly interesting for the study of turbomachinery: **flow factor φ** and **enthalpy factor φ ψ.**

By definition (equations (3.3.18) and (3.3.19)):

$$\varphi = \frac{(M_a)_c}{(M_a)_u} = \frac{4\dot{m}\sqrt{rT_a}}{\pi D^2 P_a\sqrt{\gamma}}\frac{60\sqrt{\gamma r T_a}}{\pi DN} = \frac{240\dot{m}rT_a}{\pi^2 D^3 P_a N} = \frac{240}{\pi} \qquad \delta = \frac{240\dot{V}}{\pi^2 D^3 N} \div \frac{\theta}{(M_a)_u}$$

$$\psi = \frac{|\Delta h_s|}{1/2U^2} = 2\mu = \frac{7,200|\Delta h_s|}{\pi^2 D^2 N^2} = \frac{\Omega}{(M_a)_u^2}$$

For the design and representation of turbomachinery performance, one can also use reduced performance maps defined by a combination of similarity factors φ and ψ. Moreover, given the importance of off-design operation, Mach numbers of flow are introduced, usually the wheel Mach number $(Ma)_u$.

Among the most commonly used reduced performance map parameters, we have:

the power factor Λ:

$$\Lambda = \varphi \times \psi = \frac{8}{\pi^4}\frac{\dot{V}|\Delta h_s|}{N^3 D^5} \tag{9.1.7}$$

the specific diameter of the machine Δ:

$$\Delta = \frac{\psi^{1/4}}{\varphi^{1/2}} \tag{9.1.8}$$

the specific speed σ:

$$\sigma = \frac{\varphi^{1/2}}{\psi^{3/4}} \tag{9.1.9}$$

Combining (9.1.7) and (9.1.8) with equations (3.3.18) and (3.3.19), we can also express these last two quantities in terms of the performance map of the machine:

$$\Delta = 1,054D\frac{|\Delta h_s|^{1/4}}{\dot{V}^{1/2}} \tag{9.1.10}$$

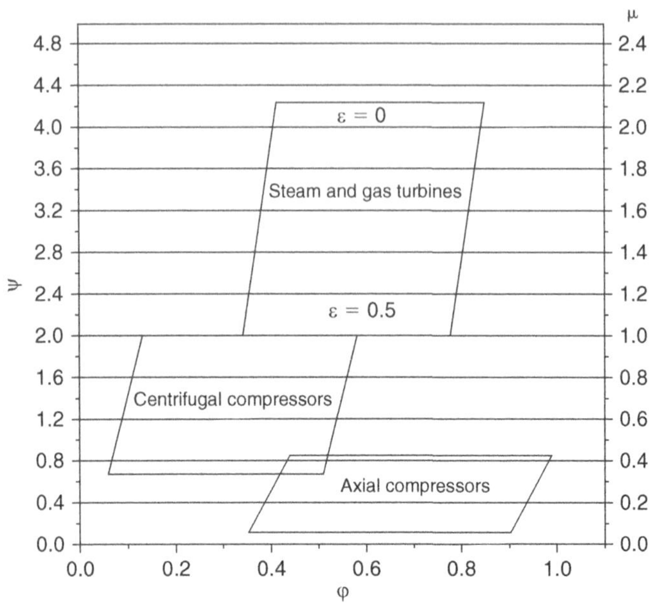

FIGURE 9.1.13 Similarity and turbomachinery stage.

$$\sigma = 2{,}108N\frac{\dot{V}^{1/2}}{|\Delta h_s|^{3/4}} \tag{9.1.11}$$

Similarity factors are often useful to guide the design of turbomachinery. Technological constraints imply that, depending on values of similarity factors, the choice is limited: for a given flow factor, the enthalpy factor or Rateau coefficient of a turbomachinery stage can be obtained from charts of the type shown in Figure 9.1.13.

Indeed, part of the interest of similarity theory comes from the fact that, in off-design operation, actual machines are such that, because of constructive possibilities, the ranges of variation of characteristic factors are relatively narrow.

Besides their applications for the selection of machines, these similarity factors prove also very interesting for studying the off-design behavior of a dynamic compressor, as we shall see.

9.2 PUMPS AND FANS

The mechanical and thermodynamic study of flows in pumps and fans is to a very large extent similar to that of dynamic compressors presented in Section 3.3.3, but with an important simplification: for them, it is perfectly legitimate to assume that the **fluid is incompressible**.

In particular, the expression for ψ is considerably simplified, because $\Delta h_s = v\Delta P$ $\Delta h_s = \Delta P/\rho$. $\Delta h_s = gH$. For a pump, the data provided is generally the head H, and with g being the acceleration due to gravity, $\Delta h_s = gH$.

Let us recall that similarity factors are always defined up to a proportionality factor. Those we have identified for turbines and dynamic compressors were developed by referring them to Mach numbers, which are not expected to be used in the case of pumps and fans. Therefore, their expressions given in the literature are generally different.

Similarity factors φ, ψ, Δ, and σ introduced in Section 9.1.5 are very useful for studying turbomachinery off-design behavior. It is possible to plot the performance map either with φ as the abscissa and ψ as the ordinate or σ as the abscissa and Δ as the ordinate. Figure 9.2.1 shows the performance map of a Gebhardt RZR 315 **fan** as provided by the manufacturer, giving, as a function of the intake flow, first the pressure supplied and also the efficiency at three speeds (1,600, 2,400, and 3,200 rpm).

When expressed in reduced coordinates in the (φ,ψ) and (φ,η) coordinate systems, they overlap perfectly as shown in Figure 9.2.2.

This simplicity is also found for pumps, for which manufacturers generally provide a single set of performance maps, including, as a function of intake flow, first the pressure supplied and second the efficiency.

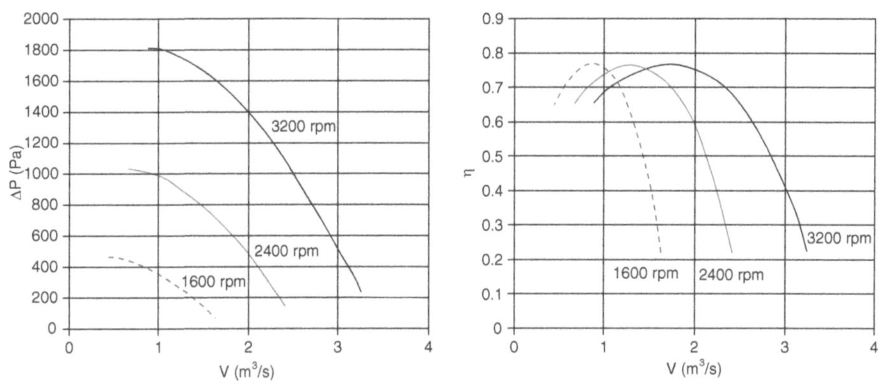

FIGURE 9.2.1 Performance map of a fan.

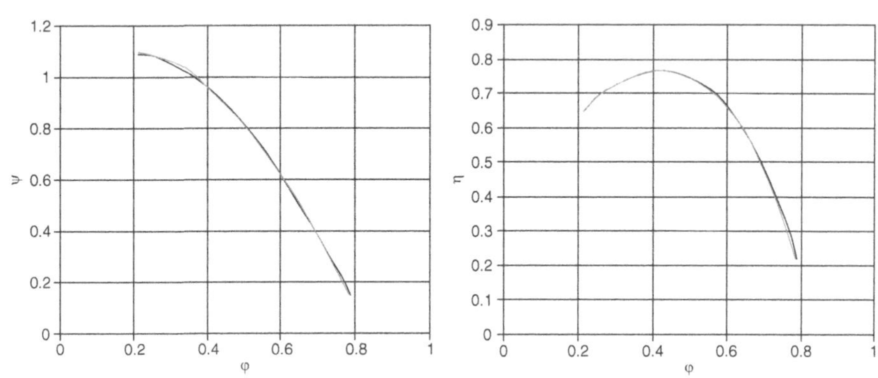

FIGURE 9.2.2 Reduced performance map of a fan.

Figure 9.2.3 shows the performance map of a Grundfos centrifugal pump (model LP 100-200/183 rotating at 2,940 rpm).

By expressing them in reduced variables in the (φ, ψ) and (φ, Λ) coordinate systems, with the power factor being representative of the efficiency, we obtain those in Figure 9.2.4, which eliminates the need for speed data.

The performance maps of **positive displacement pumps** are generally drawn in a frame where the abscissa is the head and the ordinate is the flow or efficiency. They have the form of lines with low slope.

These developments explain why the fan or pump off-design modeling can be performed by simply providing two curves giving, as a function of the flow factor φ, the enthalpy factor ψ and efficiency η or power factor Λ. Such curves can be approximated by polynomials of order 2–4, as appropriate.

As we have seen, reduced performance maps of pumps and fans are relatively simple. It is not the same with those of dynamic compressors and turbines, which we now study.

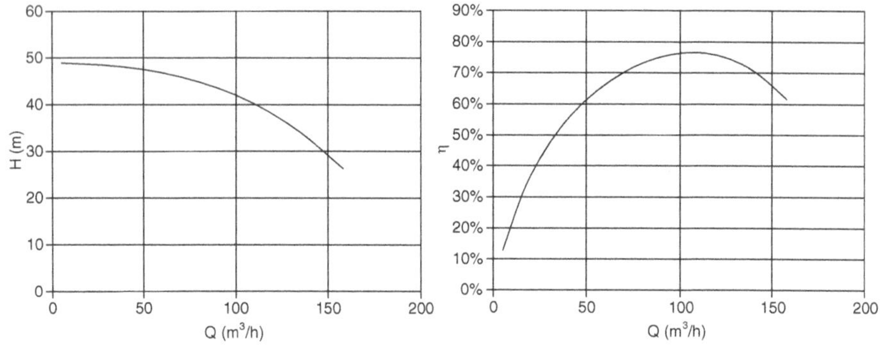

FIGURE 9.2.3 Absolute performance map of a centrifugal pump.

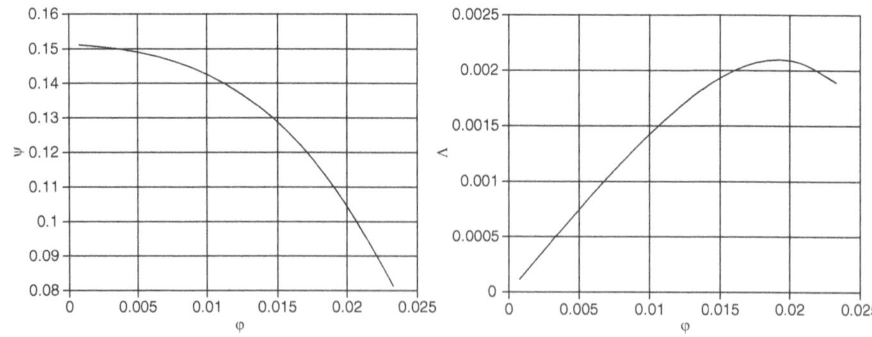

FIGURE 9.2.4 Performance map of a centrifugal pump.

9.3 DYNAMIC COMPRESSORS

9.3.1 Performance Maps of Dynamic Compressors

Models of dynamic compressors that we will consider here are based on those in Section 3.3.3. These numerical models are generally expressed in terms of the similarity factors presented in Section 9.1.5.

In practice, performance maps are usually established for constant values of the corrected rotation speed N_c, and are expressed in terms of the corrected flow $\dot{m}_c$:

$$N_c = \frac{N}{\sqrt{T_a}}; \quad \dot{m}_c = \frac{\dot{m}\sqrt{T_a}}{P_a}; \quad \frac{P_r}{P_a} = f(\dot{m}_c); \quad \eta_s = f\left(\frac{P_r}{P_a}\right) \quad \text{or} \quad f(\dot{m}_c)$$

Curves of equal efficiency can be directly plotted on charts: $P_r/P_a = f(\dot{m}_c)$.

Sometimes, curves are presented in relation to reference values of the corrected mass flow rate or corrected speed.

For dynamic compressors (Figures 9.3.1–9.3.3), the corrected mass flow is used as abscissa. On the ordinate, we plot the pressure ratio or isentropic efficiency. The corrected rotation speed is the curve parameter.

The corrected rotation speed strongly influences the performance of dynamic compressors. This is easily explained by considering that in such a machine, energy is transmitted to the fluid by the rotor in the form of kinetic energy. At most, this energy is $1/2U^2$, that is to say, is proportional to N^2. It is therefore natural that the sensitivity of these machines to regime changes is significant.

9.3.1.1 Axial Compressors

The compression ratio per stage that axial compressors can provide (Figure 9.3.2) is relatively low, generally between 1.2 and 2.

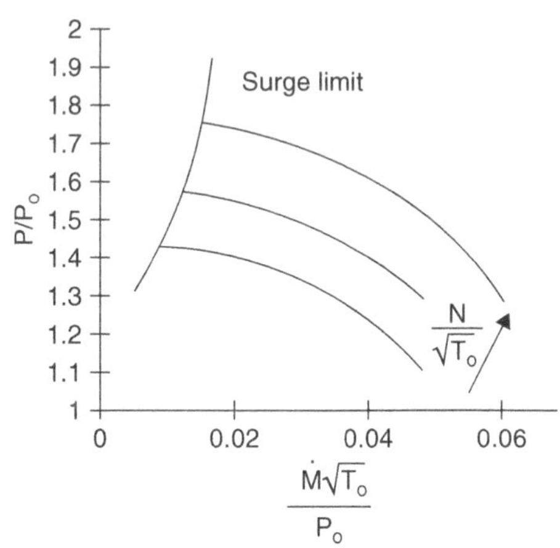

FIGURE 9.3.1 Dynamic compressor performance map.

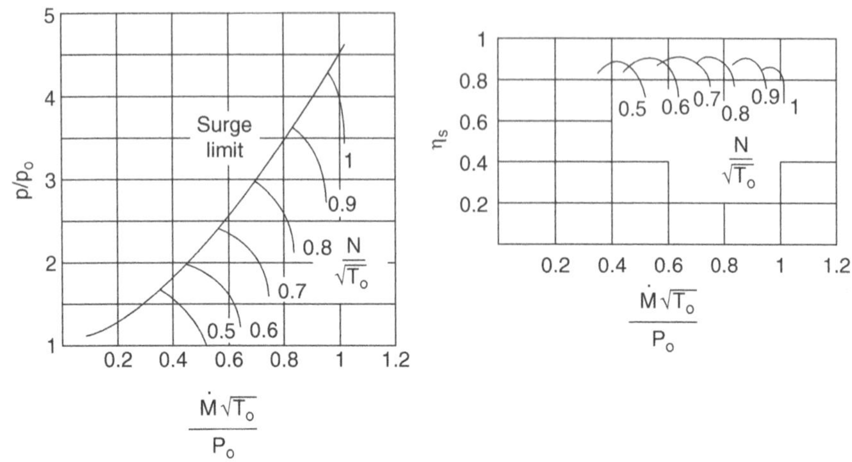

FIGURE 9.3.2 Performance map of an axial compressor.

9.3.1.2 Centrifugal Compressors

Performance maps of centrifugal compressors (Figure 9.3.3) are steeper than those of axial compressors, making them more stable.

Due to the acceleration that the fluid receives during passage through the rotor, the compression ratio per stage they can provide is far greater, between 2.5 and 9.

The analysis of these performance maps shows that for the isentropic efficiency, the effect of speed essentially results in a shift of the curves along the abscissa axis. For the pressure ratio, there is an additional amplification along the ordinate axis.

It is also common to use on the ordinate, not the compression ratio, but a coefficient Ω proportional to isentropic work Δh_s, and on the abscissa a coefficient θ

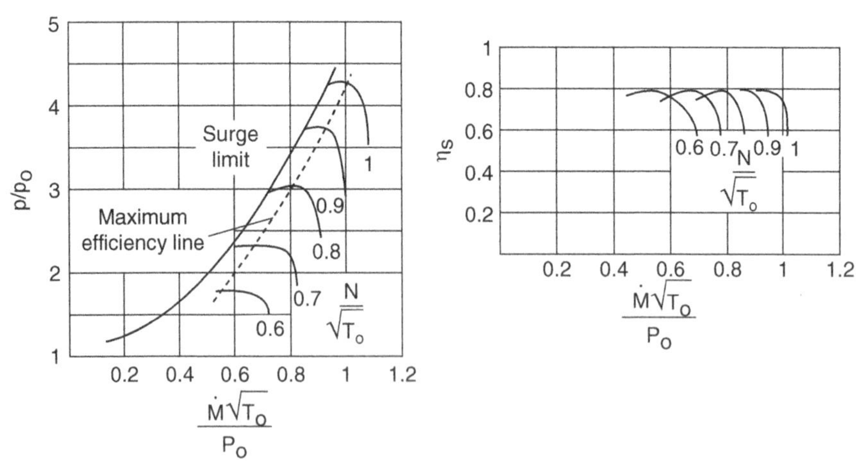

FIGURE 9.3.3 Performance map of a centrifugal compressor.

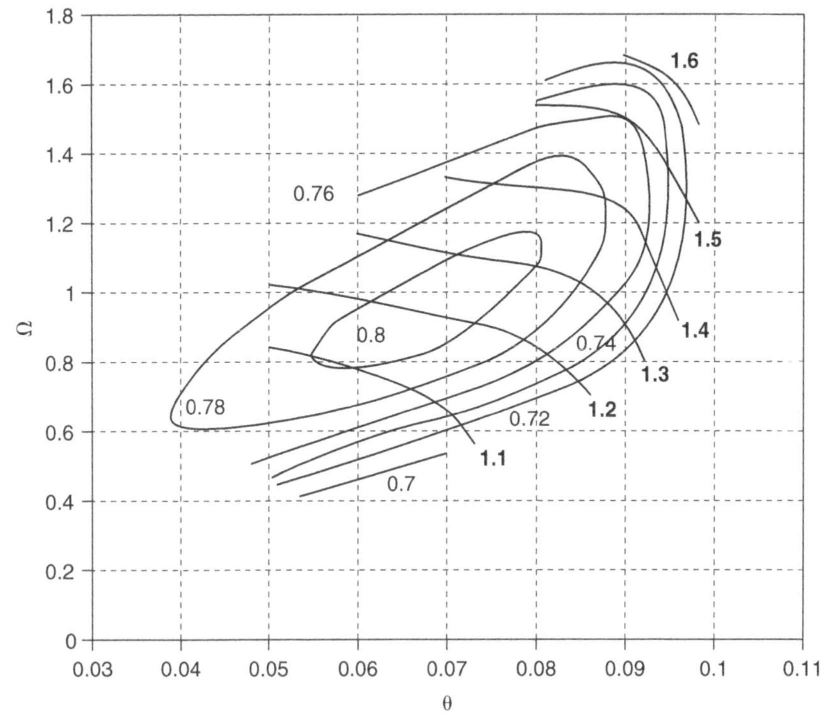

FIGURE 9.3.4 Performance map of an industrial centrifugal compressor.

proportional to corrected mass flow $\dot{m}_c$. You can shift from one coordinate system to another relatively easily.

$$\Omega = \frac{\Delta h_s}{C_s^2} \quad \theta = \frac{\dot{V}}{C_S D^2}$$

For example, Figure 9.3.4 provides the performance map of an industrial centrifugal compressor in a (θ, Ω) coordinate system, with lines in bold representing iso-speeds and Mach number, and thin lines representing iso-efficiencies.

9.3.2 Analysis of Performance Maps of Dynamic Compressors

Figure 9.3.5 shows the reduced performance map, in the (φ,ψ) and (φ,η) coordinate systems, of the industrial centrifugal compressor of Figure 9.3.4.

Iso-speed curves are much more tightly clustered than in Figure 9.3.4, but the differences between them remain significant, especially for efficiencies at extreme speeds: it is far from the simplicity highlighted in the pump and fan study.

However, as shown in Figure 9.3.6, the dispersion of the power-flow performance map is drastically reduced when using the (θ, Δ) coordinate system.

The performance map of Figure 9.3.6 can be approximated with reasonable accuracy by a cubic or a power function: $\Delta = a\sigma^b$ with $a = 2$ and $b = -0.61$, the regression coefficient r^2 being equal to 0.966.

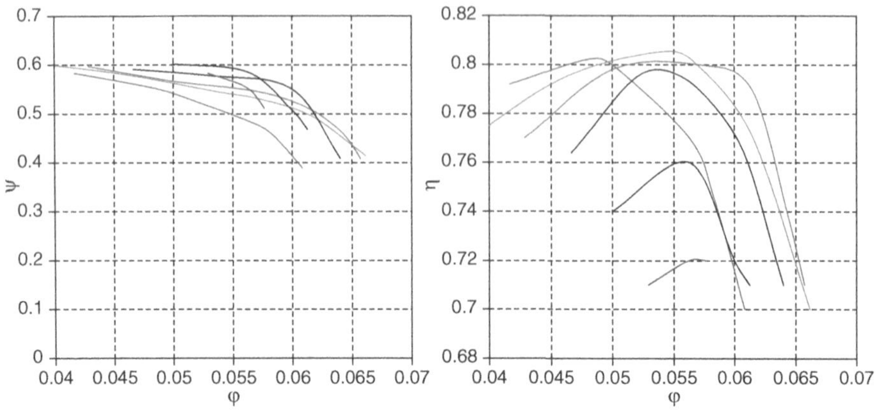

FIGURE 9.3.5 Centrifugal compressor reduced performance map.

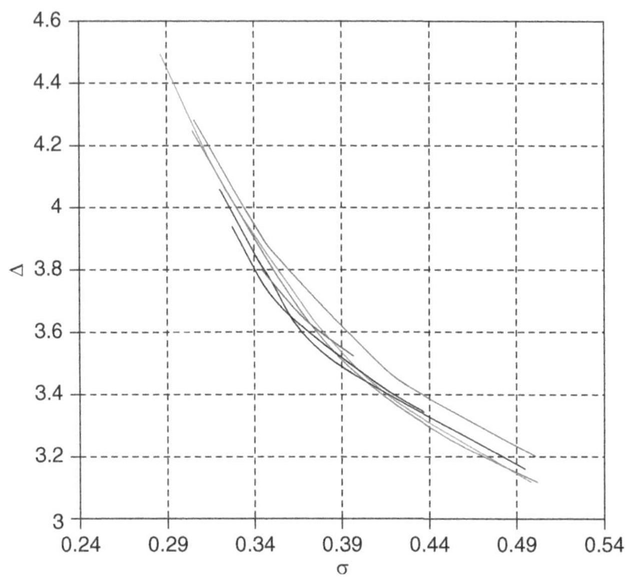

FIGURE 9.3.6 Performance map in the (θ, Δ) coordinate system.

This provides a simple relationship between Δ and σ, which can be used both in design (determination of the diameter of the wheel), and to study off-design operation. If you have a compressor efficiency map, you can either search for the pair (speed, diameter) that maximizes efficiency, or determine efficiency if you know the speed.

These results, drawn here for a particular compressor, are quite general, as shown in the figures below, drawn for different axial and centrifugal compressors.

Figure 9.3.7 shows the centrifugal compressor cell used by M. Pluviose to illustrate some of the exercises suggested in his book (Pluviose, 1988), and Figure 9.3.8

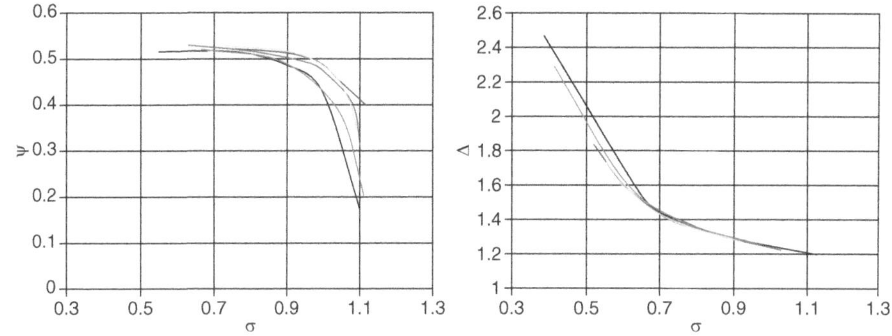

FIGURE 9.3.7 Performance map of a centrifugal compressor cell.

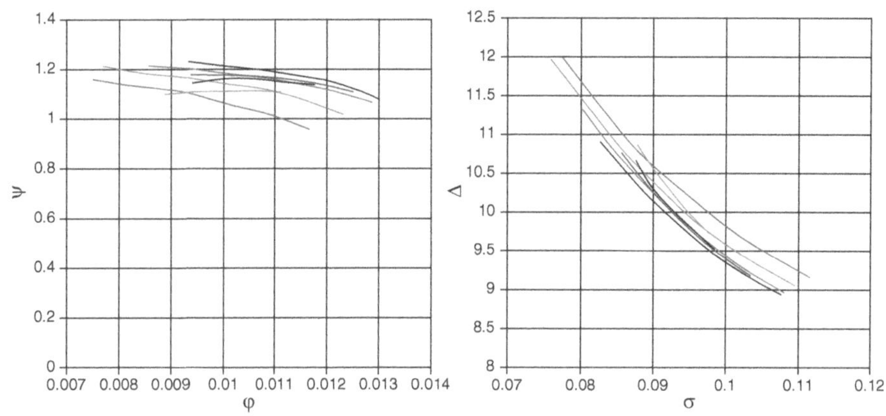

FIGURE 9.3.8 Performance map of a centrifugal compressor.

shows the centrifugal compressor presented by Conan on page 240 of his book (Conan, 1988).

Figures 9.3.9–9.3.11 correspond to axial compressors used in jet engines or gas turbines, the first two representing the same compressor.

In all cases, representation in the (θ, Δ) coordinate system allows us to group the cloud of data points. This is particularly remarkable in the case of Figure 9.3.9. However, the dispersion is not completely reduced, and only detailed adjustments allow the curves to be digitized.

The conclusions we can draw for our purposes are the following: if one has a precise map of the dynamic compressor, such as that of Figure 9.3.4, it is much better to use it because the shapes of efficiency islands can vary significantly with speed. When there are only a few experimental points, one can seek an approximate law of the type that can be identified in the (θ, Δ) coordinate system.

Examination of all these curves shows that, even if the efficiency islands are very interesting visually and allow an engineer to see the machine operating areas, it is much easier to build computer models using two sets of performance maps, one

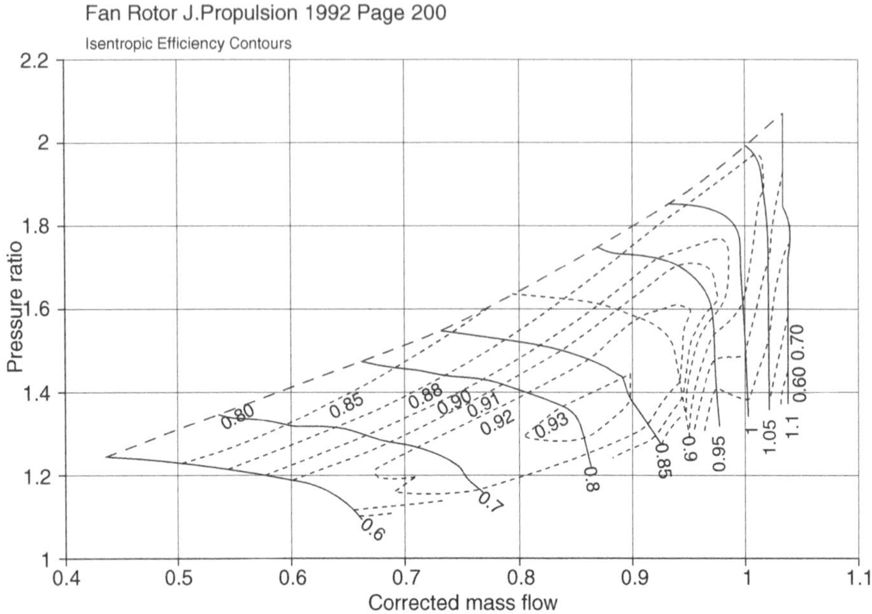

FIGURE 9.3.9 Reduced performance map of an axial compressor, according to the calculator GasTurb by J. Kurzke (www.gasturb.de).

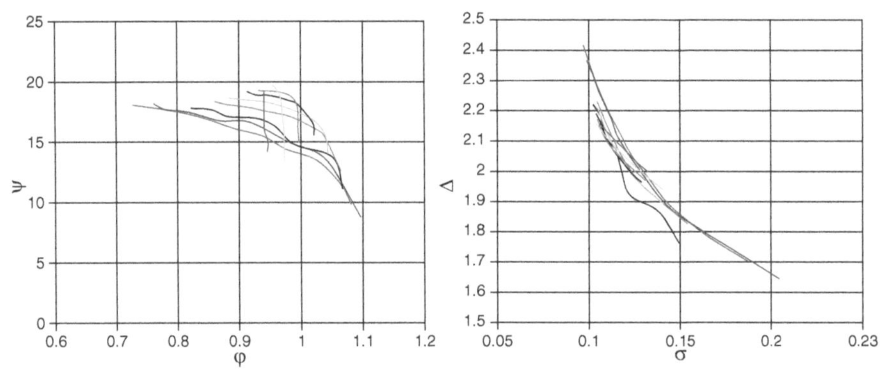

FIGURE 9.3.10 Performance map of an axial compressor.

dedicated to the pressure ratio or ψ, and the other to the machine isentropic efficiency η.

If we choose to use the complete map, various methodological problems may arise, including the fact that the iso-speed curves can be either very steep (Figure 9.3.2), or very flat (Figure 9.3.3).

The interested reader can refer to Kurzke (1996) for complementary explanations.

The digitization of detailed performance maps is an exercise that needs to be done carefully if they are to be reproduced accurately. Given the wide operating range of

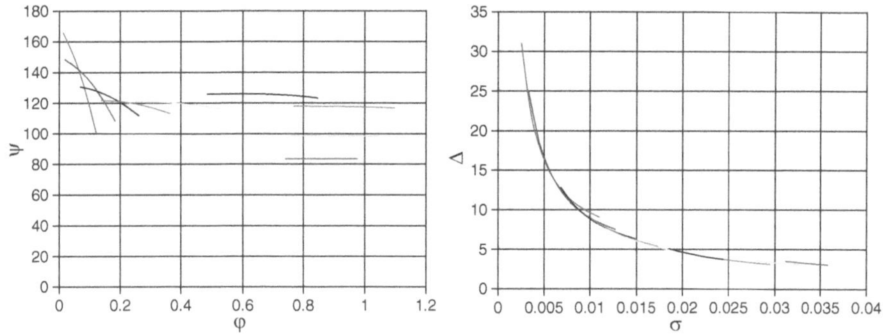

FIGURE 9.3.11 Reduced performance map of an axial compressor.

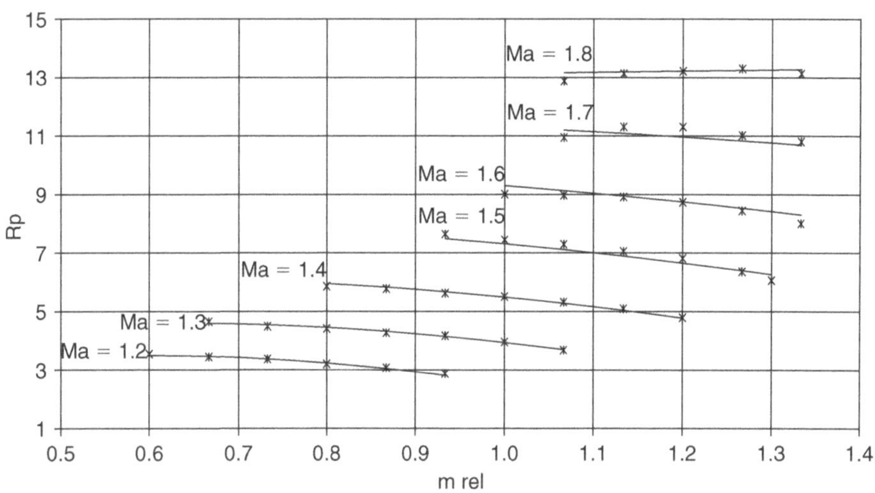

FIGURE 9.3.12 Reconstituted performance map.

dynamic compressors, it is particularly preferable to normalize all curves so that they are uniformly adjusted. Several techniques are presented in the literature (Kurzke, 1996; El-Gammal, 1991; Tirnovan et al., 2008).

Figure 9.3.12 shows the results we can obtain with a six-parameter model: initial performance maps (those of Figure 9.3.8) are represented by asterisks and curve adjustments by continuous dark lines.

The adjustment is given by a degree 3 polynomial function of corrected flow and compression ratio, whose parameters are quadratic polynomials of Mach number. The equation is: $y = a_1 + a_2 y(x + 1/x + y)$, with $a_i = \alpha_i + \beta_i Ma + \gamma_i Ma^2$.

The isentropic efficiency can also be similarly represented, with an adjustment by a cosine function of the type: $y = a_1 \cos\left(a_2 x + a_3 x^2\right)$ leading however to a better estimate (Figure 9.3.13).

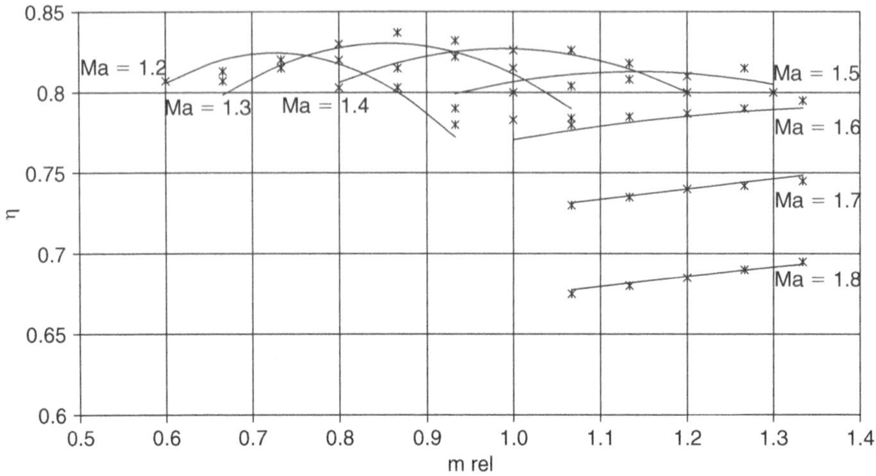

FIGURE 9.3.13 Reconstituted performance map (cosine function).

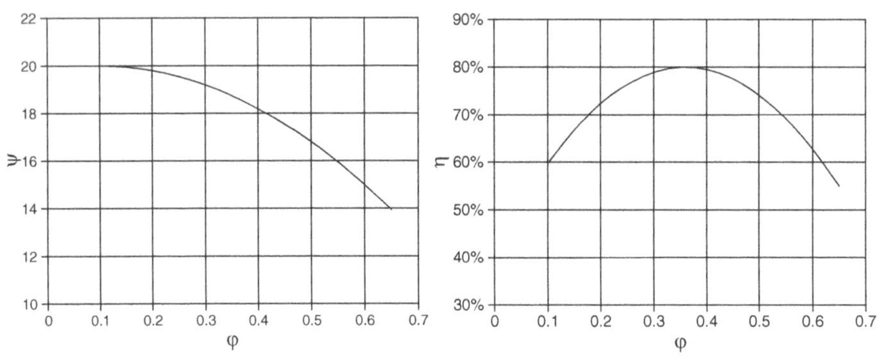

FIGURE 9.3.14 Simplified reduced performance map/corresponding performance map.

9.3.3 Technological Screen of Dynamic Compressors

For a jet engine compressor, we opted for the representation that was justified in Section 9.2 for pumps and fans, but which remains simplified for dynamic compressors. We assumed that a single curve was sufficient to represent the performance map in the (φ,ψ) coordinate system, and another in the (φ,η) coordinate system, as shown in Figure 9.3.14.

Relating these performance maps to corrected flow and replacing ψ by the compression ratio, they take the shape of Figure 9.3.15.

Although it is simplified, this approach allows for a first reasonable modeling of the behavior of a dynamic compressor.

The corresponding technological screen is given in Figure 9.3.16. It uses a series of coefficients describing the two performance maps.

The parameters to define are:

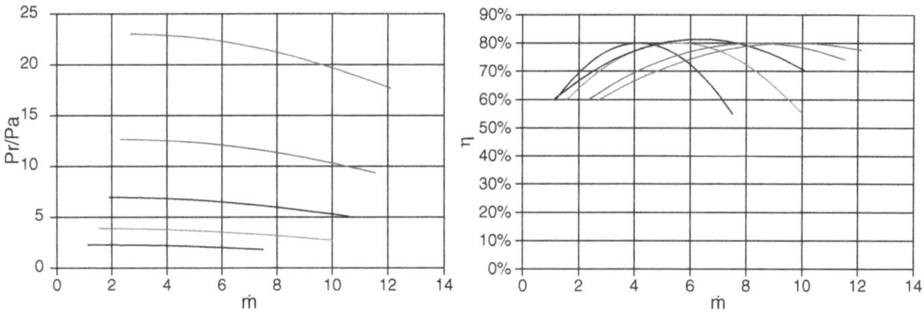

FIGURE 9.3.15 Corresponding performance map.

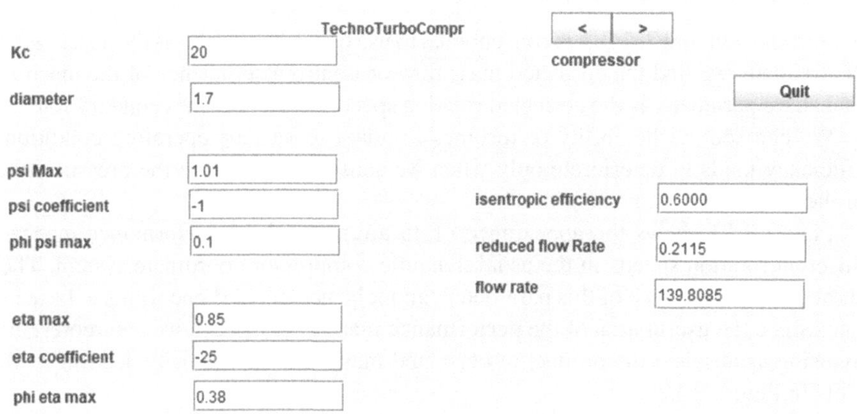

FIGURE 9.3.16 Simplified dynamic compressor sizing screen.

- Kc, compressor characteristic;
- compressor diameter;
- ψ is represented by a branch of a parabola, defined by ψ_{max}, maximum value, ψ_{coeff}, and $\varphi\eta_{max}$, abscissa of the maximum corrected flow mc;
- η is also represented by a branch of a parabola, defined by η_{max}, maximum value, η_{coeff}, and $\varphi\eta_{max}$, abscissa of the maximum corrected flow mc.

The calculated values of isentropic efficiency, reduced flow, and mass flow appear on the right of the screen.

9.4 TURBINES

Turbine models that we will consider here are based on those described in Section 3.5. Very often, the models chosen are based on the assumption that the behavior of an adiabatic turbine can be represented with reasonable accuracy by two parameters: the classical isentropic efficiency η_s, and a magnitude K_0 called the cone or Stodola constant, which characterizes the design flow.

If one searches the literature on turbine off-design modeling, almost all approaches are indeed based on the Stodola and Baumann rules.

The ASHRAE Toolkit (Bourdouxhe, 1993) holds the following assumptions:

- either the flow is considered choked in the turbine, with the isentropic efficiency being assumed constant: $\dot{m}\sqrt{T}/P = \text{Constant}$;
- or the turbine is represented as a Laval nozzle without adiabatic diffuser, and modeled from the Euler relationship and the velocity triangle. The blade orientation angle and the nozzle section must then be provided. The isentropic efficiency is calculated in this case. This approach is similar to how we have chosen to account for Leaving Losses LL as explained in Section 9.4.5.

9.4.1 Performance Maps of Turbines

For turbines (Figure 3.4.4), it is conventional to use the pressure ratio as the abscissa. On the ordinate, we find the corrected mass flow or isentropic efficiency of the machine. The curve parameter is the corrected rotation speed, which plays a secondary role.

We can see the flexibility of turbines to adapt to various operating conditions: efficiency tends to deteriorate only when we significantly reduce the pressure ratio or the speed.

Figure 9.4.1 shows the appearance of an aircraft turbine performance map, for different rotation speeds in the usual dynamic compressor coordinate system. They show that the influence of this parameter cannot be neglected if one wants to be accurate. Since the useful area of the performance map is very small, we can represent it by using as abscissa the product of corrected mass flow and velocity, leading to the plots in Figure 9.4.2.

9.4.2 Isentropic Efficiency Law

For isentropic efficiency, the shape of the performance map in Figures 9.4.1 and 9.4.2 shows that a polynomial representation of curves versus flow can provide good accuracy. It is also possible to retain an equation based on the expansion ratio which is

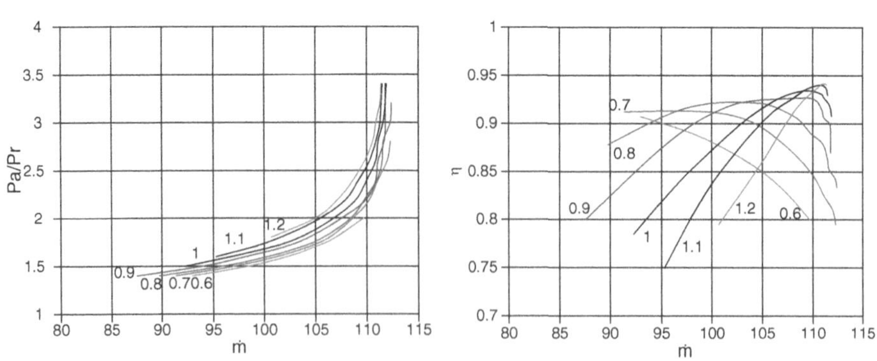

FIGURE 9.4.1 Performance map of the aircraft turbine.

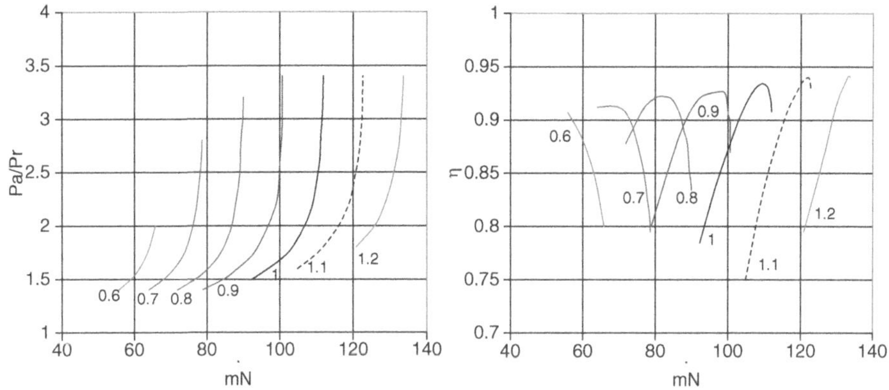

FIGURE 9.4.2 Modified performance map of the aircraft turbine.

very easy to identify, of type $y = a + bx + cx^2$, analogous to the positive displacement compressor law (3.1.2).

$$\eta_s = K_1 + K_2 \cdot \frac{P_r}{P_a} + K_3 \cdot \left(\frac{P_r}{P_a}\right)^2 \tag{9.4.1}$$

Note that it is expressed linearly as a function of the inverse of the expansion ratio and its square. Its parameters have the advantage of having a physical meaning, since it can be rewritten as follows (9.4.2):

$$\eta_s = \eta_{lim} + (\eta_{max} - \eta_{lim}) \cdot \left(2\left[\tau_{max}\frac{P_r}{P_a}\right] - \left[\tau_{max}\frac{P_r}{P_a}\right]^2\right) \tag{9.4.2}$$

η_{lim} is the asymptotic value of isentropic efficiency for high expansion ratios, and η_{max} the maximum efficiency obtained for an expansion ratio equal toτ_{max}.

Figure 9.4.3 is $\eta_{lim} = 0.6$, $\eta_{max} = 0.8$, $\tau_{max} = 3$.

We do not repeat here the developments presented in Section 8.1.1, which explain how these parameters are deduced from one another.

The estimation of K_i from the physical parameters is straightforward:

$$K_1 = \eta_{lim}$$

$$K_2 = \frac{\tau_{max}}{2}(\eta_{max} - \eta_{lim})$$

$$K_3 = -\frac{\tau_{max}}{4}(\eta_{max} - \eta_{lim})$$

This law depends on the rotation speed N, with η_{lim} being generally an increasing function of it, as is well illustrated in Figure 9.4.4, which shows the plot of the isentropic efficiency of the aircraft turbine described above, as a function of the expansion ratio.

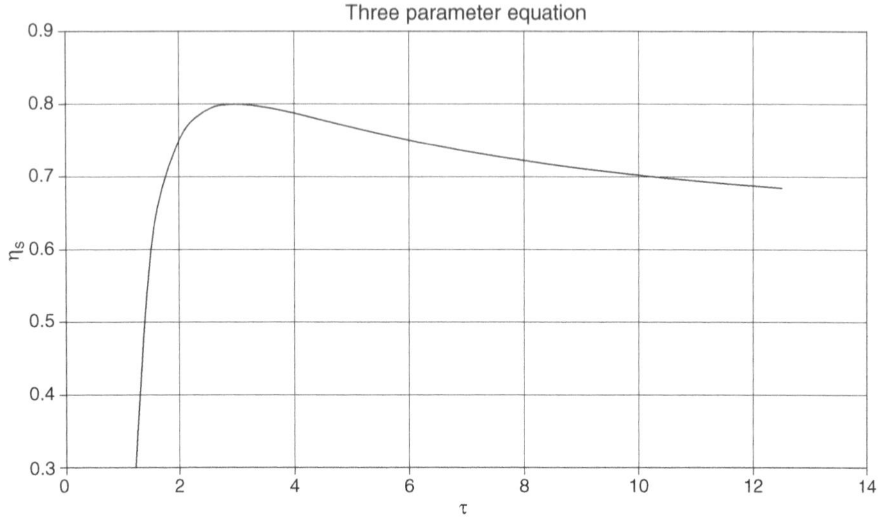

FIGURE 9.4.3 Three-parameter equation.

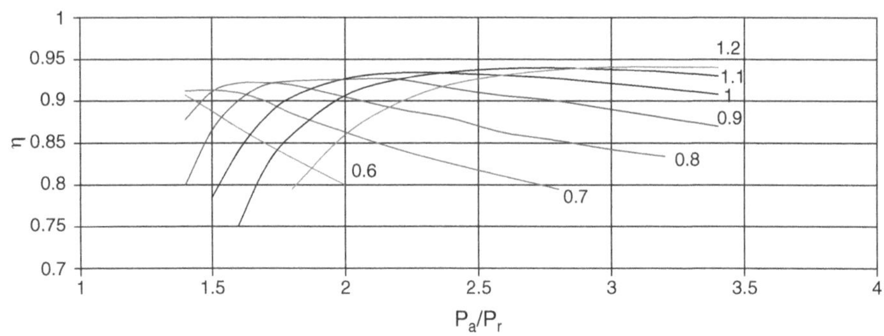

FIGURE 9.4.4 Aircraft turbine isentropic efficiency.

9.4.3 Stodola's Cone Rule

As discussed in Section 3.3.3, manufacturers design turbines respecting similarity rules for the different stages: their flow factor φ and enthalpy factor ψ can generally be considered a constant for a given turbine.

With the notations of Section 3.3.3:

- flow factor φ is the ratio of the two Mach numbers representative of flows in the machines, $(Ma)_C$ and $(Ma)_u$, i.e., C_2/U, C_2, and U being respectively the flow and circumferential velocities;
- enthalpy factor ψ is $2|\Delta h_s|/U^2$.

Moreover, the equation providing the flow rate can be obtained for a multi-stage turbine (Figure 9.4.5) by reasoning as follows (Kane, 2002).

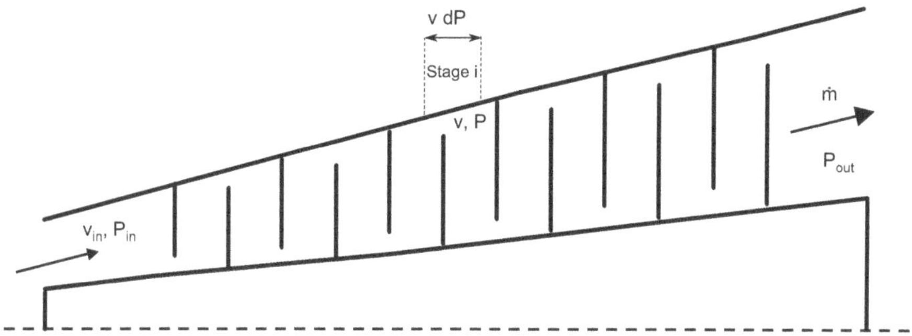

FIGURE 9.4.5 Multi-stage turbine.

The continuity equation is written, with Ai being the outlet section of stage i:

$$\dot{m} = \frac{AiCf}{v} \tag{4.4.3}$$

$$\dot{m} = \frac{AiU\varphi}{v} \tag{4.4.4}$$

$$\dot{m} = \frac{Ai\varphi\sqrt{2|\Delta h_s|}}{v\sqrt{\psi}} \tag{4.4.5}$$

Assuming a polytropic process, $|\Delta h_s|$ can be replaced by vdP/η_p. Raising both members to the square, (9.4.5) can be rewritten as:

$$\frac{\Psi}{2}\left[\frac{\dot{m}}{\varphi Ai}\right]^2 = \frac{dP}{\eta_p v} \tag{9.4.6}$$

9.4.3.1 Expression for Perfect Gases

This equation is integrated over the entire length of the turbine for a perfect gas for which the polytropic evolution follows the law Pv^m = Constant, and provides:

$$\sum_i \frac{\psi}{2}\left[\frac{\dot{m}}{\varphi Ai}\right]^2 = -\frac{1}{\eta_p P_{in}^{1/m} v_{in}} \int_{P_{in}}^{P_{out}} P^{1/m}\, dP \tag{9.4.7}$$

$$\sum_i \frac{\Psi}{2}\left[\frac{\dot{m}}{\varphi Ai}\right]^2 = \frac{m}{m+1}\frac{1}{\eta_p v_{in} P_{in}^{1/m}}\left(P_{in}^{(m+1)/m} - P_{out}^{(m+1)/m}\right) \tag{9.4.7 bis}$$

After completing all calculations, we obtain the so-called cone rule, due to Stodola:

$$\dot{m}\sqrt{\frac{v_{in}}{P_{in}}} = K_0\sqrt{\frac{m}{m+1}}\sqrt{1-\left[\frac{P_{out}}{P_{in}}\right]^{(m+1)/m}} \tag{9.4.8}$$

$$\frac{\dot{m}\sqrt{rT_{in}}}{P_{in}} = K_0\sqrt{\frac{m}{m+1}}\sqrt{1-\left[\frac{P_{out}}{P_{in}}\right]^{(m+1)/m}} \tag{9.4.9}$$

with

$$K_0 = \varphi\sqrt{\frac{2}{\psi}\frac{1}{\eta_p}\frac{1}{\sum Ai^2}} \tag{9.4.10}$$

K_0 is called the Stodola or cone constant. It depends both on structural characteristics, manufacturing quality and fluid properties, which vary slightly depending on the expansion ratio. Most of the time, it incorporates the term $m/(m+1)$.

Expressions (9.4.8) and (9.4.9) are equivalent for the perfect gas (recall that we made the implicit assumption that the fluid follows this law to calculate (9.4.7)). It is thus possible, for design conditions, to calculate the value of K_0, then, depending on operating conditions, determine the new flow rate through the turbine.

The value of the polytropic exponent k is for steam and modern turbines, approximately 1.2–1.3. Under these conditions, $(k+1)/k$ is approximately 1.7–1.8, which is close to 2. Therefore, it is common to find in the literature the Stodola rule written as (9.4.11). This is a quadratic form in space $(\dot{m}, P_{in}, P_{out})$, which justifies the name cone rule.

$$Tin\frac{\dot{m}^2}{K^2} = P_{in}^2 - P_{out}^2 \tag{9.4.11}$$

This expression is further simplified when P_{out} is negligible compared to P_{in} or the flow is choked, leading to:

$$\frac{\dot{m}\sqrt{T_{in}}}{P_{in}} = Const \tag{9.4.12}$$

Stodola's cone rule is only valid as long as the flow remains below a limit value, which is reached when sonic conditions are established in the stator nozzle throat.

For a turbine stage, the critical ratio is given (for isentropic flow) by equation (9.4.13). For n stages, it is raised to the power n.

$$\frac{P_{in}}{P_{out}} = \left[\frac{2}{\gamma+1}\right]^{(\gamma+1)/2(\gamma-1)} \tag{9.4.13}$$

Stodola's cone rule generally applies quite well when the turbine is running at constant speed and with an inlet temperature which varies little. When the speed varies significantly, experimental results generally show that its influence on the characteristics cannot be neglected, so that if one wants to be accurate, it should be taken into account, using a turbine map. When the inlet conditions differ significantly from those for which K_0 was determined, Stodola's cone rule should be used with caution.

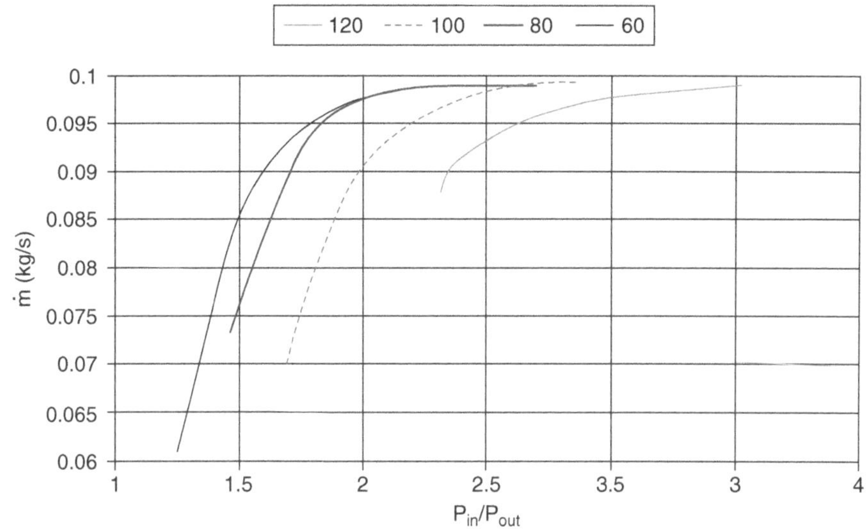

FIGURE 9.4.6 Performance map of a radial turbine.

Figure 9.4.6 shows the shape of a radial turbine performance map based on test results provided by (Rogers, 1966), for different speeds, as a percentage of design rotation speed. The influence of speed for low flows is rather significant.

In the wet zone, however, to proceed with the integration of (4), it is no longer possible to assume that the fluid is an ideal gas. We can establish an appropriate equation of state along a polytropic defined as $|\Delta h| = |\Delta h_s| / \eta_p$, which can be quite complicated but can be done numerically for different values of η_p and upstream and downstream pressures. It is also possible to look for a relationship between P and v in the wet zone of interest and to use it to integrate the expression:

$$\int_{P_{in}}^{P_{out}} \frac{dP}{v}$$

9.4.3.2 Generalized Expression in the Wet Zone

One can also establish a relation of the type $Pv^k = \text{constant}$, with k in this case not having the same meaning as if the gas were perfect. It is only this type of relation that connects the upstream and downstream points. The relation is then integrated as before and leads to a generalized expression of Stodola's rule:

$$\dot{m}\sqrt{\frac{(k+1)v_{in}}{kP_{in}}} = K_0\sqrt{1 - \left[\frac{P_{out}}{P_{in}}\right]^{(k+1)/k}} \tag{9.4.14}$$

It is formally identical to equation (9.4.8), but the k that appears there is not the usual polytropic exponent. While the term $(k/(k+1))$ is most often included in K_0, which does not pose a serious problem for a perfect gas, it must be taken into account in the humid zone, as k varies greatly from one turbine stage group to another and depends on the moisture content.

We can thus use this generalized relationship to study the partial load operation of a turbine. This is what we will do with the data from the Evolutionary Power Reactor (EPR) of Flamanville in case study 10.5.

9.4.4 Baumann Rule

The Baumann rule takes into account the efficiency degradation of a turbine operating in a humid zone.

Relationship (9.4.15) provides the isentropic efficiency of a humid stage η_{hum}, where α is the Baumann coefficient, close to 1, and η_{dry} the isentropic efficiency for dry steam.

$$\frac{\eta_{hum}}{\eta_{dry}} = 1 - \alpha(1 - x) \tag{9.4.15}$$

It is clear that the calculations of x and η_{hum} are related and cannot be made independently. We can however note that in the vapor-liquid equilibrium zone, the enthalpy of fluid leaving the turbine is a linear function of x, which allows us to write:

$$\Delta H_{is} = H_a - H_{is} = H_a - (H_l + x_{is}L_0)$$

$$\Delta H_{real} = H_a - H_r = H_a - (H_l + xL_0)$$

$$\eta_{hum} = \frac{\Delta H_{real}}{\Delta H_{is}} = \eta_{dry}(1 - \alpha(1 - x)) \tag{9.4.16}$$

$$\eta_{hum} = \frac{H_a - (H_l + xL_0)}{\Delta H_{is}} = \eta_{dry}(1 - \alpha(1 - x))$$

Equation (9.4.16) is a first-degree equation in x which is solved easily and avoids iterative calculations.

When the expansion takes place partly in the dry vapor zone and partly in the wet steam zone, the Baumann rule is to be applied only in the latter. One way to do this is to replace x in equation (9.4.16) by the ratio of the enthalpy drop in the humid zone to the total enthalpy drop.

9.4.5 Leaving Losses

When the downstream pressure drops, for a given free flow area, the flow velocity increases, so that the kinetic energy lost may not be negligible. Physically, this means that some of the available power is not recovered on the turbine shaft and is converted downstream into heat. The corresponding loss assessment can be made on the basis of the velocity triangle, which assumes that the downstream flow section is known, as well as the circumferential speed U and outlet angle β. In the equation below, C_2 represents the flow velocity C_f with our usual notation.

These LL are equal to $C_2^{2/2}$, with (Figure 9.1.1):

$$C_t = Cr \operatorname{cotg}\beta - U$$

$$C_2^2 = Cr^2 + Ct^2 = Cr^2 + (Cr \operatorname{cotg}\beta - U)^2$$

In practical terms, if one does not want to modify the isentropic efficiency expression, the calculation of LL can be performed as follows:

- we first determine the turbine outlet conditions in the absence of LL;
  - knowledge of the specific volume and the exhaust section allows us to determine Cr and thus LL;
  - work done by the turbine is equal to that in the absence of LL minus LL. Knowing the isentropic work (in the absence of LL because we assume the machine is perfect), we deduce a new value of isentropic efficiency;
  - recalculation of the turbine with this isentropic efficiency value gives the exit point state taking into account LL.

9.4.6 Technological Screen of Turbines

The screen of Figure 9.4.7 represents the turbine model presented above.

The meaning of the fields is as follows:

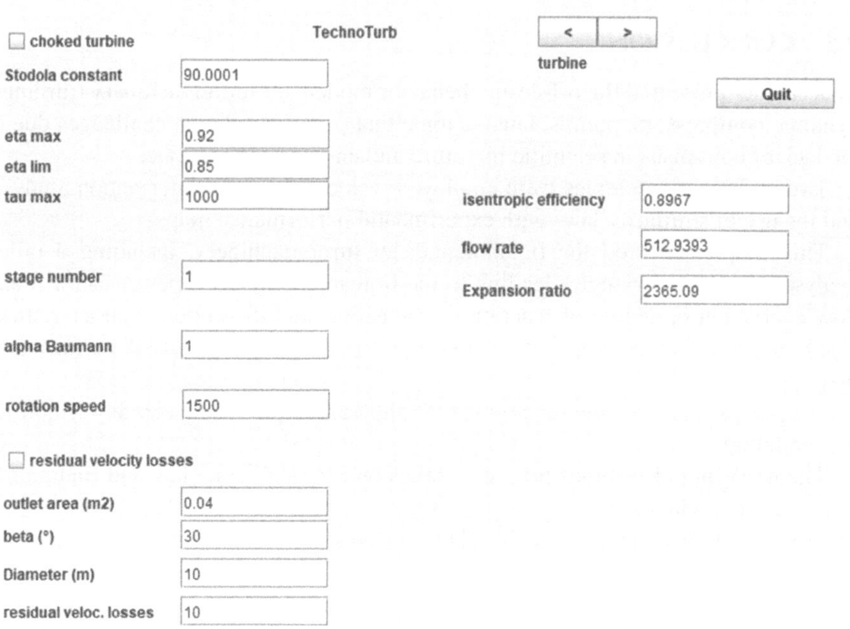

FIGURE 9.4.7 Turbine sizing screen.

- Stodola constant is K_0, the Stodola coefficient. If "choked turbine" is checked, the flow is considered choked, and the formula used is (9.4.12). Otherwise, it is (9.4.11);
  - when η is represented by a three-parameter equation, η_{max} is the maximum value, η_{lim} is the limit for very high expansion rates, and τ_{max} is the abscissa of the maximum. If the first two values are equal, the isentropic efficiency is constant;
- alpha Baumann is coefficient α of equation (9.4.16);
- rotation speed is the turbine rotation speed;
- for calculating LL, the outlet section and the turbine diameter as well as the outlet angle β must be provided.

The values of the isentropic efficiency, mass flow rate, and expansion ratio appear on the right of the screen.

9.4.7 Identification of Turbine Parameters

The identification of turbine parameters generally poses many difficulties because very few experimental data are available in practice and there is uncertainty about them. It is more appropriate to use *a priori* laws based on physical phenomena than to attempt to identify very precise models.

In particular, when the exit point of the turbine is two-phase, we usually do not know its exact quality, which poses a real problem.

9.5 CONCLUSION

This chapter presented the off-design behavior models for turbomachinery (turbines, dynamic compressors, pumps, fans), a topic that poses significant challenges due to the lack of consensus in scientific literature and engineering practice.

Two main methodologies were employed: velocity triangle deformation analysis and the use of similarity laws with experimental performance maps.

This chapter covered the fundamentals of turbomachinery, including detailed analysis of velocity triangles leading to the Rateau coefficients (power factor μ and flow coefficient δ, degree of reaction relationships, and theoretical characteristics). The similarity factors (ϕ,ψ,Δ,σ) enable the construction of reduced performance maps.

For pumps and fans, the incompressible fluid assumption considerably simplifies the modeling.

The modeling of turbines primarily relies on Stodola's cone law and Baumann's efficiency degradation rule.

Practical implementation via the Thermoptim screens was shown.

BIBLIOGRAPHY

Boehm, R.F. (2000). "Pumps and fans." In Kreith, J.F. (ed.), *The CRC handbook of thermal engineering*. CRC Press, Boca Raton, FL, pp. 4-170–4-188. ISBN 0-8493-9581-X.

Bourdouxhe, J.P., Grodent, M., Lebrun, J.J., Saavedra, C. & Silva, K. (1993). "A toolkit for primary HVAC system energy calculations." Draft prepared for ASHRAE TC 4.7, University of Liège, Belgium.

Boyce, M.P. (2012). *Gas turbine engineering handbook*, 4th ed. Elsevier, San Diego, CA.

Cohen, R., Groll, E., Harden, W.H., Hickman, K.E., Mistry, D.K. & Muir, E. (2000). "Compressors." In Kreith, J.F. (ed.), *The CRC handbook of thermal engineering*. CRC Press, Boca Raton, FL. ISBN 0-8493-9581-X.

Conan, J.G. (1988). *Réfrigération industrielle*. Eyrolles, Paris, pp. 216–244.

El-Gammal, A.M. (1991). "An algorithm and criteria for compressor characteristics real-time modeling and approximation." *Transactions of the ASME, Journal of Engineering for Gas Turbines and Power*, 113(1), 112–118.

El-Sayed, A.F. (2008). *Aircraft propulsion and gas turbine engines*. CRC Press, Boca Raton, FL. ISBN 978-0-8493-9196-5.

Giampaolo, T. (2010). *Compressor handbook: Principles and practice*. The Fairmont Press, Lilburn/CRC Press, Boca Raton, FL. ISBN 0-88173-615-5.

Gorla, R.S.R. & Khan, A.A. (2003). *Turbomachinery design and theory*. Marcel Dekker, New York. ISBN 978-0824740651.

Kane, M. (2002). "Intégration et optimisation thermoéconomique et environomique de centrales solaires hybrides." Doctoral thesis, École Polytechnique Fédérale de Lausanne.

Kurzke, J. (1996). "How to get component maps for aircraft gas turbine performance calculations." ASME Paper 96-GT-164.

Park, S. & Lee, S. (2022). "An improved off-design prediction model based on similarity laws for S-CO_2 compressors and turbines." *Proceedings of the 2022 sCO_2 Symposium*. Available at: https://sco2symposium.com/proceedings2022/059-paper.pdf

Rogers, C. (1966). "Efficiency and performance characteristics of radial turbines." SAE Paper 660754.

Stodola, A.B. (1903). Die Dampfturbine und ihre Aussichten als Wärmekraftmaschinen (The steam turbine and its prospective use as a mechanical engine). Berlin.

Stodola, A.B. (1927). *Steam and gas turbines*. English translation by L.C. Loewenstein. McGraw-Hill Book Company, New York.

Tirnovan, R., Giurgea, S., Miraoui, A. & Cirrincione, M. (2008). "Surrogate modelling of compressor characteristics for fuel-cell applications." *Applied Energy*, 85, 394–403.

Vivier, L. (1965). *Turbines à vapeur et à gaz*. Albin Michel, Paris.

de Vlaminck, M. & Wauters, P. (1988). *Thermodynamique et turbines*. Ciaco, Louvain-la-Neuve.

10 Case Studies

INTRODUCTION

In this chapter, we present several case studies on component sizing and off-design operation. Our goal is primarily, as indicated in the general introduction of this book, to demonstrate methodologies for users who wish to develop their own models. Although we have attempted to select models and settings as realistic as possible, the choices we made cannot be considered fully validated. In particular, the sensitivity analyses provided should be considered indicative only: the values given depend on the settings chosen and the equations selected. The numerical tools we consider use either specific Thermoptim classes to perform simple searches by dichotomy, or generic libraries such as minPack, which enables nonlinear optimization and, in particular, solving systems of equations.

We focus mainly on modeling assumptions, controller implementation, and result presentation. For details on the computer implementation of external component sizing classes for the various components involved in these examples, please refer to Volume 4 of the Thermoptim reference manual as well as their code.

In any case, since external classes are distributed as open-source code, they can be customized quite easily to match specific requirements.

Various resources related to these case studies are provided in this page of this book's website.[1] Reminders will be indicated in the text.

We will explore four complementary cases of increasing difficulty:

- The first case involves an air piston compressor with exchanger cooling that charges a compressed air storage of given volume. It helps familiarize users with creating a controller and studying off-design operation of a simple system. We present first the model of the exchanger alone, and then the cooled compressor.
- The second case is the one used in Chapter 6 to illustrate the component sizing and off-design operation methodology. It is a refrigeration machine involving a displacement compressor, a thermostatic valve, and two two-phase heat exchangers whose pressures vary with external conditions. The resolution is performed using minPack.
- The third case corresponds to a simplified steam power plant (without extraction or reheat). It involves a turbine and two heat exchangers, and particularly allows studying the evolution of cycle performance when varying the cooling water temperature, maximum pressure, or superheat temperature. It is also solved numerically with minPack.
- The fourth case study differs somewhat from the first three. It presents data analysis on the partial load operation of the Flamanville 3 EPR turbine, for which very detailed official data have been provided by EDF to the

 DOI: 10.1201/9781003606017-11

French Nuclear Safety Authority (ASN) for licensing approval. This data made it possible to develop the NUSCLE software, a simplified model of the thermodynamic cycle of a water-cooled nuclear power plant (WCR type: Water-Cooled Reactor), such as PWRs, BWR, Russian RBMK (graphite-moderated pressure tube reactors), or CANDUs.

10.1 COMPRESSOR FILLING A STORAGE OF COMPRESSED AIR

10.1.1 Modeling of the Heat Exchanger

Let us consider a tube-and-fin heat exchanger cooling approximately 0.0117 kg/s of air leaving a compressor at 5 bar and 275°C with a flow rate of 0.02 kg/s of cold water passing through a coil of two parallel tubes arranged in several layers.

The synoptic view of the heat exchanger is given in Figure 10.2.1.

This is a tube-and-fin heat exchanger with an effectiveness epsilon equal to 0.84.

The tube diameter is 15 mm, and its thickness is 1.5 mm. The spacing between the fins is 3 mm.

Inside the tubes, dh is given, and the calculation of Ac is straightforward: it equals the product of the number of tubes by the cross-section of one tube. With a tube outside diameter of 15 mm and a thickness of 1.5 mm, the inner diameter is 12 mm. A simple calculation shows that for two tubes, Ac is 0.000226195 m^2.

Outside the tubes, the calculations are more complex. The sketches in Figure 10.2.2 show the arrangement of fins around the water tubes in bold lines: rectangular fins with half-length l = 1 cm and depth h = 2 cm, spacing e = 3 mm, pressed perpendicular to the tubes. Their number must be determined. We begin by taking a value of 100.

The determination of different dimensions poses no particular problem and can be done in a spreadsheet[2] (Table 10.2.1).

Two versions of controllers are available for this case study. The first uses the generic controller, which allows configuration and simulations without writing a single line of code, while the second uses a specific controller, which is simple in this case.

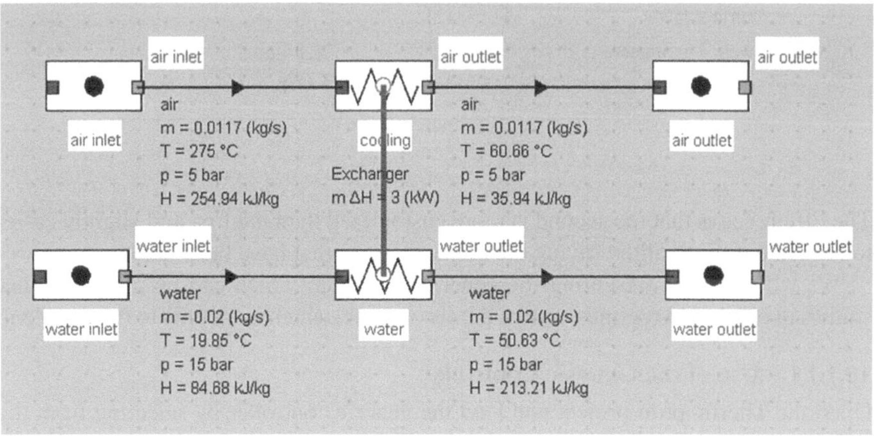

FIGURE 10.2.1 Air exchanger synoptic view.

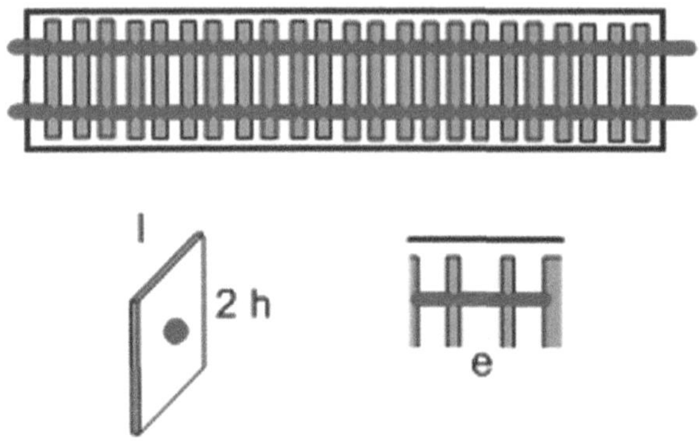

FIGURE 10.2.2 Schematics of the heat exchanger.

TABLE 10.2.1
Heat Exchanger Parameters

Parameter	Value	Formula
spacing e	0.0030	
half-height h	0.01	
depth l	0.02	
nb of fins	100	
unit Ac	0.000015	(2h − d)e
Ac total	0.003	
fin thickness a	0.0003	
exchanger length	0.33	nb(e + a)
extended surface	0.00057381	$2(2lh - \pi d^2/4) + \pi d(e - a)$
initial surface	0.00014137	πde
surface factor	4.05884242	
wetted perimeter	0.046	2(2h + e)
dh	0.00130435	4 Ac/p

The difference is that the second is more customized than the first and slightly easier to implement. Both allow saving the configurations that have been made.

We will start by presenting the generic controller, which can be used for other studies in off-design regimes, unlike the second one, which is specific to this problem.

10.1.1.1 Use of the Generic Controller

Open the Thermoptim project and load the generic controller by selecting from the list of controllers the one named "generic sizing controller," and then click "Set the sizing screens" (Figure 10.2.3). The list of components for which sizing screens exist

FIGURE 10.2.3 Generic sizing controller.

is displayed in the table with their name, type, and name of the sizing screen class instantiated by default: here TechnoHx for the exchanger.

In this example, this initial setting is suitable. If necessary, to change a component TechnoDesign class, select its line and double-click. A message asks you to confirm your choice, and then displays a list of available classes. Choose the one you want and confirm.

You can then access these sizing screens either from their own screens ("tech. design" button) or from the upper part of a global window that opens from the simulator screen by typing Ctrl T (Figure 10.2.4).

10.1.1.2 Sizing the Heat Exchanger

Once the sizing screen is created, you can proceed to size the heat exchanger. This step must be done carefully, because it involves making a series of choices about internal configurations, geometrical dimensions, etc. Some clarification on this point is given in this section, but refer to Chapters 6 and 7 for more details.

First choose the type of configuration: "airCoil_wcc|Wang, Chi & Chang…" for air, with the settings shown in Figure 10.2.5, and "int_tube|Mac Adams…" for the heating of water.

We have used a correlation specific to tube-and-fin exchangers, that of (Wang et al., 2000). Its parameters can be deduced from the spreadsheet: spacings Fp = 3 mm between the fins, Pt = 20 mm between the tubes and Pl = 55 mm between the layers, outer diameter of the tubes Dc = 15 mm, and the number of layers N = 3.

We assume that the fluid cross-sectional flow area is minimum at the center of the tubes, arranged supposedly in the same plane, and the thickness of the fins equals one-tenth of the spacing, 0.3 mm.

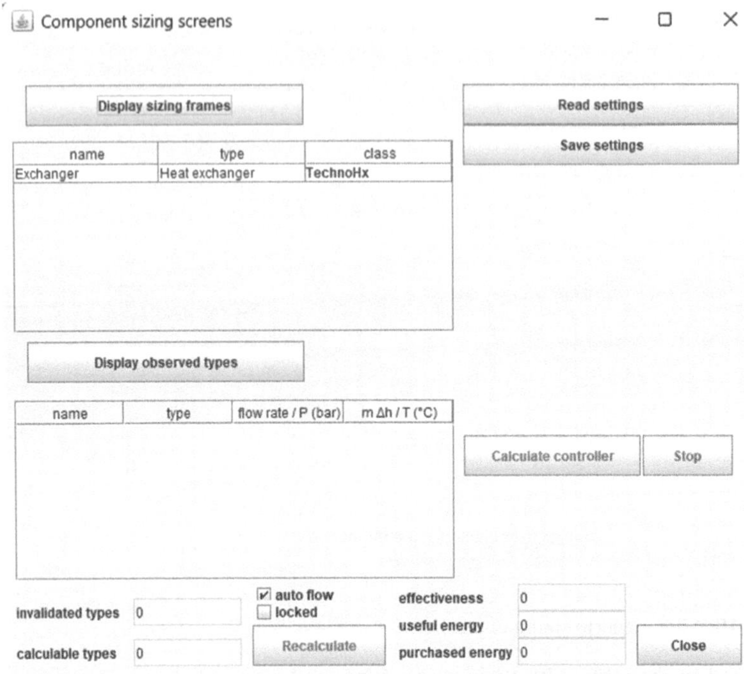

FIGURE 10.2.4 Sizing screen window.

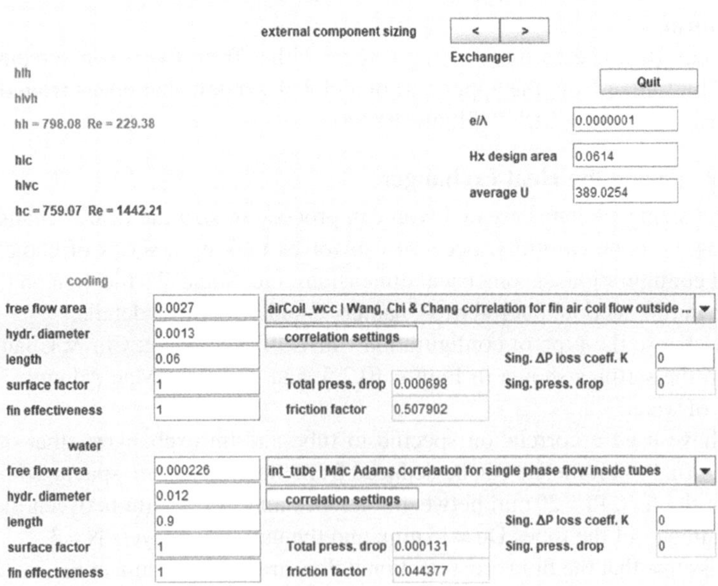

FIGURE 10.2.5 Component sizing screen.

The length of the heat exchanger on the air side is 3 cm × 2 = 6 cm, and that of water tubes is 3 times 0.297 cm = 0.9 m. These are used to calculate the pressure losses, which are negligible here.

As explained above, on the air side, we considered a fluid free flow area of 0.003 m^2 and a hydraulic diameter of 1.3 mm with a length of 0.06 m, and on the water side a free flow area of 2.26 cm^2 and a hydraulic diameter of 12 mm, with a tube length of 0.9 m.

We now size the heat exchanger from the generic controller by selecting it in the list and clicking the button "Design the selected components."

By entering this setting in the exchanger, the calculated area is 0.0684 m^2, leading to a requirement for two layers of 2.8 tubes.

By reducing the number of fins from 100 to 90, the size calculated in the spreadsheet changes slightly, with the air flow area becoming 0.0027 m^2.

After entering this value in the sizing screen, the calculated area is 0.0614 m^2, and the number of layers equals 3.

There are therefore six tubes of 0.297 m, approximately 1.80 m of tubing total, with 540 fins.

Exchange coefficients are 798 W/m^2/K on the air side and 759 W/m^2/K on the water side, resulting in an overall coefficient of 389 W/m^2/K.

The heat exchanger will ultimately consist of two tubes of 90 cm length arranged in coils on three layers and passing through 540 iron plates 2 cm^2, with a total area of 0.31 m^2, separated from each other by 3 mm.

10.1.2 Design of the Controller

Let us now move on to the presentation of the controller dedicated to this exchanger.

The controller class is called AirExchangerController.java.[3] Its graphical interface is simple and conventional, so we do not elaborate here.

10.1.2.1 Initializations

For the name of the exchanger, one can either enter it directly in the code or use the Project method getHxList(), but this has the disadvantage of requiring that the diagram associated with the project be opened.

For the other initializations, we use PointThopt instances, which are like clones of the core Thermoptim points, allowing easy access to their values. Four of these objects are instantiated, representing the inlets and outlets of both fluids. Here we have entered their names in the code, but it would be possible to recognize them automatically.

```
//initializations for the simulation
//watch out: the names of points and components must be correct, otherwis
hxName="Exchanger";
hotUpstream=new PointThopt(proj,"air inlet");
hotDownstream=new PointThopt(proj,"air outlet");
coldUpstream=new PointThopt(proj,"water inlet");
coldDownstream=new PointThopt(proj,"water outlet");

//instantiation of the TechnoDesign in the external class
technoExchanger=new TechnoHx(proj, hxName, hotUpstream, hotDownstream, co
addTechnoVector(technoExchanger);

//initialization of the TechnoDesign in Thermoptim
setupTechnoDesigns(vTechno);
```

The last line initializes the TechnoDesign when opening an existing project file or initially loading the controller. Without it, the TechnoDesign does not appear in the list of sizing screens.

Then, the heat exchanger is initialized in the controller based on the information available in the Thermoptim project. Using the Project method getProperties(), we obtain the values of UA and ΔH:

```
if(!hxName.equals("")){//initialization of the heat exchanger
    args=new String[2];
    args[0]="heatEx";
    args[1]=hxName;
    vProp=proj.getProperties(args);
    Double f=(Double)vProp.elementAt(15);
    UAech=f.doubleValue();

    hotUpstream.getProperties();
    hotDownstream.getProperties();
    coldUpstream.getProperties();
    coldDownstream.getProperties();
```

10.1.2.2 Calculations

The calculations here are straightforward: the exchanger is sized by the method makeDesign() and the results are displayed on the screen:

```
    //TechnoDesign initializations
    technoExchanger.UA=UAech;
    technoExchanger.makeDesign();

    //display
    U_value.setText(Util.aff_d(technoExchanger.getU(),4));
    UAech_value.setText(Util.aff_d(UAech,4));
    AechReel=Util.lit_d(technoExchanger.ADesign_value.getText());
    AcalculatedEch_value.setText(Util.aff_d(AechReel,4));
}
```

This controller is not intended to perform simulations. It is used solely to size the heat exchanger. In the following case study, we will use this model to simulate the cooling of air exiting a compressor.

10.1.3 Analysis of the Cooled Compressor

To illustrate Thermoptim's capability to perform off-design calculations, we will study the behavior of an air piston compressor filling a compressed air storage of given volume at variable pressure. The compressed air is cooled before storage by a water heat exchanger of the type presented in the previous section.

The system can easily be modeled in Thermoptim and produces a diagram of the type shown in Figure 10.2.6, which differs from that studied in the previous example by the addition of the compressor.

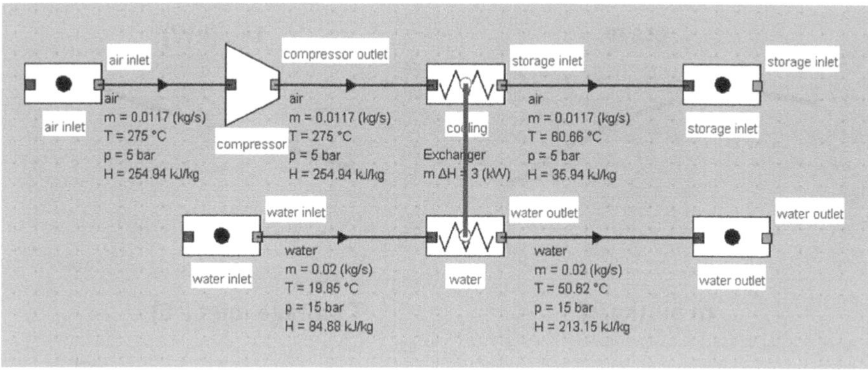

FIGURE 10.2.6 Cooled compressor example.

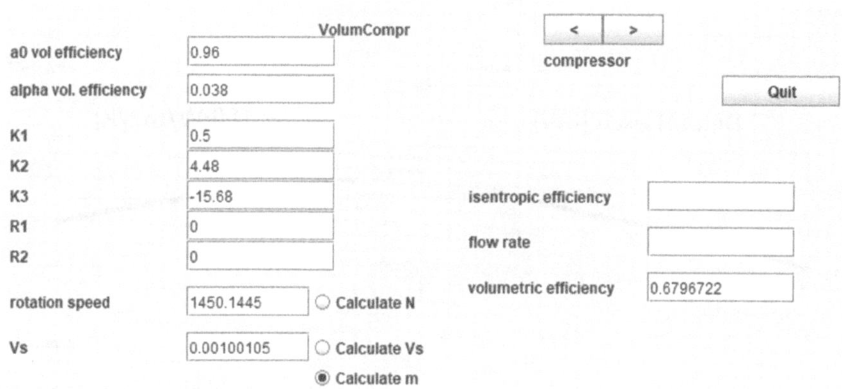

FIGURE 10.2.7 Compressor sizing screen.

The compressor model is that presented in Section 3.1.1. Its volumetric and isentropic efficiencies are given by equations (3.1.1) and (3.1.2). Its sizing screen is shown in Figure 10.2.7. That of the heat exchanger is similar to the previous one.

10.1.3.1 Results

Once the controller is built (the screen will be presented in the next section, Figure 10.2.9), it is straightforward to vary the compression ratio to obtain the evolution of the main variables as the tank pressure varies. Simply save the project file under a different name after each simulation, then load these files into the macro, and extract the relevant values.

Figure 10.2.8 shows, as a function of storage pressure, changes in the compressor isentropic efficiency ηs, work consumed, intake air flow, temperature of air entering the tank, exchanger load, and heat transfer coefficient U.

Between 5 and 12 bar, ηs and U vary by 22% and 7%, respectively. As the curves are in some cases far from linear, overly simplistic approaches are insufficient, which demonstrates the usefulness of our approach.

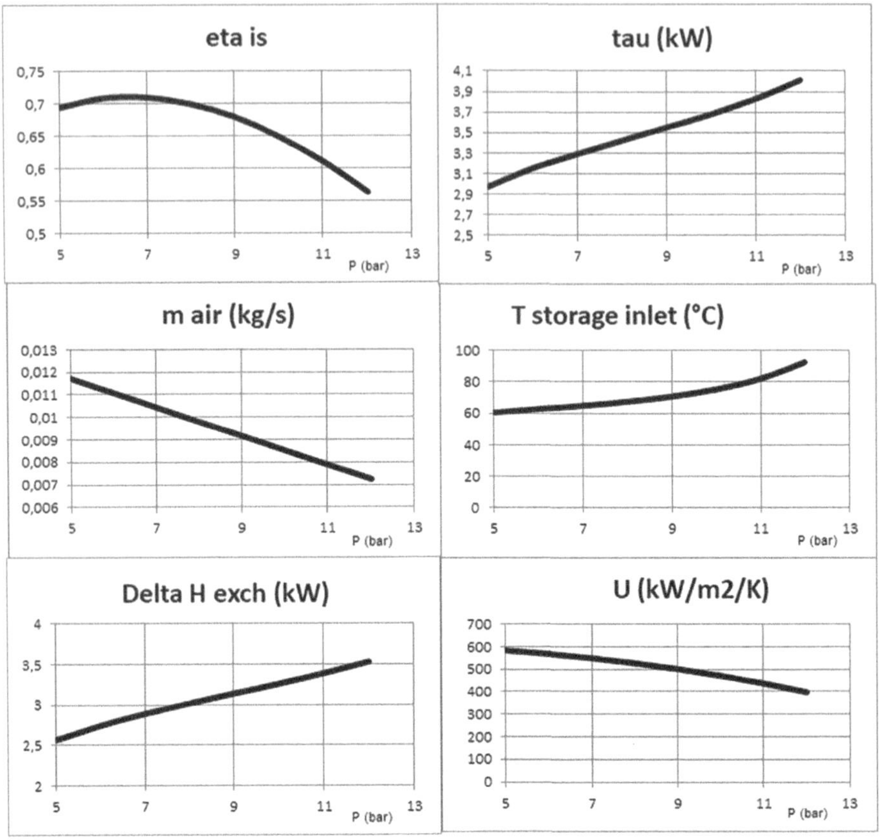

FIGURE 10.2.8 Cooled compressor simulation.

10.1.3.2 Controller Design

The controller class is VsCompressorController.java,[4] type "Vs compressor controller."

The controller screen is shown in Figure 10.2.9. Its graphical interface is simple and conventional, so we do not elaborate here.

To create the relevant class, we take the exchanger controller class and complete it in two ways:

- First by instantiating and initializing the compressor sizing screen, analogous to that of the exchanger.
- Second, by introducing a "Calculate" button and configuring the calculations and simulator configuration changes that should be performed when the downstream pressure is varied.

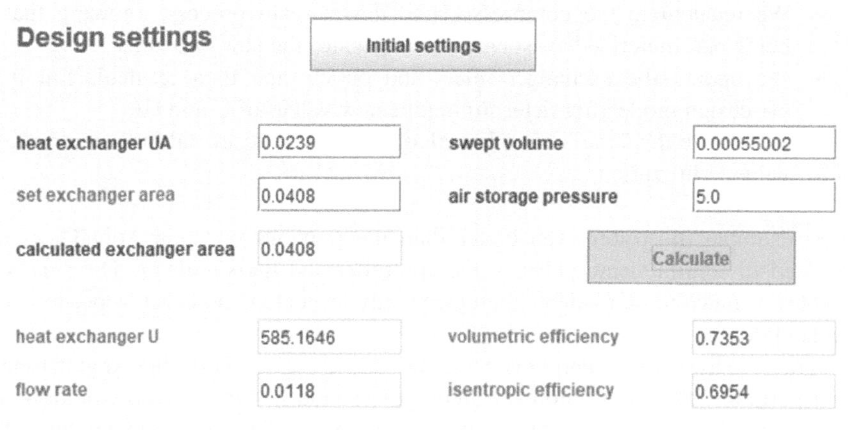

FIGURE 10.2.9 Controller screen.

10.1.3.2.1 Initializations

During initialization, after both PointThopt and TechnoDesign have been constructed, the displacement of the compressor needed to obtain the desired flow can be determined based on the sizing screen settings.

```
//initializations for the simulation
//watch out: the names of points and components must be correct, otherw
hxName="Exchanger";
compressorName="compressor";
hotDownstream=new PointThopt(proj,"storage inlet");
hotUpstream=new PointThopt(proj,"compressor outlet");
comprUpstream=new PointThopt(proj,"air inlet");
coldUpstream=new PointThopt(proj,"water inlet");
coldDownstream=new PointThopt(proj,"water outlet");

//instantiation of the TechnoDesign in the external classes
technoExchanger=new TechnoHx(proj, hxName, hotUpstream, hotDownstream,
addTechnoVector(technoExchanger);
technoCompr=new VolumCompr(proj, compressorName, comprUpstream, hotUpst
addTechnoVector(technoCompr);

//initialization of the TechnoDesign in Thermoptim
setupTechnoDesigns(vTechno);
```

10.1.3.2.2 Calculations

The calculations proceed as follows:

- We start by updating the pressure at the compressor outlet.
- We calculate its volumetric and isentropic efficiencies.

- We recalculate the compressor and downstream process, knowing that being parameterized as isobaric, it propagates the new pressure.
- We update the exchanger inlets and outlets and then recalculate it in off-design mode after determining the new values of U and UA.
- We update the simulator and recalculate the project several times to ensure value stabilization.

This example illustrates even better than the previous one how PointThopt can be used to communicate between the controller and the simulator. The syntax of updates is much more readable than using only Project methods getProperties() and updatePoint().

The compressor calculation is performed using two methods named getRisentr() and getLambdaVol(). Method updateprocess() modifies the compressor flow and isentropic efficiency, and then recalculates the process. The downstream point, called here "hotUpstream" because it corresponds to the upstream point of the exchanger hot fluid, is then updated.

```
void bCalc_actionPerformed(java.awt.event.ActionEvent event)
{
    Vs=Util.lit_d(Vs_value.getText());//reads the compressor swept volume
    technoCompr.setVs(Vs);
    Preservoir=Util.lit_d(P_value.getText());
    double UA_ech=Util.lit_d(UAech_value.getText());

    hotUpstream.P=Preservoir;// updates the compressor outlet pressure
    hotUpstream.update(!UPDATE_T,UPDATE_P,!UPDATE_X);
    hotUpstream.getProperties();

    massFlow=technoCompr.getMassFlow(comprUpstream.V);
    double eta_is=technoCompr.getRisentr();// calculates compressor isentropic
    lambdaVol=technoCompr.getLambdaVol();

    //recalculates the compressor and the downstream process
    updateprocess(compressorName, "Compression",RECALCULATE,IS_SET_FLOW, UPDATI
    hotUpstream.getProperties();
    updateprocess("refroidissement", "Exchange",RECALCULATE,IS_SET_FLOW, UPDATI
    updateprocess("entree air", "Exchange",RECALCULATE,IS_SET_FLOW, UPDATE_FLO

    hotUpstream.getProperties();//updates heat exchanger inlets and outlets
    hotDownstream.getProperties();
    coldUpstream.getProperties();
    coldDownstream.getProperties();
```

The heat exchanger calculation proceeds as follows: we perform a first calculation using method updateHx() by setting the value of UA read on the screen, UA_ech (the exchanger is configured for an off-design calculation, so the UA value is taken into account). This calculation initializes the exchanger for the new operating conditions.

We then use the method makeDesign() of TechnoDesign, which updates the value of U. The actual UA is obtained by multiplying the new U by the initial A value (AechReel), which allows correct recalculation of the exchanger.

```
        //calculates the heat exchanger (several iterations)
        for(int i=0;i<5;i++){
            updateHx(hxName, RECALCULATE, UPDATE_UA, UA_ech, !UPDATE_EPSI, 0, !UPDAT
            coldDownstream.getProperties();
            hotDownstream.getProperties();

            technoExchanger.UA=UA_ech;//calculates U
            technoExchanger.makeDesign();
            double U=technoExchanger.getU();

            UAech=U*AechReel/1000.;//updates UA and recalculates the heat exchanger
            UAech_value.setText(Util.aff_d(UAech,4));
            updateHx(hxName, RECALCULATE, UPDATE_UA, UAech, !UPDATE_EPSI, 0, !UPDATE

            coldDownstream.getProperties();
            hotDownstream.getProperties();
            U_value.setText(Util.aff_d(U,4));
            UA_ech=UAech;
        }

        //updates the simulator and displays
        for(int j=0;j<3;j++)proj.calcThopt();
        flow_value.setText(Util.aff_d(massFlow,4));
        eta_is_value.setText(Util.aff_d(eta_is,4));
        lambdaVol_value.setText(Util.aff_d(lambdaVol,4));
        AcalculatedEch_value.setText(technoExchanger.ADesign_value.getText());
    }
```

Since U depends on the average temperatures of fluids, and therefore on outlet point temperatures, we iterate the calculations five times to ensure proper stabilization, resetting UA_ech each time.

10.1.4 Use of the Model to Simulate the Filling of a Compressed Air Storage

The results in Figure 10.2.8 can be used to simulate the filling of a compressed air storage. To determine the flow rate, power consumed by the compressor, and outlet temperature, simply perform polynomial regressions based on the previous Thermoptim simulations.

$$\tau = 0.0019x^3 - 0.0485x^2 + 0.5328x + 1.2794, R^2 = 1$$

$$\dot{m} = -0.0006x + 0.0149, R^2 = 1$$

$$\text{Tair} = 0.1135x^3 - 2.2814x^2 + 17.361x + 16.564, R^2 = 0.9997$$

Then, develop a second model, solved with Excel or a Python script, whose files are available here,[5] with the following equations:

- The mass M of air contained in storage equals the initial mass plus the integral of the flow.
- Internal energy U equals initial internal energy plus the integral of the product of flow by the enthalpy of air leaving the cooler, minus the integral of losses by convection with ambient air.
- Storage temperature is inferred from its internal energy.
- Pressure is determined by the ideal gas law.

The appearance of the results is shown in Figure 10.2.10, for a 0.5 m^3 storage (diameter 80 cm, length 1 m) supplied by a compressor with displacement Vs = 0.55 l.

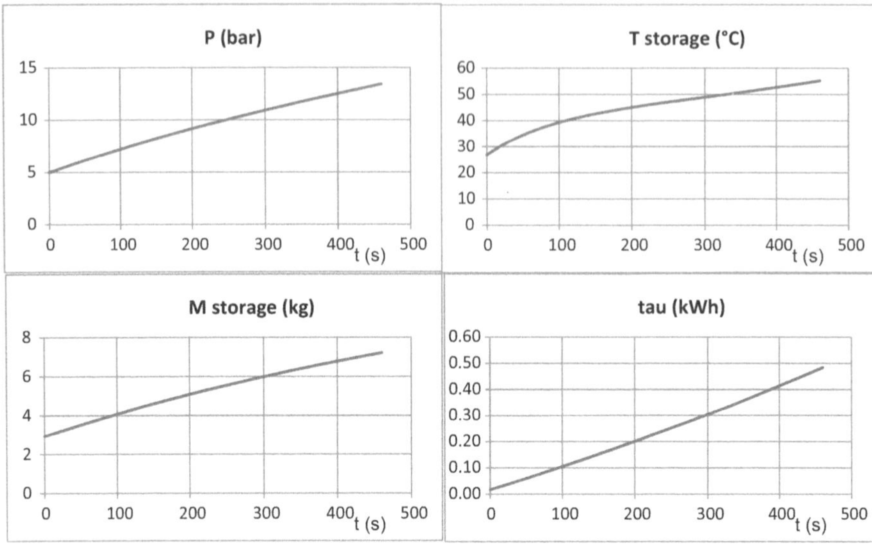

FIGURE 10.2.10 Filling of the compressed air storage.

The pressure and temperature in storage, its mass, and the compressor workload are displayed versus time in seconds.

10.2 OFF-DESIGN BEHAVIOR OF A REFRIGERATION MACHINE

The refrigeration machine on which we focus in this case study corresponds to that used to illustrate the introduction of Chapter 6.

This is a refrigeration machine with a power of 15 kW producing chilled water at −10°C. In Chapter 6, a detailed presentation of the model and its configuration was given. A complete file is available in the Thermoptim model library.

This refrigeration machine uses a compressor and a finned condenser for which we have retained the same modeling as for the air–water heat exchanger in the first case study.

10.2.1 Introduction, Results

Figure 10.3.1 shows a synoptic view of the refrigerator model we wish to control.

Figure 10.3.2 displays the controller that we have built to coordinate the calculations of the various components. It allows changing input variables such as compressor rotation speed or outside air temperature.

To calculate an operating point, we must first initialize calculations by clicking the button "Initial settings," which displays:

- Compressor rotation speed corresponding to the simulator flow.
- UA values for the evaporator and condenser.

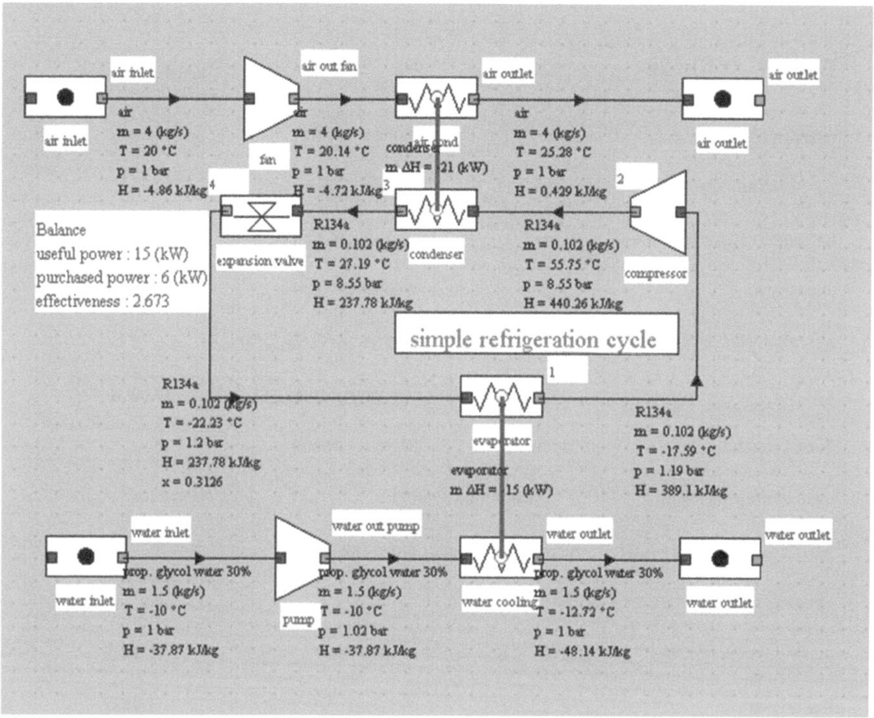

FIGURE 10.3.1 Model of the refrigeration cycle.

- Temperature of the air used to cool the condenser.
- Heat rates of fluids in the two exchangers.

If you change values in the editable fields, those are the values taken into account in simulation calculations.

Figure 6.3.1 shows the simulation results when varying the temperature of the cooling air, for the settings of sizing screens selected in Section 6.2.4.2.

10.2.2 Principle of Resolution

Unlike other examples, we will not discuss here the code of the Java class RefrigControllerMinPack.java, referring the reader to a programming note available here.[6]

We will only explain the principle of resolution that was adopted, which has the advantage of being generic. The problem we are trying to solve is the following:

$V_{s,fl}$ being the swept volume, the refrigerant volumetric flow is given by equation (10.3.1):

$$\dot{V} = \lambda \frac{N}{60} V_{s,fl} \tag{10.3.1}$$

FIGURE 10.3.2 Refrigeration cycle controller screen.

with λ volumetric efficiency, given by (10.3.2):

$$\lambda = \alpha_0 - \alpha_1 \frac{P_{ref}}{P_{asp}} \tag{10.3.2}$$

The mass flow rate thus depends on the specific volume at suction and the compression ratio.

The evaporation temperature is set by the evaporator heat balance, which depends mainly on the temperature and flow of coolant (brine water here), and on the refrigerant flow. Equation (10.3.3) provides the overall exchanger balance, but in reality it is an evaporator with a two-phase part and slight superheating, so it must be modeled as multizone, knowing that only the total area is known, with the boundary between the two zones varying according to operating conditions.

$$Q_e = \dot{m}\left(h_v\left(T_i\right) - h_l\left(T_c\right)\right) = UA_{evap}\left(T_e - T_c\right) = U_e A_e\left(T_e - T_c\right) \tag{10.3.3}$$

Note that the exchange coefficient U_e also depends on many parameters, including temperature and flow of fluids through the exchanger, so solving this equation is particularly complex. Assuming the value of U_e is known, it can be solved as indicated below.

The compressor balance also depends on many parameters:

$$\tau = \dot{m}\frac{\Delta h_s\left(T_e, P_{ref}\right)}{\eta_s\left(T_e, P_{ref}\right)} \tag{10.3.4}$$

Similarly to the evaporator, the condensation pressure and temperature is set by the thermal equilibrium of the condenser, which depends mainly on the temperature and flow of cooling air, on the refrigerant flow, and on the condenser outlet temperature.

The latter depends on the compression ratio and compressor isentropic efficiency, which itself also depends on this ratio.

Equation (10.3.5) provides the overall balance of the condenser, but as it includes a desuperheater, a two-phase condensation zone, and slight subcooling, it must be modeled as three zones knowing that only the total area is known, with the boundaries between zones varying according to usage conditions.

$$Q_c = \tau + Q_e = UA_{cond}\left(T_c - T_a\right) = U_c A_c\left(T_c - T_a\right) \tag{10.3.5}$$

In addition, we want the surface areas of the heat exchangers to equal the design values, and we impose an additional constraint: that the refrigerant charge remains constant, allowing us to calculate the subcooling after condensation. Therefore, there are six variables to determine.

The search for the values of these variables (refrigerant flow, evaporation temperature, condensation temperature, evaporator, and condenser UA, subcooling) corresponds to solving a set of relatively complex nonlinear equations. The interest of the external controller is that it can implement an algorithm for solving this system of equations, to which must be added all those necessary to determine the thermophysical properties of fluids and values of heat exchange coefficients.

The solution we have adopted is to configure an external controller that provides the resolution of this system of equations and updates Thermoptim once they are determined.

According to the remark made earlier about the difficulties of calculating U, it uses two nested algorithms: one searching for a solution at constant UA for the two exchangers, and one adjusting U according to the solution found previously.

We use the Marquardt-Levenberg method with algorithms developed in Fortran as the minPack 1 package and translated to Java. This method combines the Gauss-Newton method and gradient descent. Its main advantage is robustness and requiring only an approximate solution as initialization.

Its implementation in Java uses an interface called optimization.Lmdif_fcn, which requires the calling class (in this case our controller) to have a function called fcn().

This function fcn() receives as key arguments a vector (array x[n]) containing the variables and a vector (array fvec[m]) referring to residuals of the functions we seek to set to zero. Their number may exceed that of variables, but in our case it is the same.

The vector of variables here is: Tcond, Tevap, massFlow, UAevap, UAcond, and DTssrefr. The six fvec methods are functions resEvap(), resCond(), resFlow(), resAevap(), resAcond(), and resLoad() returning these residuals. To estimate the residuals, we must first update Thermoptim variables corresponding to vector x, and then calculate the other variables.

We presented simulation results in Figures 6.3.1 and 6.3.2 when either the ambient temperature or the rotational speed varies.

10.3 OFF-DESIGN BEHAVIOR OF A STEAM POWER PLANT

This section introduces external class SimpleSteamPlantControllerOptim.java,[7] which is used as a steam power plant controller to perform the study of its off-design operation.

10.3.1 Introduction, Results

Figure 10.4.1 shows a synoptic view of the steam power plant model which we wish to control.

This is a 660 MW coal-fired plant cooled by river water. The technological settings selected here for both exchangers were conducted in a manner analogous to that described in Section 10.2.1. For the evaporator, we consider that water flows inside tubes and retain the Borishanskij correlation. For condensation, we consider that the steam circulates outside the tubes and choose the correlation proposed by Dhir-Lienhard (see Section 7.3.3.5).

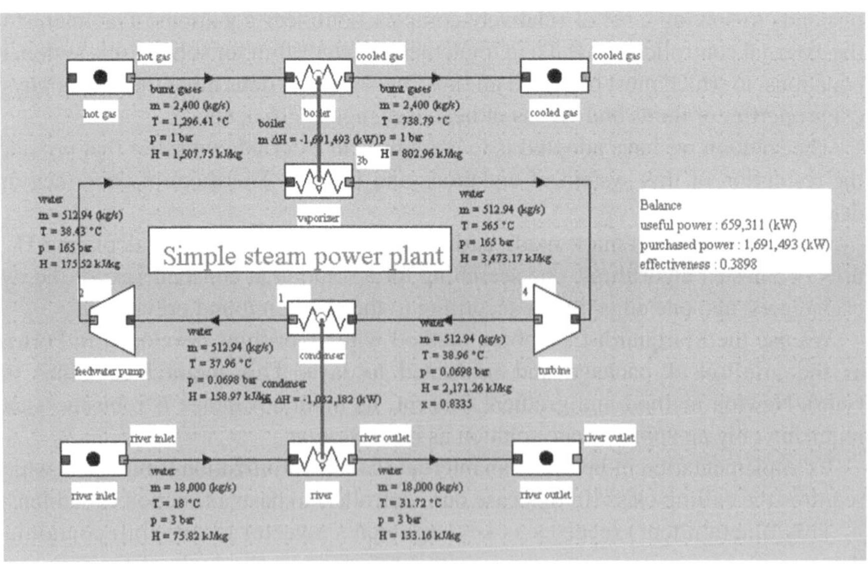

FIGURE 10.4.1 Steam power plant model.

The problem to solve is the following (we have assumed that the pressure and temperature at the steam boiler outlet are known, defined in the controller screen).

The steam mass flow is given by formula (9.4.11), K being the Stodola constant:

$$T_{in}\frac{\dot{m}^2}{K^2} = P_{in}^2 - P_{out}^2 \tag{9.4.11}$$

The condensing temperature is set by the condenser thermal balance, which depends mainly on the temperature and flow of cooling water, on the steam flow, and on the condenser inlet temperature.

The latter depends on the expansion ratio and turbine isentropic efficiency.

$$Q_c = \tau + Q_b = U_c A_c \Delta T_{ml\_cond}$$

Note that the exchange coefficient U_c also depends on many parameters, including temperature and flow of fluids through the exchanger and the distribution of areas between the condensation zone and the subcooling, so the resolution of this equation is particularly complex (Figure 10.4.2).

The turbine balance also depends on many parameters. It is given by:

$$\tau = \dot{m}\eta_s(P_b, P_c)\Delta h_s(T_b, P_b, P_c)$$

Finally, the boiler heat balance sets the value of hot gas outlet temperature.

Searching for a pair (steam flow, condensation temperature) already corresponds to solving a set of relatively complex nonlinear equations, but since the surfaces of the two exchangers are fixed, we also need to find the correct values for their UA products.

The algorithm determines the load of the boiler assumed to be controlled at the selected temperature and pressure, deduces that of the condenser after having determined the power produced by the turbine, and seeks the residual of the condenser heat balance, which serves as a convergence criterion when iteration is performed by varying the condensation pressure.

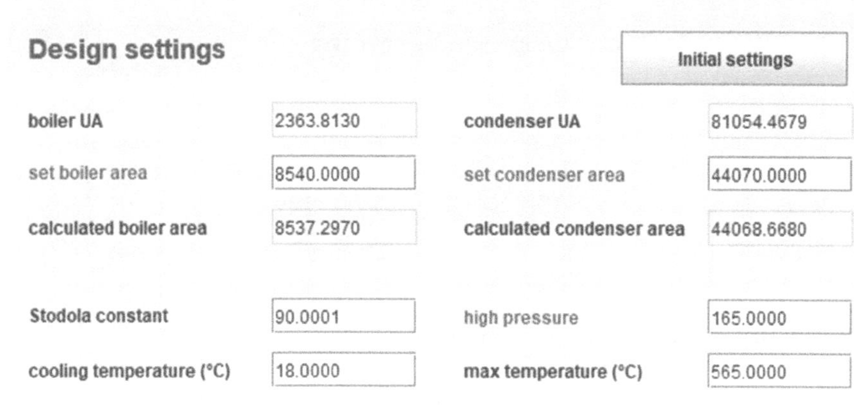

FIGURE 10.4.2 Steam power plant controller screen.

The strong coupling between fluid thermophysical properties and U value makes it very difficult to directly search for the exact solution of this system of equations, particularly since the exchangers are multizone, which does not allow changing their UA as easily as we did in the first case study. There are therefore four main variables to determine here, with the others being deduced easily afterwards. The solution is also obtained here using the minPack library.

To calculate an operating point, we must first initialize calculations by clicking the button "Initial settings," which displays:

- UA values for the evaporator and the condenser.
- Desired values for exchange surfaces.
- Calculated values for exchange surfaces.
- Turbine Stodola constant.
- Steam evaporation pressure.
- Temperature of the water used to cool the condenser.
- Superheat temperature.

If you change the editable values, those are the values taken into account in calculations. Once a simulation is performed, the results are displayed (Figure 10.4.3), and it is advisable to save the project file by renaming it appropriately to facilitate exploitation of multiple simulations, for example, through Excel post-processing macros.

Figure 10.4.4 shows the results provided by the model when the temperature of the cold source varies from 1°C to 24°C.

Figure 10.4.5 shows results obtained when varying the evaporation pressure from 40 to 165 bar. We call this type of regulation sliding pressure control. The steam flow is quasi-proportional to the evaporation pressure, with the influence of the downstream pressure being very small (see equation (4.4.11)).

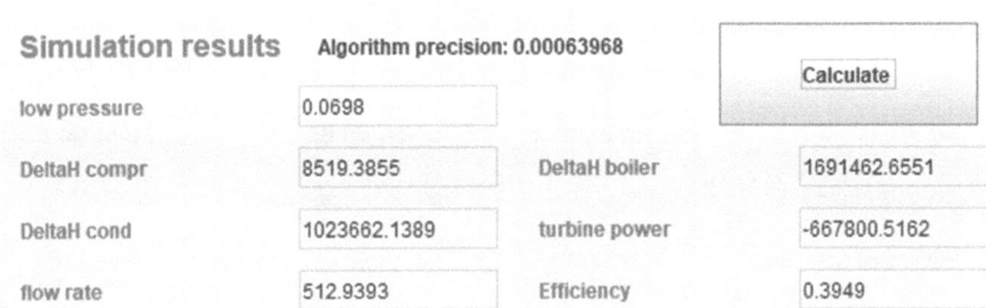

FIGURE 10.4.3 Steam power plant simulation results.

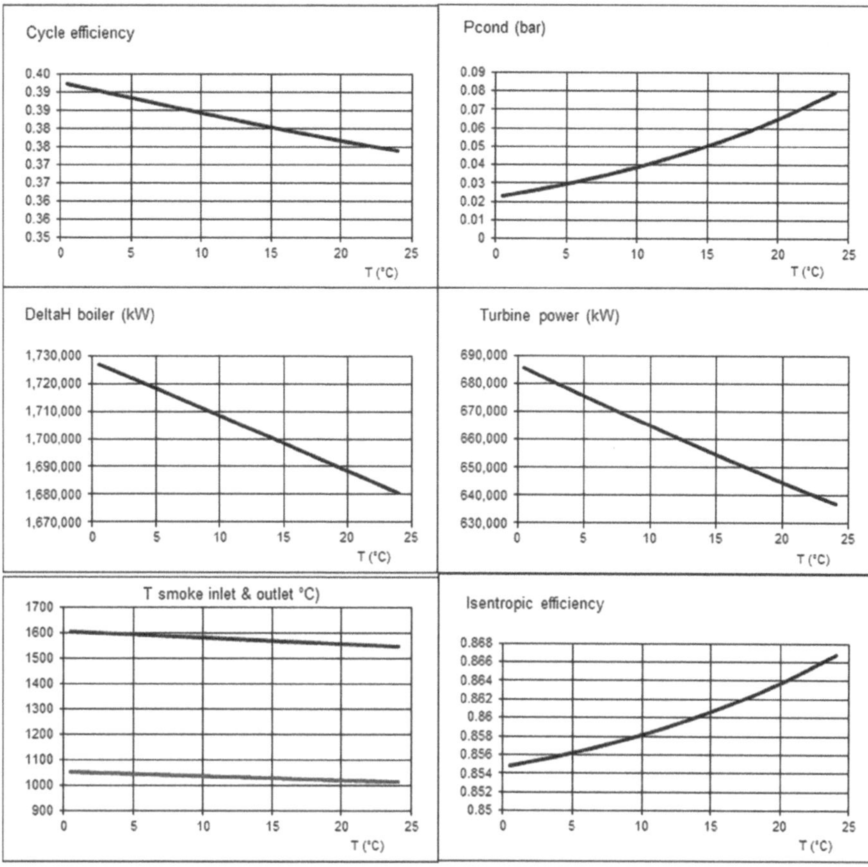

FIGURE 10.4.4 Effect of cooling water temperature.

10.3.2 Presentation of the External Class

The controller class (SimpleSteamPlantControllerOptim.java) is an extension of extThopt.ExtPilot, which derives from rg.thopt.PilotFrame.

10.3.2.1 Overall Organization

The controller begins with initializations:

- To facilitate code development, it creates eight external points (class PointThopt) for points existing in Thermoptim, and instantiates the three required TechnoDesigns.
- It then calculates the values of the various parameters that will be used later in calculations.

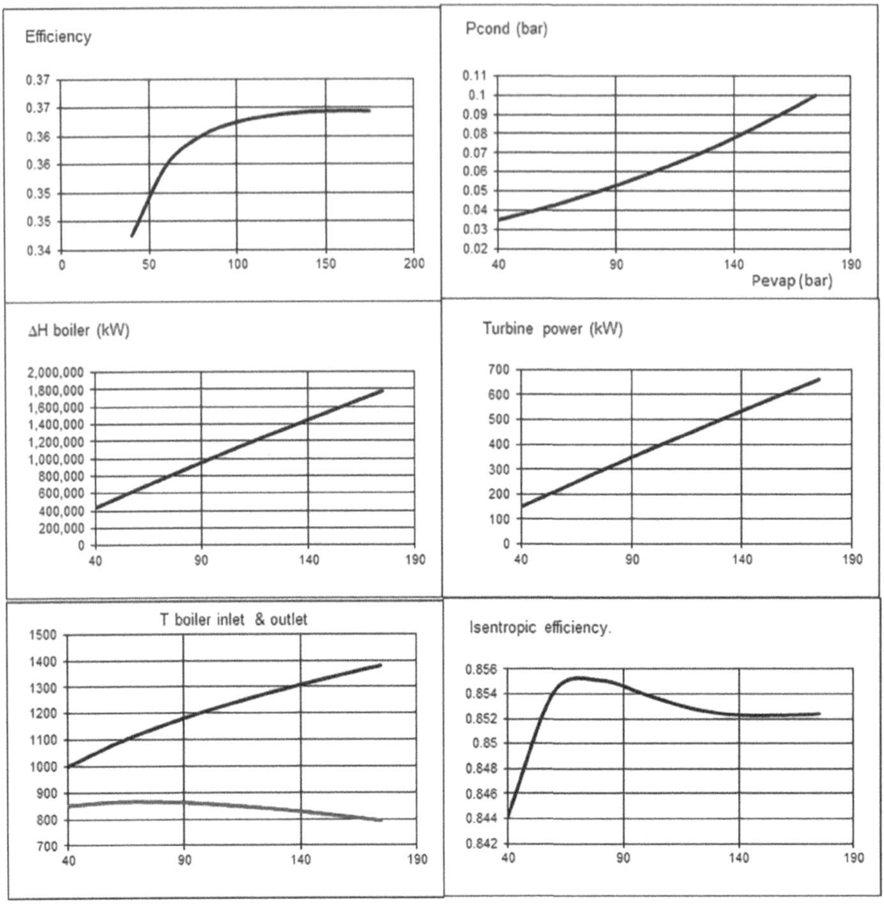

FIGURE 10.4.5 Effect of evaporation pressure.

It then launches a search for a solution by iterating on four variables: the condensation temperature Tcond, the steam mass flow, and the two UA values of the boiler and the condenser, UAevap and UAcond.

Once convergence is obtained, the Thermoptim project is completely updated and recalculated, and main results are displayed in the controller screen (Figure 10.4.3).

10.3.2.2 Initializations

We will not present here the construction of the interface, which poses no particular problem and whose code is entirely conventional.

The sequence of initializations and calculations is as follows:

1. Creation of PointThopt and TechnoDesign

```
//initialisations des PointThopt et des TechnoDesign
//initializations for simulation
boilerUpstream=new PointThopt(proj,"2");
boilerDownstream=new PointThopt(proj,"3b");
condUptream=new PointThopt(proj,"4");
condDownstream=new PointThopt(proj,"1");
hotGases=new PointThopt(proj,"hot gas");
coldGases=new PointThopt(proj,"cooled gas");
inRiver=new PointThopt(proj,"river inlet");
outRiver=new PointThopt(proj,"river outlet");
condDownstream.getProperties();
subst=(rg.corps.Corps)condDownstream.lecorps;

//noms des composants / component names
boilerName="boiler";
condenserName="condenser";
turbineName="turbine";

//instanciation des TechnoDesign dans les classes externes
//instanciation of TechnoDesign in external classes
technoBoiler=new TechnoEvaporator(proj, boilerName, hotGases, coldGases, boilerU
addTechnoVector(technoBoiler);
technoCond=new TechnoCondensor(proj, condenserName, condUptream, condDownstream,
addTechnoVector(technoCond);

technoTurbine= new TechnoMultiStageTurb(proj, turbineName, boilerDownstream, cor
addTechnoVector(technoTurbine);

//initialisation des TechnoDesign dans Thermoptim / TechnoDesign initialisation
setupTechnoDesigns(vTechno);
```

2. We use update methods for external points to access the properties of those of Thermoptim, for both exchangers and the turbine, to determine the quantities in their calculation. Here is the code for the turbine:

```
if(!turbineName.equals("")){//initialization of the steam turbine
    args=new String[2];
    args[0]="process";
    args[1]=turbineName;
    vProp=proj.getProperties(args);
    String amont=(String)vProp.elementAt(1);
    String aval=(String)vProp.elementAt(2);
    Double f=(Double)vProp.elementAt(3);
    massFlow=f.doubleValue();
    f=(Double)vProp.elementAt(11);
    eta_is=f.doubleValue();

    Kh=technoTurbine.getStodolaConstant();
    Kh_value.setText(Util.aff_d(Kh,4));
    RVL=0;
}
```

10.3.2.3 Simulation

1. We initialize the simulation by updating some settings from the values entered on the screen, and then the calculations themselves are launched:

```
//parameters and variables are read on the controller screen
  AdesignBoiler=Util.lit_d(AdesignBoiler_value.getText());
  AdesignCond=Util.lit_d(AdesignCond_value.getText());
  Kh=Util.lit_d(Kh_value.getText());
  technoTurbine.setKh(Kh);
  UAevap=Util.lit_d(UAevap_value.getText());
  UAcond=Util.lit_d(UAcond_value.getText());

  Pevap=Util.lit_d(Pevap_value.getText());
  boilerDownstream.P=Pevap;
  boilerDownstream.T=Util.lit_d(Tmax_value.getText())+273.15;
  boilerDownstream.update(UPDATE_T,UPDATE_P,!UPDATE_X);
  boilerDownstream.getProperties();

  boilerUpstream.P=Pevap;
  boilerUpstream.update(!UPDATE_T,UPDATE_P,!UPDATE_X);
  boilerUpstream.getProperties();
  Twater=Util.lit_d(Teau_value.getText())+273.15;
```

2. The search for the solution is carried out using the minPack library. There are four variables: Tcond, massflow, UAevap, and UAcond.

```
x[1] = Tcond;
x[2] = massFlow;
x[3] = UAevap;
x[4] = UAcond;

double residu0;
double residu1;

fcn(m,n,x,fvec,iflag);
residu0 = optimization.Minpack_f77.enorm_f77(m,fvec);

nfev2 = 0;
njev2 = 0;

double epsi=0.0005;//précision globale demandée sur la convergence / overall a
double epsfcn = 1.e-6;//précision calcul des différences finies finies / preci

if(twoStepAlgorithm.isSelected()){
    epsi=0.01;
}

//appel modifié de lmdiff avec modification précision calcul des différences f
//modified call to lmdiff with precision change in the calculus of finite diff
optimization.Minpack_f77.lmdif2_f77(this, m, n, x, fvec, epsi, epsfcn, info);
residu1 = optimization.Minpack_f77.enorm_f77(m,fvec);
```

3. Four residuals to minimize correspond to these four variables: resCond(), resFlow(), resAevap(), and resAcond(), presented here:

```
double resCond(){//équation 17.4.5
    //renvoie la différence entre la température de condensation
    //recalculée à partir de la méthode du NUT et Tcond=x[1]
     // Returns the difference between the condensation temperature
     // recalculated using the NTU method and Tcond = x [1]

    double Toutlet_refrig=0;
    mCpRefrigCond=DeltaHcond/(condUptream.T-(Tcond+DTssrefr));
    double[] res=Util.epsi_NUT(mCpRefrigCond,mCpCalopCond,UAcond);
    epsilon=res[0];
    mCpmin=res[1];
    double Tinlet_refrig=Twater+DeltaHcond/epsilon/mCpmin;
    Toutlet_refrig=Tinlet_refrig-DeltaHcond/mCpRefrigCond;

    //le résidu est l'écart entre les deux températures de condensation
  // The residual is the difference between the two condensing temperatures
    double z= Tcond-(Toutlet_refrig-DTssrefr);
    z= 2*z/(Tcond+Toutlet_refrig-DTssrefr);
    System.out.println("resCond: " + z + ", "+ Tcond + ", " + Tevap+ ", "  +
    return z;
}
```

10.3.2.4 Project Update and Result Display

1. When the four residuals are close to 0, the solution is found and the project is updated with the new values:

```
        //appel modifié de lmdiff avec modification précision calcul des diffé
        //modified call to lmdiff with precision change in the calculus of fin
        optimization.Minpack_f77.lmdif2_f77(this, m, n, x, fvec, epsi, epsfcn,
        residu1 = optimization.Minpack_f77.enorm_f77(m,fvec);

        System.out.println();
        System.out.println(" Initial L2 norm of the residuals: " + residu0);
        System.out.println("Final L2 norm of the residuals: " + residu1);
        System.out.println("Number of function evaluations: " + nfev);
        System.out.println("Number of Jacobian evaluations: " + njev);
        System.out.println("Info value: " + info[1]);
        System.out.println("Final approximate solution: " + x[1] + ", " + x[2]
        System.out.println(); /**/
        algorithmResults.setText("Algorithm precision: " + Util.aff_d(residu1,
    }
    eff_value.setText(Util.aff_d(-tauTurb/DeltaHevap,4));
    DeltaHevap_value.setText(Util.aff_d(DeltaHevap,4));
    massFlow_value.setText(Util.aff_d(massFlow,4));
    DeltaHcond_value.setText(Util.aff_d(DeltaHcond,4));
    Pevap_value.setText(Util.aff_d(Pevap,4));
    Pcond_value.setText(Util.aff_d(condDownstream.P,4));
    tauTurb_value.setText(Util.aff_d(tauTurb,4));
    DeltaHcompr_value.setText(Util.aff_d(DHcompr,4));
}
```

10.4 PART LOAD OPERATION OF THE FLAMANVILLE 3 EPR TURBINE

The results presented in this case study are based on data provided by EDF to the Nuclear Safety Authority to obtain authorization to commission the Flamanville EPR power plant, which is a third-generation nuclear plant with a capacity of 1,650 MW.[8]

This data comes from the knowledge models possessed by the constructor and operator for the design and operation of their installations.

Although some questions remain concerning certain equations used in these models,[9] they can be regarded as the most reliable representations available until operational data become accessible—a process that will take several years, especially since there is no guarantee that such data will ever be made public.

All of this information therefore constitutes a rare and high-quality educational resource that should be leveraged as much as possible, justifying the focus of this case study on it.

The insights gained from their analysis may be applicable to many steam thermodynamic cycles coupled with nuclear reactors—not just the EPR. In particular, they will help better model certain small reactors under development: the Small Modular Reactors.

The original format of this data consists of medium-quality copies of flowsheets representing the plant's secondary cycle, showing the thermodynamic state of the fluid (steam or liquid water) and the main flow rates and extractions between the different Turbine Stage Groups (TSG).

The available data corresponds to six complementary operating flowsheets[10] for two turbines, labeled A and B. We used those for turbine B, in regime R, at five load rates: 100%, 90%, 70%, 50%, and 30%.

They were used to build Thermoptim models for the various cases, which were then used to generate the data for studying the evolution, as a function of the fraction of nominal power (also called load rate), of the most relevant variables: main flow rates and pressures, Stodola constants, polytropic efficiencies, and so on.

Figure 10.5.1 shows the detailed flowsheet of the full load cycle, divided into two parts for better readability.

10.4.1 Steam Circuit

The nuclear reactor receives 2,429 kg/s of feedwater at 84.5 bar and 226.6°C, outputting steam at 78 bar and 293.2°C (quality 0.9976). The flow splits: 2,288 kg/s to the high-pressure turbine (HP) turbine and 140.3 kg/s to the second reheater.

The HP turbine has three TSGs with extractions at each stage: 96.97 kg/s to supply the first reheater and 100.6 kg/s after TSG1, 119.6 kg/s after TSG2, and 92.22 kg/s at the outlet, feeding the HP feedwater heaters (FWHs) and feedwater tank.

The main flow enters the moisture separator reheater. The liquid phase (275.2 kg/s) joins the feedwater tank, while vapor passes through two reheaters (reaching 275.1°C at 10.62 bar) before entering the IP turbine (two TSGs) with extractions of 68.22 and 111.6 kg/s for LP4 and LP3 FWHs.

The LP turbine comprises six TSGs. After TSG1, 12.34 kg/s condensate and 95.52 kg/s steam extraction feed LP2 FWH. After TSG2, 24.38 kg/s condensate and 6.52 kg/s steam partially feed LP1. After TSG3, 33.32 kg/s condensate and 48.79 kg/s steam also feed LP1 FWH. The flow exits at 0.0442 bar with quality 0.8989.

10.4.2 Feedwater Circuit

The extraction pump compresses 1,316 kg/s to 19 bar. The flow is progressively heated through LP1 (54°C), LP2 (92.2°C), LP3 (133°C), and LP4 (157°C) FWHs (pumped forward type), with condensed steam repressurized and mixed at each stage.

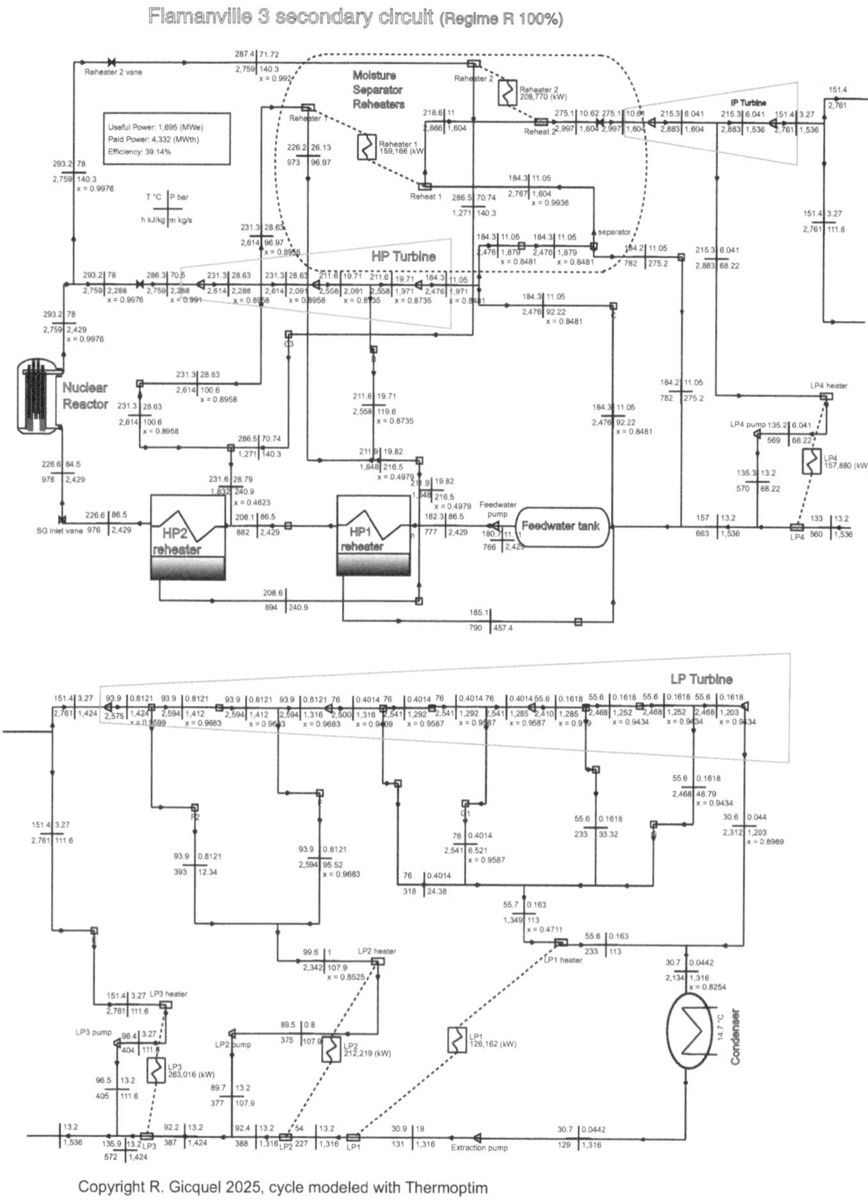

FIGURE 10.5.1 Detailed flowsheet of the full load Flamanville 3 EPR cycle.

The feedwater tank receives all flows, outputting 2,429 kg/s at 180.7°C, which is compressed to 86.5 bar and heated through HP1 and HP2 (cascaded drains type) to 226.6°C before entering the steam generator.

This page of this book's website gives you access to the flowsheets of the five cycles, created in SVG format.[11]

In the following charts, the load rate always appears on the x-axis. On the y-axis are the different variables, either as relative values compared to nominal power or as absolute values.

A spreadsheet with the raw data presented here is available here.[12]

10.4.3 Flow Rate Variation

The chart of Figure 10.5.2 shows the relative evolution of the main flow rates in the cycle as a function of load rate.

They can be represented by linear relations. Extraction flow rates are determined as residuals to ensure flow conservation.

10.4.4 Pressure Variation

In the secondary circuit, the pressure at the SG point at the outlet of the steam generator is set by plant operators depending on both the requested power and the operational constraints of the primary circuit. The steam moisture content is very low, about 0.5%. The SG temperature and enthalpy can thus be deduced. This SG pressure decreases as the power decreases.

At full load, the control valves upstream of the HP are almost fully open, as are the diaphragms that modulate the flow entering the machine.

The pressure in its inlet torus, often called the first-stage turbine pressure (here P1RT), is thus, at nominal power, very close to the pressure at the SG point. Since this value is not provided in the diagrams, we will use 70.5 bar for nominal power.

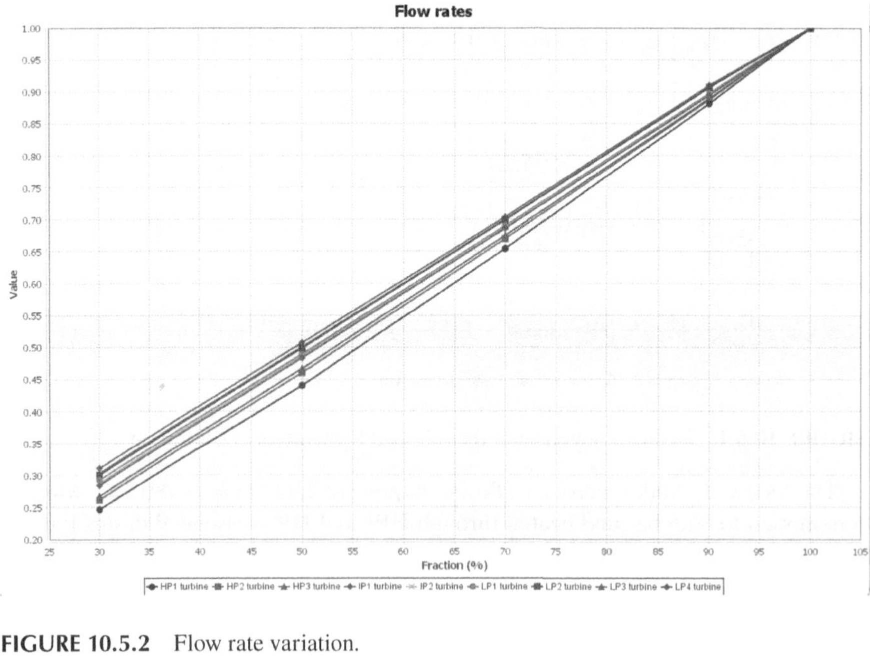

FIGURE 10.5.2 Flow rate variation.

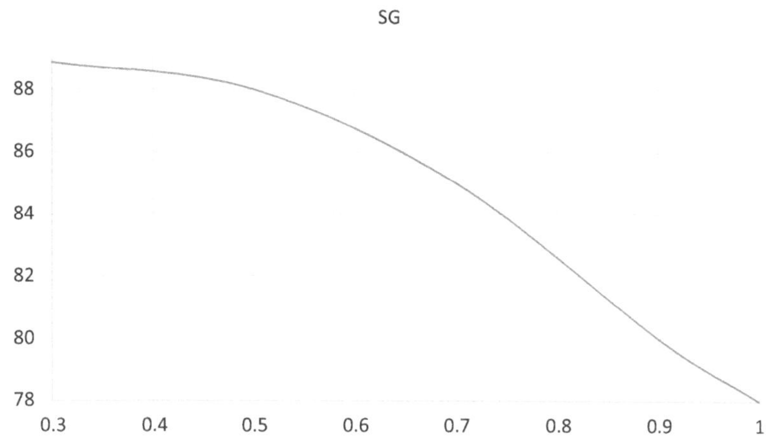

FIGURE 10.5.3 Outlet SG pressure.

The enthalpy at P1RT equals that at SG, as the expansion in the valves is isenthalpic.

When load decreases, the thermal power transferred by the reactor must be reduced. For this, it is necessary to increase the secondary side temperature to reduce the temperature difference with the primary. Since the steam is two-phase, this can be done by increasing its pressure.

Flamanville's data show that outlet SG pressure can be represented by a cubic function of load rate (Figure 10.5.3); other HP pressures can be deduced by fixed proportionality coefficients.

Downstream from the diaphragms, the flow once again becomes unconstrained and must satisfy the Stodola law if it is to operate under optimal conditions. Determining P1RT at partial load is, however, quite complex, because this pressure affects both the Stodola law and the turbine's polytropic efficiency.

Like the other variables, P1RT follows an approximately linear relation with load rate.

The changes in cycle pressures at different points are shown in Figure 10.5.4, which demonstrates that most pressures increase with load rate—except, of course, those between the feedwater pump and the turbine inlet, which evolve oppositely.

The Stodola law, which we will analyze now, makes it possible to go even further by revealing an almost invariant relationship between the upstream and downstream states of the turbine and the flow that passes through it.

10.4.5 Influence of the Stodola Law Used

As discussed in Chapter 9, the Stodola law can be formulated in several ways. The influence of the chosen model is addressed in this section. These results show the evolution of the Stodola constants for the nine TSGs shown in the ASN diagrams: three TSGs for the HP turbine (HP1, HP2, HP3), two TSGs for the IP turbine (IP1 and IP2), and four TSGs for the LP turbine (LP1–LP4).

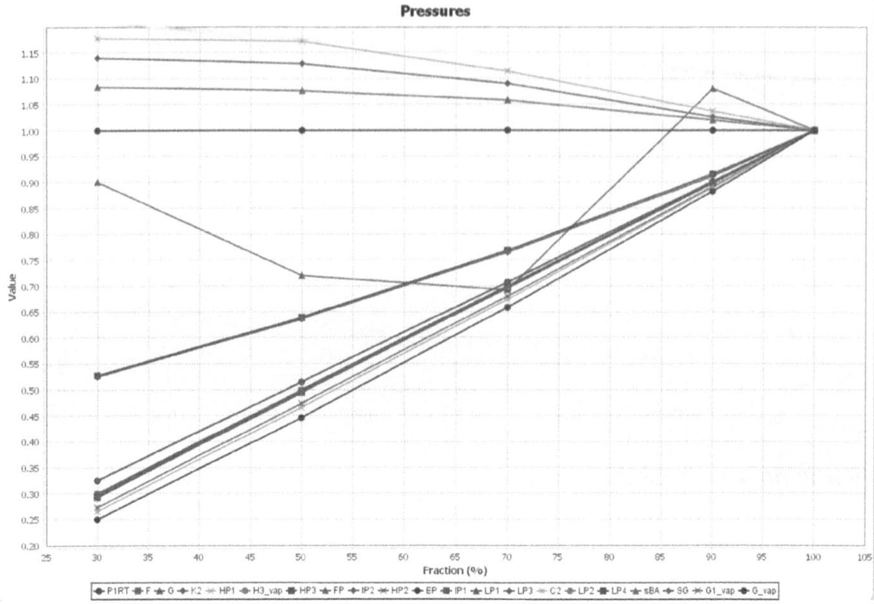

FIGURE 10.5.4 Cycle pressures.

We defined the quality criterion for estimating the Stodola constant for each TSG as the square root of a quarter of the sum of squared errors between different load rates and nominal power, and as a global criterion, the mean of these deviations.

1. If the Stodola law is expressed as a function only of upstream temperature and pressure, the relative variation of the calculated value for different stage groups exceeds 8%, and the global criterion is 2.98% (Figure 10.5.5).

$$\frac{\dot{m}\sqrt{T_{in}}}{P_{in}} = \text{Const} \tag{9.4.12}$$

2. Taking into account the downstream pressure by quadratic correction, the relative variation is slightly reduced, with a global criterion of 1.88%.

$$T_{in}\frac{\dot{m}^2}{K^2} = P_{in}^2 - P_{out}^2 \tag{9.4.11}$$

3. Using a relation based on volume rather than temperature, without downstream correction, the variation in the Stodola constant decreases significantly and becomes less than 4% except for turbine LP4, where it remains high. The global criterion decreases to 1.1%.

$$\dot{m}\sqrt{\frac{V_{in}}{P_{in}}} = K$$

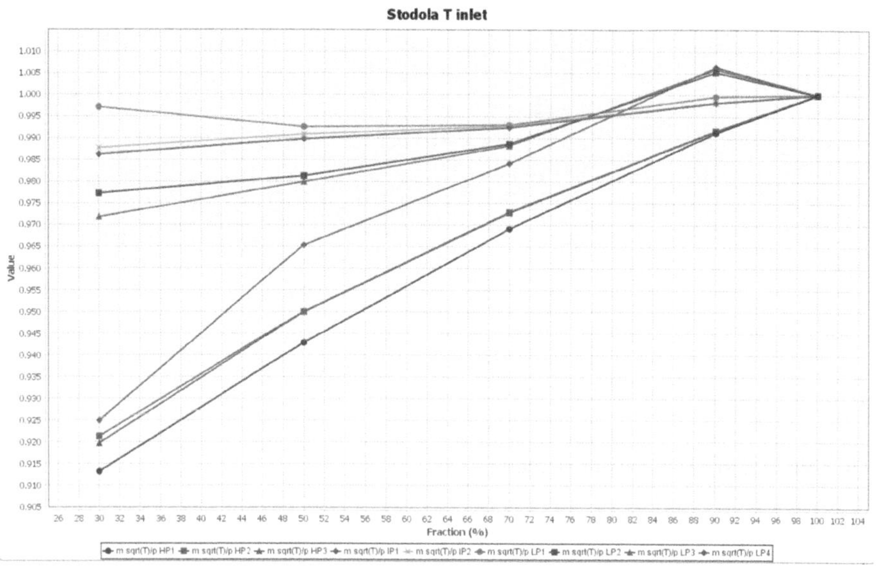

FIGURE 10.5.5 Stodola Law 1.

4. Using the volume relation with an upstream multiplicative factor (k + 1)/k, the Stodola constant's variation further decreases and becomes less than 1.3%, with the global criterion dropping to 0.46%.

$$\dot{m}\sqrt{\frac{(k+1)v_{in}}{kP_{in}}} = K$$

5. However, the best match is achieved with the v/P formulation with quadratic correction, suggesting this is the model used, even if it lacks solid theoretical justification. The relative error is always below 1%, and the global criterion is 0.23%.

$$\frac{v_{in}}{P_{in}}\frac{\dot{m}^2}{K^2} = P_{in}^2 - P_{out}^2$$

   A comparison of these two best matches is shown in Figure 10.5.6.

6. Using the volume-based relation without the upstream multiplicative factor and with downstream correction in (k + 1)/k, the Stodola constant's variation is slightly larger for HP2 and HP3 TSGs. The global criterion is 0.5%.

$$\dot{m}\sqrt{\frac{v_{in}}{P_{in}}} = K_0\sqrt{1-\left[\frac{P_{out}}{P_{in}}\right]^{(k+1)/k}}$$

7. Using the volume-based relation with both upstream and downstream multiplicative factors in (k + 1)/k, the Stodola constant's variation is greater, especially for TSG LP4, with a global criterion of 1.05%. Nevertheless, this model is likely the most theoretically sound.

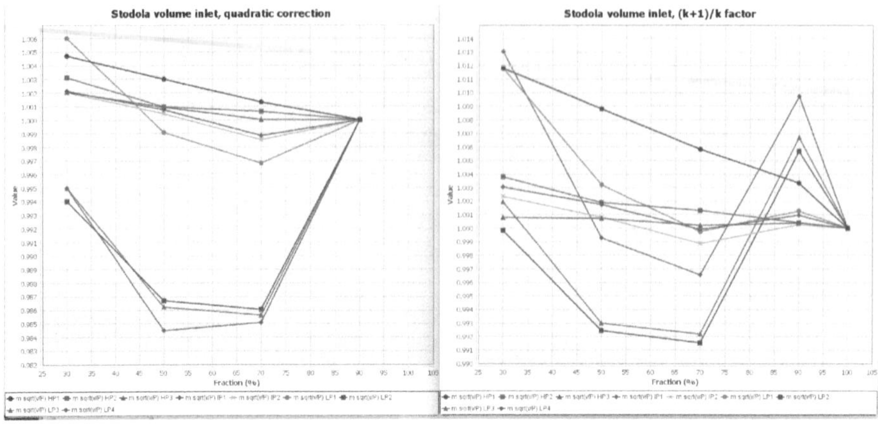

FIGURE 10.5.6 Comparison of Stodola Laws.

$$\dot{m}\sqrt{\frac{(k+1)v_{in}}{kP_{in}}} = K_0\sqrt{1-\left[\frac{P_{out}}{P_{in}}\right]^{(k+1)/k}}$$

Regardless, the discrepancies remain quite small as soon as the Stodola law is formulated based on volume rather than temperature. Given the extreme difficulty of accurately measuring the steam state in the two-phase zone in actual power plants, it will at this time unfortunately not be possible to definitively validate any of the formulations just discussed.

10.4.6 Polytropic Efficiencies

The polytropic efficiencies of the various TSGs are shown in Figure 10.5.7, given in absolute values, not relative. They are referred to as "wet" as they take into account the potential efficiency reduction due to steam moisture.

To determine the polytropic efficiency values for TSGs with dry steam, they must be corrected for the influence of the steam quality and a factor often called Baumann's factor.

$$\frac{\eta_{hum}}{\eta_{dry}} = 1-\alpha(1-x)$$

Taking a value of α as 0.9 yields results exceeding one for LP4, which is physically impossible, and the other values are also very high.

To bring these to more plausible values, α must be significantly lowered. More plausible results are obtained for $\alpha = 0.4$.

In any case, the phenomena related to droplet condensation at the outlet of low-pressure turbines are still very poorly understood, and there is no guarantee that Baumann's rule is truly applicable.

Generally, values for these efficiencies obtained by identifying turbine models on TSG data are quite high, even though they include corrections for moisture.

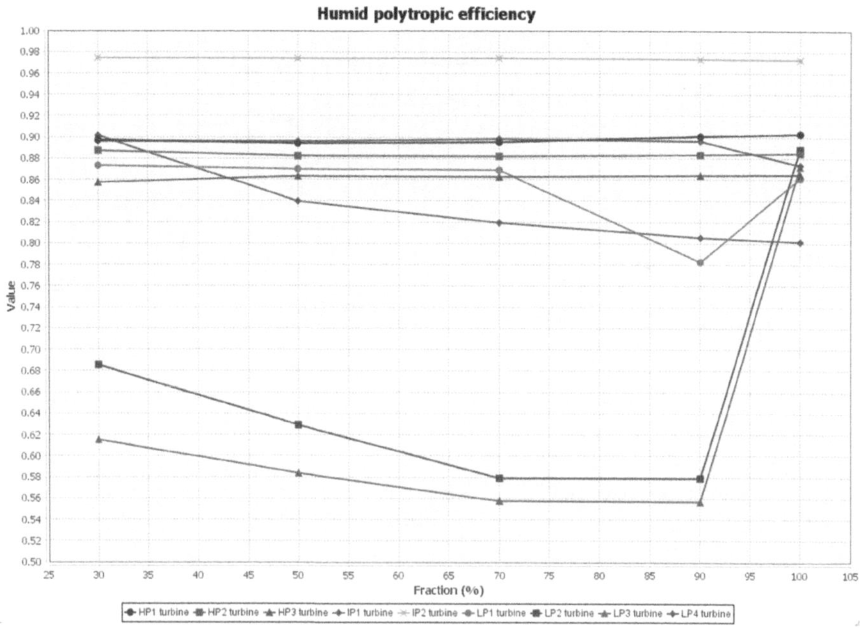

FIGURE 10.5.7 Variations of polytropic efficiencies.

Returning to "dry" efficiencies by taking a Baumann coefficient close to one sometimes yields values exceeding one, which is physically impossible, whereas a value of 0.4, much lower than what is usually used in the literature, would yield more plausible results.

10.4.7 Polytropic Exponents

The evolution of the polytropic exponent k, used for downstream pressure correction, is shown in Figure 10.5.8. It is identified between the inlet and outlet of each TSG from the law $Pv^k = \text{Constant}$.

The most remarkable point is the very large variation for LP4 TSG, which is further analyzed later.

Another way to present these results is to show the evolution of the corrective terms related to downstream pressure in the equation:

$$\dot{m}\sqrt{\frac{v_{in}}{P_{in}}} = K_0\sqrt{1-\left[\frac{P_{out}}{P_{in}}\right]^{(k+1)/k}}$$

The lower their value, the greater the correction. These values are quite stable, except for LP4 (Figure 10.5.8).

Only the correction for TSG LP4 changes significantly with the load rate. Figure 10.5.9 shows it is strongly correlated with condensation pressure, which

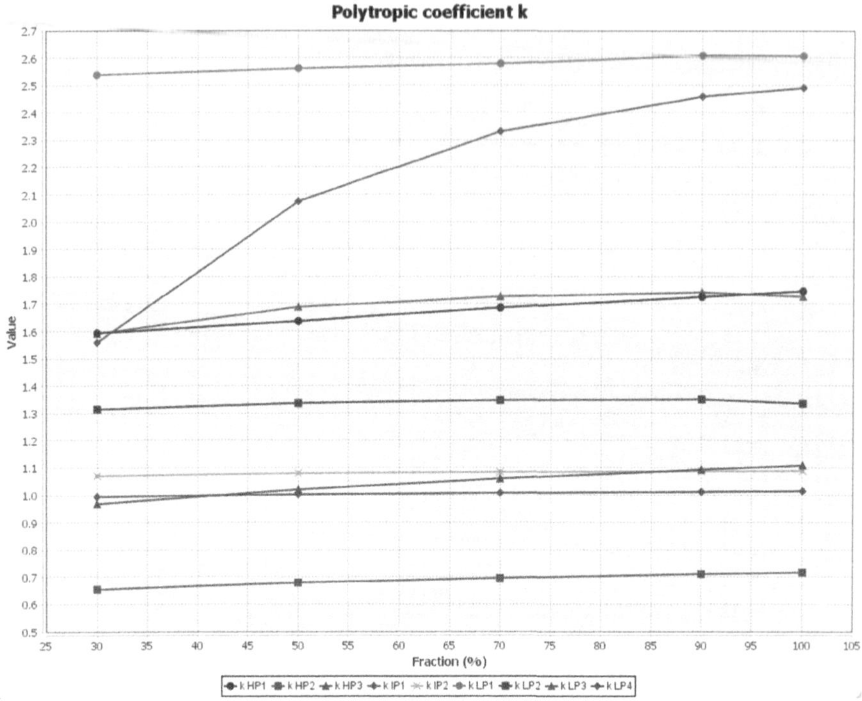

FIGURE 10.5.8 Polytropic efficiency correction factor.

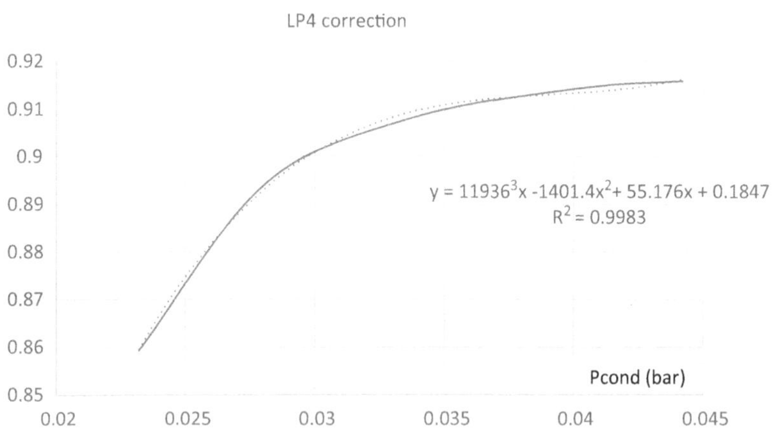

FIGURE 10.5.9 LP4 correction.

decreases as the load rate drops—the temperature difference in the condenser is smaller due to decreased heat to evacuate.

10.4.8 Separator

The separator is characterized by a drying efficiency that varies little with load rate (a relative decrease of 1.5% at 30% nominal power).

10.4.9 NUSCLE

As we mentioned in the introduction, all of this data constitutes a rare and exceptionally high-quality pedagogical resource that should be valorized to the fullest extent possible.

It has enabled us to validate a number of empirical correlations that we have implemented in a simplified thermodynamic cycle model for water-cooled nuclear power plants, which we have named NUSCLE. We are distributing the entire code following an open-source approach, so that other users can modify it if they wish to meet their specific needs.

This independent tool comes with complete documentation: the full equation set, solution methodology, and operational data for about ten reference plants.

NUSCLE allows the study of partial load operation of many WCRs, such as such as PWRs, BWRs, Russian RBMK (graphite-moderated pressure tube reactors), or CANDUs.[13]

With four extractions and five TSGs, it enables much more detailed modeling than the models typically used for training purposes, even though there are respectively 11 and 9 in the Flamanville 3 EPR cycle.

This compromise choice aligns with a pedagogical logic: to offer an intermediate tool between a complete industrial model and an elementary model.

10.4.9.1 Capabilities and Educational Value

NUSCLE enables users to:

- Explore how key parameters affect the performance of various WCR types (PWR, BWR, Vodo-Vodyanoi Energetichesky Reaktor or Russianwater-water energy reactor (VVER), RBMK, CANDU).
- **Analyze part load behavior** using Stodola law for TSG modeling and Number of Transfer Units (NTU)-effectiveness models for heat exchangers.
- Compare different reactor secondary cycles with reasonable accuracy.
- Access and modify the complete set of equations for deeper understanding (a key advantage of the open-source approach).

The tool is coupled with the IF97 water property library for high accuracy. Its graphical interface (Figure 10.5.11) displays the cycle diagram with interactive value readouts and enables sensitivity studies as well as off-design condition analysis—capabilities few educational software achieve with such accessibility (Figure 10.5.10).

The simulation results can be exported in several formats:

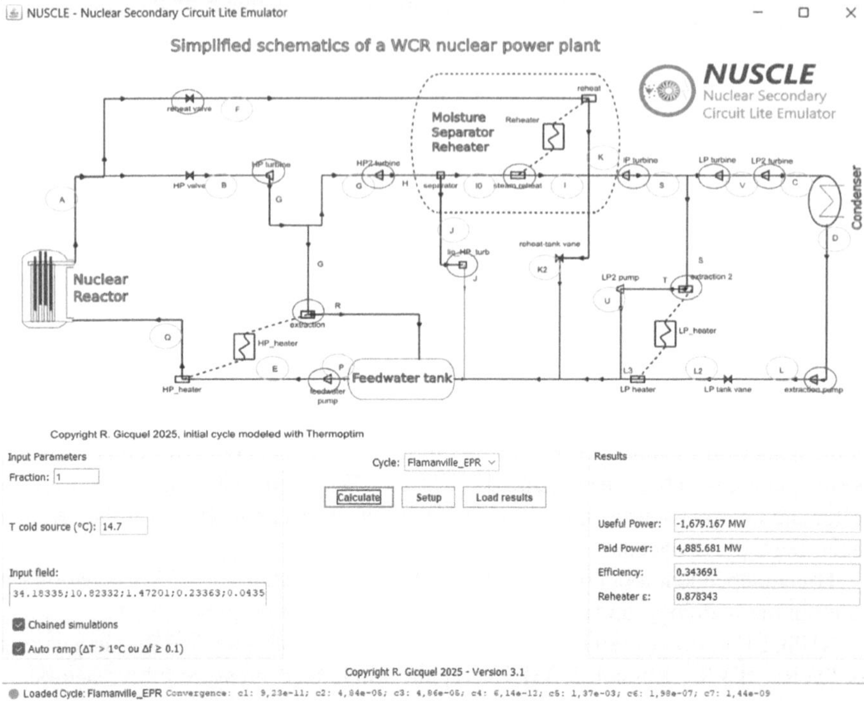

FIGURE 10.5.10 NUSCLE interface.

- a detailed text file containing almost all calculated quantities;
- a summary window displaying the thermodynamic states of the main points, the mass flow rates and power of the turbines, the overall performance indicators, and the global convergence criterion of the model;
- a graphical window showing the expansion line of the turbines on the Mollier (*h*–*s*) diagram (Figure 10.5.11).

10.4.9.2 Technical Foundation

The main heat exchangers, the steam reheater and the condenser, are calculated in off-design conditions according to the principles presented in Chapters 4 and 7, and the TSGs are modeled using either of Stodola's equations presented in Chapter 9.

The model requires solving several hundred highly nonlinear equations. Given the load factor and condenser inlet cooling water temperature as inputs, seven principal variables must be determined: the condensing pressure, five TSG inlet pressures, and the high-temperature reheater outlet temperature. These principal variables allow all other cycle quantities to be derived.

The key challenge lies in equation hierarchy. Generic solvers treating all equations indiscriminately fail beyond one hundred equations. NUSCLE's solution separates unknowns into three categories: pre-calculable parameters, principal variables recalculated at each iteration, and dependent variables computed once the solution

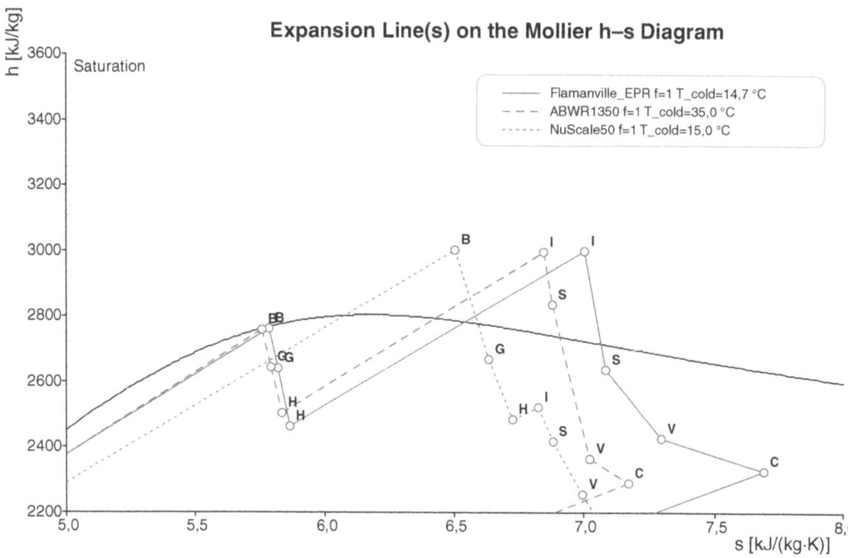

FIGURE 10.5.11 Expansion line(s) on the Mollier h–s diagram.

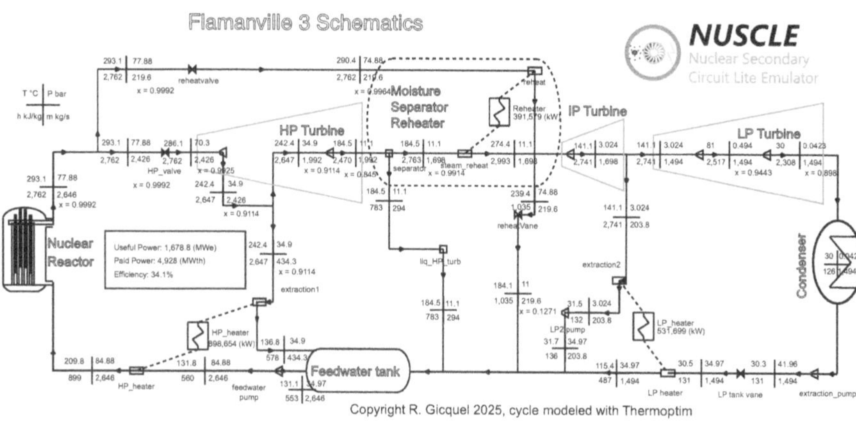

FIGURE 10.5.12 Flamanville 3 EPR NUSCLE model.

converges. This organization ensures convergence when properly initialized despite the equation count.

The model originated from Thermoptim, which generated 256 equations (about ten redundant). These were then supplemented to implement off-design calculations and the solution algorithm.

10.4.9.3 Application to Flamanville 3 EPR

Configured to represent the Flamanville 3 EPR, NUSCLE's thermodynamic diagram is shown in Figure 10.5.12. Although simplified, the model successfully captures

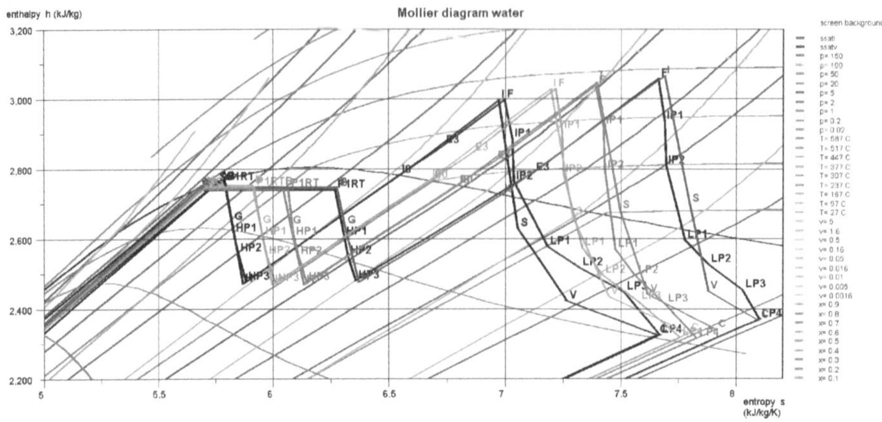

FIGURE 10.5.13 Comparison of detailed and NUSCLE models in the Mollier diagram.

the main cycle characteristics: maximum pressure, electrical power output, separator pressure, superheating temperature, and feedwater temperatures.

The detailed Flamanville cycle model was presented in the previous section. It is interesting to compare the results it provides with those of NUSCLE.

Figure 10.5.13 illustrates simulated cycles plotted in the Mollier diagram for power levels of 30%, 50%, 70%, and 100%. The simulation curves closely match actual data for HP and IP turbines. Discrepancies in the LP expansion line arise because the simplified model includes only two stage groups instead of the real system's six.

Detailed documentation is available online at: https://server.s4e2.com/crc/amtc/nuscle/index.html.

10.5 CONCLUSION

This chapter presented four case studies of increasing complexity regarding sizing and off-design operation, demonstrating the methodologies for users developing their own models.

1. The first case involves an air piston compressor with intercooling feeding a compressed air storage system, introducing the creation of controllers and simple off-design analysis.
2. The second case examines the refrigeration machine from Chapter 6, involving a positive displacement compressor, thermostatic expansion valve, and two two-phase exchangers, solved using minPack for nonlinear equation systems.
3. The third case studies a simplified steam power plant, analyzing the evolution of performance by varying cooling water temperature, maximum pressure, or superheat temperature. It is also solved with minPack.
4. The fourth case differs by presenting the data analysis of the partial load of the Flamanville 3 EPR turbine using official EDF data. The results demonstrate methodological implementation, revealing optimal formulations of

Stodola's law, variations in polytropic efficiency, and separator performance across the 30%–100% load range. This data made it possible to develop the NUSCLE software, a simplified model of the thermodynamic cycle of a water-cooled nuclear power plant (WCR type: Water-Cooled Reactor), such as PWRs, BWRs, RBMKs, or CANDUs.

Additional examples illustrating how the concepts presented in this chapter can be applied using Thermoptim are available on this book's website.[14]

NOTES

1 https://server.s4e2.com/crc/amtc/chapters/chapter10/_resources/index.html.
2 The spreadsheet is available at: https://server.s4e2.com/crc/amtc/chapters/chapter10/_resources/index.html.
3 The controller is available at: https://server.s4e2.com/crc/amtc/chapters/chapter10/_resources/index.html.
4 The controller is available at: https://server.s4e2.com/crc/amtc/chapters/chapter10/_resources/index.html.
5 https://server.s4e2.com/crc/amtc/chapters/chapter10/_resources/index.html.
6 https://server.s4e2.com/crc/amtc/chapters/chapter10/_resources/index.html.
7 https://server.s4e2.com/crc/amtc/chapters/chapter10/_resources/index.html.
8 https://www.asn.fr/content/download/194137/file/10.%20Circuit%20vapeur%20et%20transformation%20de%20l%E2%80%99%C3%A9nergie%20-%20Rapport%20de%20s%C3%BBret%C3%A9.pdf.
9 We have communicated our questions to the constructors, but despite numerous follow-ups, we have never received responses.
10 We would like to thank Mr. J. M. Monteil, senior expert at EDF now retired, for his valuable help in interpreting certain subtleties of these flowsheets.
11 https://server.s4e2.com/crc/tcnpp/resources/epr/.
12 https://server.s4e2.com/crc/amtc/chapters/chapter10/_resources/index.html.
13 In volume 2 of this set, you will find in-depth developments on the thermodynamic cycles of Water-Cooled Reactors.
14 Available at: https://server.s4e2.com/crc/amtc/chapters/chapter10/_resources/index.html.

Index

M

N

O

P

Q

R

S

T

U

V

W